SPECIAL RELATIVITY
THE FOUNDATION OF MACROSCOPIC PHYSICS

T0296719

SPECIAL RELATIVITY

THE FOUNDATION OF
MACROSCOPIC PHYSICS

W. G. DIXON

Fellow of Churchill College, Cambridge

CAMBRIDGE UNIVERSITY PRESS

CAMBRIDGE

LONDON   NEW YORK   NEW ROCHELLE

MELBOURNE   SYDNEY

CAMBRIDGE UNIVERSITY PRESS
Cambridge, New York, Melbourne, Madrid, Cape Town, Singapore, São Paulo, Delhi

Cambridge University Press
The Edinburgh Building, Cambridge CB2 8RU, UK

Published in the United States of America by Cambridge University Press, New York

www.cambridge.org
Information on this title: www.cambridge.org/9780521272414

© Cambridge University Press 1978

First published 1978
First paperback edition 1982
Re-issued in this digitally printed version 2009

A catalogue record for this publication is available from the British Library

Library of Congress Catalogue Card Number: 77–83991

ISBN 978-0-521-21871-9 hardback
ISBN 978-0-521-27241-4 paperback

Contents

Preface

The special theory of relativity is often considered as irrelevant to the macroscopic physics of ordinary material systems. The range of velocities, pressures and temperatures encountered under terrestrial conditions is such that the differences between the Newtonian and relativistic theories are negligibly small. Either theory can thus be used, and as the Newtonian theory is usually considered to be the simpler, it is the one usually adopted. But *is* the Newtonian theory actually the simpler of the two? This depends on what one is trying to do. Ballistic calculations are undoubtedly made more complicated by the use of relativistic formulae in place of the corresponding Newtonian ones, but special relativity has more to offer than nuisance factors of $\sqrt{(1 - v^2/c^2)}$. The aim of this book is to show that an understanding of the basic laws of macroscopic systems can be gained more easily within relativistic physics than within Newtonian physics. The speed of the systems concerned is irrelevant. Even equilibrium thermodynamics gains by being seen from a relativistic viewpoint.

The book is not directed towards any particular university course. It tries to show the unity of dynamics, thermodynamics and electromagnetism under the umbrella of special relativity, and it should be accessible to any second year undergraduate in mathematics or physics. The emphasis throughout is on the extraction by systematic development of a maximum of information from a minimum of assumptions. With this in mind, the first chapter lays minimal physical foundations for the special theory of relativity and explores its relationship to Newtonian physics. The assumption that the speed of light is independent of the motion of the observer is found to be unnecessary. A prior knowledge of special relativity is not essential but an acquaintance with its basic ideas will be helpful. The second chapter lays the mathematical foundations needed for the subsequent development. The remaining three chapters develop the foundations of particle and continuum dynamics, and the thermodynamics and electrodynamics of fluids, within this relativistic framework. Fluids are studied in preference to solids as they are conceptually simpler. As four-dimensional spacetime techniques are used throughout, much of the theory can be taken over into general relativity with little alteration.

The S.I. system of units that is now widely adopted for the presentation of formulae in electromagnetism does not combine naturally with the four-dimensional tensor formulation that is used in special relativity. For this reason Gaussian (c.g.s.) units have been used instead for the development of electrodynamics in Chapter 5.

Chapters are divided into sections, and equations are numbered consecutively within each section. These numbers run continuously through the subsections into which some sections are divided, thus §4 of Chapter 5 has subsections labelled 4a to 4c and equations numbered (4.1) to (4.42). Sections and equations within the current chapter are referred to simply by these numbers. References to sections and equations of another chapter are prefixed by the chapter number and a hyphen, thus §4a means subsection a of section 4 of the current chapter but §4-1 means section 1 of Chapter 4. Reference to publications is by author and year. Details of these publications are given in a list at the end of the volume.

This book was begun during leave of absence from Churchill College, Cambridge, for the academic year 1974/75. I am grateful to Churchill College for financial support during that period, and to the Department of Physics and Astronomy, University College London, for its hospitality during it. I would also like to thank my wife for her patience and constant encouragement during my writing and typing of the book.

March 1978   W. G. Dixon

1

The physics of space and time

1 Introduction

The special theory of relativity has its historical origin in a study of electromagnetic phenomena. It takes its name from its denial of the concept of absolute motion and the consequent recognition that only relative motion has any physical significance. However, it does recognize a preferred class of observers who are in uniform motion relative to one another, even though it denies that it is meaningful to ask which of them is at rest in any absolute sense. Hence the qualification 'special', the hope being that it would ultimately be superseded by a theory in which all observers are treated as equivalent.

At the time that the special theory was being developed, around the beginning of this century, it was believed that all forces in nature would ultimately be reducible to electromagnetism and gravitation. With the success of the special theory in resolving the conflicts that had existed between Newtonian dynamics and Maxwell's electromagnetic theory, it became natural to try to fit gravitation into this new physical framework. That this proved so difficult seems perhaps more surprising now than it did at that time. It is now realized that the ultimate structure of matter is considerably more complicated than was suspected seventy years ago, when the quantum theory was still in its infancy and even the Bohr theory of the atom was still in the future. Although the forces that occur within the atomic nucleus are not yet fully understood, tremendous progress has been made, and underlying it all is the basic framework provided by the special theory of relativity. This is indeed the main strength of the theory. The fact that it predicts modifications of Newtonian dynamics for particles whose speeds are comparable with that of light is important, but its real achievement has been in providing a foundation on which almost the whole of modern physical theory has been built. However, this increasing scope of the special theory has also seemed to increase the apparent perversity of gravitation in refusing to be fitted into this growing structure.

A study of the foundations of the special theory should reveal the origins of its limitations as well as of its successes. Gravitation must thus be expected to play a distinctive part in such a study, inasmuch as the reason for its exclusion from the theory should become clear. But although it is excluded from the theory, it cannot be excluded from the laboratories in which terrestrial physical experiments are performed. To understand the validity of the special theory in such circumstances, some knowledge is required of the modifications which are required to allow for the presence of gravitation. These modifications form the basis of the *general theory of relativity*, so named because Einstein considered that these same modifications also place all observers on an equal footing.

The programme of the present chapter is to give a physical basis for the mathematical models of space and time used in relativity theory. For the reason given above, both the special and general theories will be considered. The mathematical and physical developments of the subsequent chapters will however be confined to the special theory. The physical results used will be ones which hold also in Newtonian theory. Consequently no details will be given of the experimental evidence in their support – the success of the Newtonian theory over a wide range of conditions is sufficient evidence in itself. When such results are particularly simple and have far-reaching implications, they may be dignified with the description 'principle'. This is not intended as a claim that they are 'obvious', but instead that they are firmly supported by the success of Newtonian theory. It will not be necessary to assume the constancy of the speed of light. This speed, 'c', is a fundamental constant of nature which is not primarily connected with electromagnetic phenomena, and a development based on properties of light gives electromagnetism an unnecessary prominence.

2   Frames of reference

It is quite impossible to make any physical statement at all without some implicit assumptions about the nature of space and time. The best that can be done in an investigation of these fundamental concepts is to try to be as explicit as possible about the assumptions that are being made. Our first task must thus be to provide ourselves with a language with which we can discuss the physical world, and which is as free as possible from undefined terms. This involves setting up frames of reference, and hence requires an examination of the concept of such a frame.

Until the advent of the theory of relativity, space and time were believed to be independent and absolute. In the historical development of special relativity, absolute time was the first of these to fall. By assuming the constancy of the speed of light, and examining the practical process of synchronizing clocks using light rays, Einstein showed in 1905 that simultaneity is not an absolute concept – it depends on the motion of the observer. The concept of an absolute space with a fixed three-dimensional Euclidean geometry survived for a further three years, although it was necessary to ascribe rather peculiar behaviour to rods and clocks in motion in order to retain it. But in 1908, Minkowski (1908a) showed that the natural framework within which to express special relativity is to consider space and time united to form a single four-dimensional continuum. To quote him in translation: 'Henceforth space by itself, and time by itself, are doomed to fade away into mere shadows, and only a kind of union of the two will preserve an independent reality.'

This union will be taken as our starting point. It may not seem much of an assumption, as no particular geometry is yet being ascribed to this spacetime continuum. So it is worth a pause to consider just what are the assertions about the physical world that are hidden within it. There are essentially two. The first is that space and time are continuous, which may be questioned in the light of the quantum nature of so much of physics. The second is that if two events appear coincident in both space and time to one observer, then they appear so to every other observer. It is difficult to envisage the implications of this being false, but it is not logically impossible. No attempt will be made here to justify these assumptions, but it is good to make clear that they are there as the basis of our subsequent development.

Since our everyday language and experience is based on a separate space and time, it is necessary to begin by considering in general terms how this separation is made by a scientific observer. The qualitative nature of this separation as perceived by our senses will be accepted without analysis. But a scientific observer must make this quantitative. His first step must be to make a clock. It is convenient to envisage this as a hand turning continuously around a graduated dial, together with a counter to count complete revolutions. The hand should turn smoothly (an intuitive concept based on the assumed continuity of time), but until some dynamics has been put into our theory, we cannot ask that it should turn uniformly as this is a concept that needs further definition. With this clock he can 'time' events in his immediate locality, but before he can 'time' distant events, he must decide on an operational definition of simultaneity for widely separated events.

Having done so, he can unambiguously say 'when' any event occurs, but not 'where' it occurs. To do this, he needs also to decide what is meant by the same point of space at different times, i.e. he needs a standard of 'rest'. The 'where' of an event then becomes meaningful, but for him to be able to communicate this information to anyone else, he must also set up a spatial coordinate system. Again, such coordinates are naturally required to vary smoothly from place to place (also supposed intuitive), but apart from this, all that can be said is that three coordinates will be required to specify a location uniquely. (It is perhaps worth noting that one could get away with only a single spatial coordinate if the smoothness requirement were dropped, e.g. by interleaving the decimal expressions of three smooth coordinates so that $(0\cdot114, 0\cdot225, 0\cdot336)$ becomes $0\cdot123123456$, but since physical measurements cannot be made with infinite precision, non-smooth coordinate systems are useless for physical purposes.) When this has been done, he will have set up a complete coordinate system for the spacetime continuum which enables every event to be uniquely specified by four coordinates, three being spacelike and one timelike.

To clarify the procedure, we give an example of a way in which these various constructions may be made. It is not intended, however, to be any more fundamental than any other method. This is the radar method. Suppose the observer sends out a pulse of light at time t_1, which is reflected by a distant object and arrives back at the observer at time t_2. Then the instant of reflection is allocated a time coordinate $\tau = \frac{1}{2}(t_1 + t_2)$ and a radial distance $r = \frac{1}{2}(t_2 - t_1)$. The direction of the reflected pulse may be specified by two angular coordinates (θ, ϕ) which together with r make up the three spatial coordinates of the object at the instant of reflection. The rest state is then characterized by the constancy of the spatial coordinates. If one wishes to envisage the measurement of the angular coordinates, one can think of the observer as being surrounded by a transparent sphere with a grid of latitude and longitude lines marked on it.

In this example the coordinate system is constructed first and the definitions of simultaneity and of rest then follow in the obvious way.

Although this is likely to happen in practice, it is conceptually preferable to think of simultaneity and rest being defined before the coordinates are constructed. For these are clearly physical concepts, while coordinates belong to mathematics. It will be useful to try to keep track of what belongs to mathematics and what to physics in the initial development of the theory, and for this purpose a distinction will be drawn between the physical concept of a frame of reference and the mathematical one of a coordinate system. This will be abstracted from common usage, which makes such a distinction even though it is seldom made explicit.

The coordinate system concept is simple, although coordinates will be allowed which are more general than those used in elementary physics. All that is essential in a coordinate system for spacetime is that there should be four coordinates, each of which varies smoothly, and independently of the other three. This degree of generality is necessary for the time being as so little has so far been assumed about the physical world. The preferred coordinate systems usually used in special relativity and in Newtonian physics can only be introduced after further physical assumptions have been made. But more of this later.

One other feature of these generalized coordinate systems is that it will not necessarily be assumed that the whole of spactime is covered by a single nondegenerate coordinate system. Sometimes it is simply convenient to use coordinate systems with degenerate points, e.g. plane polar coordinates (r, θ), where the origin is degenerate as θ is indeterminate there. In this case degeneracies could be avoided by the use of Cartesian instead of polar coordinates. But in other circumstances one may have no choice in the matter. On the surface of a sphere, for example, there is no coordinate system which covers the whole surface without degeneracy. Hence, to avoid any implicit assumptions about the global topological structure of spacetime, coordinate systems will be allowed which cover only a portion of spacetime. If one considers operational definitions of coordinate systems such as the radar method described above, in a finite time it is possible to survey only a finite volume of space, and so such coordinates are naturally restricted in this way.

If, in this same example, the observer decided to transform from the polar type of coordinate system that he has constructed by direct measurement to a rectangular type of coordinate system by the mathematical transformation $x = r \sin \theta \cos \phi$, $y = r \sin \theta \sin \phi$, $z = r \cos \theta$, this would not usually be considered as a change of reference frame. But if he set his transparent sphere, with its angular grid, in rotation (relative to its initial state – absolute rotation has not yet been defined), one would say that he was then using a frame of reference that was rotating relative to the initial one. Viewed as coordinate transformations, the difference between these two cases is that in the latter case the transformation of the spatial coordinates is time-dependent, while in the former case it is not. If this is taken as a characterization of those coordinate transformations which are not regarded as changing the corresponding reference frame, then what is left as belonging specifically to the reference frame is just the observer together with the definitions of rest and simultaneity.

This enables us to talk meaningfully about space and time separately in a given reference frame, but that is about all it does allow. Too much has been removed, and what is left is of little use. It would be preferable to leave some structure in the reference frame which is unaffected by a change from rectangular to polar coordinates, but which, say, makes it meaningful to talk about uniform motion in a straight line. This may be achieved by giving a suitable geometric structure to space and, more trivially, also to time. This does not involve any new assumption about the physical world, as no 'reality' will be attributed to the geometry. It is just a step in the construction of a language with which to discuss physical phenomena. One possible geometric structure for time is provided by any arbitrarily constructed clock, or equivalently by the time coordinate of any spacetime coordinate system. Two time intervals are simply *defined* to be equal when such a clock measures them as equal. The clock also gives a unit of time, which when taken together with this definition of equality of interval gives time the metric structure of the real line.

An equally arbitrary construction will be used for the spatial geometry. If a suitable spacetime coordinate system is given, each of the three-dimensional spaces of constant time can be considered as having that three-dimensional Euclidean geometry in which the given spatial coordinates are rectangular Cartesian. These coordinates also provide a unit of length for this geometry. The description 'suitable' is intended to allow for the possibility that some coordinate systems may be better interpreted as, say, spherical polar than rectangular Cartesian. To include these in the procedure, they should first be transformed to a corresponding rectangular system before the geometry is abstracted.

The frame of reference associated with the coordinate system will be taken as consisting of (a) the definitions of rest and of simultaneity which are used to separate space and time, (b) the corresponding metric structure for time, which includes a unit of time, and (c) the corresponding three-dimensional Euclidean geometry for space, together with its unit of length. For conformity with our allowing coordinate systems covering only a portion of spacetime, frames of reference must similarly be allowed in which these geometric structures also only cover portions of space and time. The initial step discussed above is now complete, for all the language permitted by this rich structure can now be used unambiguously, while the underlying assumptions as to the nature of space and time have been made explicit.

The above construction, which proceeds from a coordinate system to a reference frame, raises the question of the extent to which such a frame determines the coordinate system from which its geometric structure was abstracted. Let us say that (x, y, z, t) are *natural* coordinates for a frame if:

(i)   Simultaneity of two events corresponds to equality of t,

(ii) The state of rest corresponds to constancy of (x, y, z),

(iii) (x, y, z) are rectangular Cartesian coordinates in space, which agree with the length unit of the frame, and

(iv) t measures time consistently with the metric structure given by the frame.

Then the original coordinate system is a natural one, but the frame alone does not distinguish it from any other natural coordinate system. It is convenient to write the coordinates (x, y, z) as a (3×1) column vector $\mathbf{x}$. We see that a second coordinate system $(\mathbf{x}^*, t^*)$ is natural for some frame if and only if it is related to a given natural coordinate system $(\mathbf{x}, t)$ for that frame by a transformation of the form

$$\mathbf{x}^* = A\mathbf{x} + \mathbf{a}, \quad t^* = t + k, \tag{2.1}$$

where A is a (3×3) orthogonal matrix, $\mathbf{a}$ is a (3×1) column vector and k is a scalar. For future reference it should be emphasized that A, $\mathbf{a}$ and k are constant, and thus in particular independent of time. Geometrically, $\mathbf{a}$ and k represent a change of origin in both space and time, while A describes a rotation or reflection of the spatial coordinate axes.

3 Newtonian conceptions

Let us now consider in more detail the assumptions underlying Newtonian dynamics, as only by so doing can we fully appreciate the origins of the special theory of relativity. As has already been remarked,
underlying all Newtonian thought is the concept of absolute time.
This comprises more than a belief in the meaning of absolute simultaneity for spatially separated events. It also implies a metric structure for time, so that the equality of two time intervals is also a primitive undefined concept. Once absolute simultaneity is assumed,
'space' becomes absolute in the sense of being the same for all observers. Based on the idealization of the perfectly rigid rod as a measure
of distance, Newtonian physics also implicitly assumes that the
geometry of space as surveyed with such rods is Euclidean.

For the time being, let us not question these assumptions, but
instead investigate the dynamical laws based on them. It is then
possible to restrict attention to those frames of reference which are
compatible with these natural geometries for both space and time, and
with fixed but arbitrary units of both length and time. Such frames
will be said to be *allowable*. It is easily seen that the relation between
the natural coordinates of any two such allowable frames must have
the form
$$\mathbf{x}^* = A(t)\,\mathbf{x} + \mathbf{a}(t), \quad t^* = t + k, \tag{3.1}$$
where again k is constant, but now the orthogonal (3×3) matrix A
and the (3×1) column vector $\mathbf{a}$ may be time-dependent. Conversely,
if two frames have natural coordinates which are so related, and if one
of them is allowable, then so is the other.

It is here that the explicit development of Newtonian dynamics
starts. Its first step is to pick out a subset of the allowable frames which
are dynamically privileged, by means of the *Principle of Inertia*,
otherwise known as Newton's First Law of Motion. This states that:
*There exists a family of reference frames in which any particle will
continue in its state of rest or of uniform motion in a straight line unless it
be compelled by some external force to change that state.*
Such frames are said to be *inertial*.

Suppose now that (3.1) connects two inertial frames. Then the state
of uniform motion $\mathbf{x}(t) = \mathbf{u}t + \mathbf{c}$ must correspond to uniform motion
in the second frame for all values of $\mathbf{u}$ and $\mathbf{c}$. But it follows from (3.1)
that
$$\frac{d^2\mathbf{x}^*}{dt^{*2}} = \frac{d^2A}{dt^2}\,(\mathbf{u}t + \mathbf{c}) + 2\,\frac{dA}{dt}\,\mathbf{u} + \frac{d^2\mathbf{a}}{dt^2},$$

which must vanish for all **u** and **c**. Hence A must be constant and $\mathbf{a}(t)$ must have the form $\mathbf{a}(t) = \mathbf{v}t + \mathbf{b}$ for some constant column vectors **v** and **b**. The most general transformation between the natural coordinates of two inertial frames thus has the form

$$\mathbf{x}^* = A\mathbf{x} + \mathbf{v}t + \mathbf{b}, \quad t^* = t + k, \qquad (3.2)$$

where now A is a (3×3) constant orthogonal matrix, **v** and **b** are constant column vectors and k is a constant scalar. This is called a *Galilean transformation*. It differs from (2.1) only through the extra term $\mathbf{v}t$ in $\mathbf{x}^*$, so that the two frames are in uniform relative motion.

Although this gives the relation between two inertial frames, it does not give an absolute process for the recognition of an individual inertial frame. This needs a method by which one can decide whether or not an external force is acting on a given particle. This involves an examination of our intuitive notions of force, since to avoid a circular argument in which force and inertial frames are defined in terms of one another, force must remain a primitive undefined concept.

The obvious first step towards removing all forces from a particle is to remove it from direct physical contact with other bodies. But this is only the beginning. Somehow one must decide whether any action-at-a-distance forces, e.g. electromagnetic forces, are acting. To do this, suppose that an attempt is made to influence the motion of the particle from a distance by altering its surroundings. If it succeeds, the particle is susceptible to the type of field being generated. It is then necessary either (i) to screen the particle from such fields, the effectiveness of such a screen being tested by whether or not the particle can still be influenced when it is inside the screen, or (ii) to remove the particle to a great distance from any source of the field, or (iii) to use instead a different particle which is unaffected by this type of field. There may not be a free choice of these alternatives. There is in the case of electromagnetism, but for gravitation only the second is possible as there is no known gravitational screen and neither do particles exist which are unaffected by a gravitational field. This is in fact a feature peculiar to gravitation. It will later be seen to have profound consequences for any theory of gravity.

If either the first or third alternatives above is adopted, then a test of the reference frame for inertiality can be performed anywhere, but if the second alternative is followed, the test must be carried out in a field-free region. For gravitation this means at a great distance from any other material body, which can best be achieved in intergalactic space. Although impractical, this is in principle possible, so the circular argument mentioned above has been successfully avoided. However, in practice it means that any test of a frame for inertiality, and so also any test of the principle itself (which is equivalent to the existence of at least one such frame), must rest on less direct evidence based on the whole theory which is built on the principle together with other basic assumptions.

These other assumptions are Newton's Second and Third Laws of Motion, in either their original or some equivalent form. They will be considered in some detail in Chapter 3, but their content is not important for present purposes. The point of immediate significance is that they are all invariant under a Galilean transformation, so that if they hold in one inertial frame then they hold in all such frames. Consequently, the same is true for the whole dynamical theory which is built upon them. It is thus impossible, by any dynamical experiment, to distinguish between the various inertial frames. This naturally raises the question of whether or not it is possible to distinguish between them by any other type of physical experiment. With the development of electromagnetic theory in the second half of the nineteenth century, it appeared that the answer was 'yes', since Maxwell's equations governing the electromagnetic field are *not* invariant under Galilean transformations. In particular, they predict that the speed of light (in fact, of all electromagnetic radiation) is an absolute constant, and so is independent of both the motion of the source and the direction of propagation. It is clear that this cannot hold in more than one Newtonian inertial frame, and hence that frame in which Maxwell's equations hold is uniquely distinguished.

At that time, this seemed a perfectly acceptable state of affairs. Electromagnetism was assumed to be propagated as waves in an all-pervading medium known as the aether, and hence that inertial frame which is at rest relative to the aether would be expected to be experimentally distinguishable. So, in 1887, Michelson and Morley performed an experiment which, according to these theories, would measure the velocity of the Earth through the aether. They obtained a zero result. Although it is conceiveable that at some instant in its orbit around the Sun, the Earth is at rest relative to the aether, it is inconceiveable that it remains so throughout its orbit. This was the first direct evidence that something was wrong with the theory either of dynamics or of electromagnetism.

Einstein's great achievement in setting up the special theory of

relativity was to have the courage to question not the explicit laws of
these two theories, but instead of preconceptions of space and time
that underlie them. If these are examined, it is seen that they rely
essentially on the existence, at least as idealizations that can be
approached arbitrarily closely, of perfect clocks and perfectly rigid
rods. To merit the description 'perfect', a clock must be:

(i) Reproducible, so that identical copies of it can be made and
carried throughout space, and

(ii) Consistent, in that if two such copies are brought together and
synchronized at some time, then they will still be synchronized
whenever they are brought together in the future.

If suitably calibrated, such a clock will read 'absolute time' wherever
it is placed and whatever its state of motion may be. The perfect
measuring rod must be similarly reproducible and consistent. If one
grants the existence of such rods and clocks, then verification of the
Newtonian conceptions of space and time only consists of checking
that the geometry of space surveyed with the rigid rods is Euclidean.

Relativity theory begins by questioning the reasonableness of these
idealizations of everyday rods and clocks. Absolute rigidity implies an
infinite velocity of sound, and no known material has a sound velocity
even remotely approaching the velocity of light. It is thus too much of
an abstraction from experience to assume that in practice, absolute
rigidity can be approached arbitrarily closely, for this would assume
that the sound velocity can be made arbitrarily large by a suitable
choice of material Similarly, if clocks have to be constructed out of
such imperfect materials, they cannot be expected to be good time-
keepers under all conditions. Nevertheless, science is based on idealization and abstraction from experience. We must thus seek a more
reasonable idealization on which to base a new theory.

4 Foundations of the special theory of relativity

The materials that are commonly called rigid in everyday experience
are in reality elastic solids. They are deformable under stress but they
return to their original state, which will be called their ground state,
when the stress is removed and any induced vibrations have died
away. This suggests that an acceptable alternative to the use of rigid
measuring rods would be to use elastic ones, with the proviso that they
only be used when in their ground state. But before this can be adopted,
some operational procedure must be given by which this state can be

distinguished. A method that ensures the absence of any external forces has already been discussed, but this alone is not sufficient even if it is supposed that the body has settled down into a steady state. A body can be stressed by being set in rotation, for example, and although it will settle down to a steady state, this will not be the unstressed ground state. This type of stress can be excluded if it is additionally required that each particle of the body should move along a path which is a possible path for a freely moving particle. With this interpretation, we now adopt the use of standard measuring rods of elastic material as discussed above. It is implicitly assumed both that elastic solids exist satisfying the above criteria for the absence of stress, and that such rods satisfy the same criteria of reproducibility and consistency laid down for rigid rods in the preceding section.

Note that although the above criteria are sufficient to ensure that an elastic body is in its ground state, they are not also necessary conditions. It is possible, by exerting precisely the right forces on each particle of a body, to accelerate it while keeping it throughout in its ground state. If it could observationally be ensured that this was being done, one would be part of the way back to the Newtonian rigid body. But as there seems no way of ensuring this, it will not be pursued further here.

It is also necessary to construct standard clocks of elastic material, and again only to use them in an unstressed state. It is difficult to give a precise interpretation of this for clocks, since by their very nature clocks contain moving parts, and any mechanical clock relies on the periodic oscillations of a stressed component. However, there is no difficulty in principle in defining a standard state of motion for the essential moving parts, and so the interpretation will be left to a judicious application of common sense.

Let us now consider the consequences of this relaxation of the Newtonian postulates. The most important is that the concept of absolute time must be given up. It is natural still to require the standard clocks to be reproducible and consistent, but the consistency can now only be taken as a requirement that they run at the same rate when brought together again after synchronization at some earlier time. For in the process of separating them and then bringing them back together again, at least one of them will have been subjected to stress and so may have ceased to run properly. When it is again brought back to equilibrium, free from stress, it may thus have gained or lost in comparison with a permanently unstressed clock. With the abandonment of absolute time, the conception of space as something independent of the motion of the observer is also lost. This apparently simple change away from the conception of rigidity thus has far-reaching consequences, and so there lies ahead a major process of reconstruction.

The first step, following the development of the Newtonian theory, is to set up a preferred class of inertial frames. But here again the replacement of rigidity by elasticity causes a fundamental difficulty right at the outset. In the Newtonian theory, once an inertial frame is set up in some region of space with the aid of free particles, it can be extended throughout the whole of space if rigid rods are used to extend the coordinate system and perfect clocks are transported to give the time standard. It was seen above that this procedure is actually essential if one wishes to set up an inertial frame in a gravitational field, as such a field affects all particles and hence none exist which can be used to test the inertiality of a frame directly. It is thus necessary to test the frame at such great distances from all large masses that the gravitational field is negligible, and then to extend the frame into the field in the above way. The same procedure is not needed for the electromagnetic field due to the existence of uncharged particles.

It seems that it must be necessary to adopt some such procedure for the gravitational field also in the new theory that we are seeking. This time the extension will have to be made using elastic rods in their ground state. But this too is impossible, as the test described above of their freedom from stress also uses free particles! A little thought shows that this cannot be overcome by thinking up an alternative test for freedom from stress, since any body which extends throughout a region of varying gravitational acceleration will necessarily have gravitationally induced stresses regardless of its state of motion. For the moment these additional complications will be avoided by supposing that we are dealing with a region free from gravitation. This will lead to the special theory of relativity. We shall later return to develop the above train of thought, which leads to the general theory.

To enable inertial frames to be set up in the new theory, it is necessary first to extend the Principle of Inertia, as it now has a bigger job to do. In the Newtonian case a large initial restriction could be made on the allowed class of frames simply due to the underlying hypotheses of absolute time and of an absolute space with a Euclidean geometry. The Principle of Inertia then only had to select the inertial frames from this class. But now there is nothing that can be done in advance, so the

inertial frames have to be selected from the whole class of possible reference frames described in §2. The following postulate will thus be adopted, which will be called the *Extended Principle of Inertia*: *In the absence of gravitation, there exist frames of reference in which the motion of any elastic body that is in its ground state and free from external forces is such that the particles of the body move uniformly along parallel lines, each with the same velocity. In such a frame, all such motions are so realizable, and the geometry of space as surveyed with stress-free measuring rods at rest in the frame is Euclidean. By a change of the criterion for distant simultaneity, but without alteration of the rest state or of the metric structure of time, it is possible additionally to ensure that any clock which is stress-free and in uniform motion will run at a rate, as compared with the standard time of the frame, which depends only on its speed and not on its position or its direction of motion.*
The inertial frames of the new theory are those which have both of these properties.

The first part of this principle extends to unstressed elastic bodies the same assumptions as were made for particles in the Newtonian principle. In a sense it is a converse of the criterion given above for the recognition of such bodies. For that implied that the particles of any elastic body which is in an unstressed state and free from external forces follow paths that are possible paths for free particles. It is now being assumed conversely that any path of a free particle is also a possible path for a particle of a free and unstressed body.

Consider now how to set up an inertial frame in which a given free particle will appear at rest. Take a free elastic body in its ground state, one of whose particles has the motion of the given free particle. Define the rest state of the frame as that of the particles of the body, and the metric structure of time as that given by identical clocks distributed at rest throughout the frame. Then, as was seen in §2, the construction of a frame of reference needs for its completion a definition of simultaneity and a Euclidean geometry for space. The latter is easily provided, for the second sentence of the above principle ensures that whatever definition of simultaneity is adopted, the geometry of space as surveyed with unstressed rods at rest will necessarily be Euclidean. This leaves only the definition of simultaneity, which requires an operational procedure for the synchronization of clocks at different parts of the frame. A restriction on clock synchronization is provided by the requirement that freely moving particles should move with constant speed. However, this is not sufficient as the linear time transformation $t \to t + ax + by + cz$ preserves the whole of the first two sentences of the principle. Unique synchronization can only be achieved by some additional assumption such as that contained in the third sentence.

An alternative would be to assert that the clocks can be so synchronized that space appears isotropic and homogeneous, from which the specific assumption made above follows trivially. This will indeed be asserted later on, but it is a far greater assumption than is actually necessary for present purposes. It seems preferable first to make a relatively weak assumption that enables the required synchronization to be uniquely determined, and then later to assert more specifically that this isotropy and homogeneity holds with respect to the resulting inertial frames. It will be seen below that even this relatively weak assumption has consequences that extend far beyond the basic reason for its introduction.

To complete the construction of an inertial frame, suppose that we wish to synchronize clocks at rest at points A and B. Take a third clock and set it in uniform motion from A to B. When it reaches B, send it back to A, again uniformly, so that by its own measure of time its speed is the same in each direction. If the clock at A reads t_1 when the moving clock leaves for B, and t_2 when it arrives back again, then the clock at B must be set so that it reads $\frac{1}{2}(t_1 + t_2)$ at the instant of reflection. The infinite acceleration that is apparently required for the moving clock can be avoided if instead two identical clocks are used.

This shows that given any freely moving particle, there exists an inertial frame in which that particle is at rest, and that this frame is unique to within a change of the units of length and time. This is exactly the same as the situation in Newtonian dynamics. The next task is to obtain the transformation law connecting the natural coordinate systems of different inertial frames. This will be the equivalent in the new theory of the Galilean transformation (3.2) of Newtonian theory.

5   The transformation between inertial frames

Let (x, y, z, t) and (x^*, y^*, z^*, t^*) be natural coordinate systems of two inertial frames K and K^* respectively, and as in (2.1) write the spatial coordinates as column vectors $\mathbf{x}$ and $\mathbf{x}^*$. The most general transformation between K and K^* which preserves uniform motion in a straight line can be shown to be a projectivity. This has the form

$$\mathbf{x}^* = \frac{A\mathbf{x}+\mathbf{a}t+\mathbf{b}}{\mathbf{p}'\mathbf{x}+qt+r}, \quad t^* = \frac{\mathbf{k}'\mathbf{x}+lt+m}{\mathbf{p}'\mathbf{x}+qt+r}, \tag{5.1}$$

where the nonsingular (3×3) matrix A, the (3×1) column vectors $\mathbf{a}, \mathbf{b}, \mathbf{k}, \mathbf{p}$, and the scalars l, m, q, r are constant but otherwise arbitrary. A prime denotes the matrix transpose, so that $\mathbf{k}'\mathbf{x}$ for example is the scalar product of the vectors $\mathbf{k}$ and $\mathbf{x}$. It is easy to verify that (5.1) has the desired property. The proof that it is the most general such transformation is more difficult, and will be omitted.

Had we adopted the Newtonian form of the Principle of Inertia, we could go no further. The part of the extended principle concerned with elastic bodies additionally implies, however, that if two particles have parallel motions with the same velocity in one frame, then they also have in the other frame. Now the motion given by $\mathbf{x}^* = \mathbf{u}t^* + \mathbf{c}$, for constant $\mathbf{u}$ and $\mathbf{c}$, corresponds to

$$(A - \mathbf{u}\mathbf{k}')\mathbf{x} + (\mathbf{a} - l\mathbf{u})t + (\mathbf{b} - m\mathbf{u}) = \mathbf{c}(\mathbf{p}'\mathbf{x} + qt + r).$$

If $\mathbf{c}$ is varied for fixed $\mathbf{u}$, the corresponding motions in K^* are parallel and have the same velocity. However, unless $\mathbf{p} = \mathbf{0}$ and $q = 0$, it may be seen that in K they all intersect the hyperplane $\mathbf{p}'\mathbf{x} + qt + r = 0$ of the four-dimensional $(\mathbf{x}, t)$ space in the same point, except for those special values of $\mathbf{u}$ for which they never meet it. The required condition can thus only hold for all $\mathbf{u}$ if $\mathbf{p} = \mathbf{0}$ and $q = 0$, in which case r may be taken as unity without loss of generality.

This restricts the transformation to the linear form

$$\mathbf{x}^* = A\mathbf{x} + \mathbf{a}t + \mathbf{b}, \quad t^* = \mathbf{k}'\mathbf{x} + lt + m. \tag{5.2}$$

Note that this restriction could also have been made if transformations with degenerate points had been rejected, for if either $\mathbf{p}$ or q is nonzero, (5.1) is degenerate where $\mathbf{p}'\mathbf{x} + qt + r = 0$. However, it was seen in §2 that coordinates with degenerate points are not necessarily unacceptable, and so it seems more satisfactory to seek a physical reason for rejecting them. By a change of origin of either coordinate system, both $\mathbf{b}$ and m can be reduced to zero, which further simplifies (5.2) to

$$\mathbf{x}^* = A\mathbf{x} + \mathbf{a}t, \quad t^* = \mathbf{k}'\mathbf{x} + lt. \tag{5.3}$$

At this stage a little more simplification could be produced if use were made of the freedom to rotate or reflect the two sets of axes. Instead, it is actually more convenient next to invoke the hypothesis about moving clocks. Consider a clock moving along a straight line through the origin of K with constant but arbitrary velocity $\mathbf{u}$, such that it passes through the origin at $t = 0$ and itself reads zero at that instant. If its reading at a later instant is τ, then the hypothesis applied to the frame K is that t/τ depends only on u^2, where $u = |\mathbf{u}|$ denotes the magnitude of the vector $\mathbf{u}$. Let this ratio be $f(u^2)$, so that

$$\mathbf{x} = \mathbf{u}t \quad \text{and} \quad t = \tau f(u^2). \tag{5.4}$$

Then it follows from (5.3) and (5.4) that in K^*

$$\mathbf{x}^* = \mathbf{u}^* t^* \quad \text{and} \quad t^* = (\mathbf{k}'\mathbf{u} + l)f(u^2)\tau, \tag{5.5}$$

where

$$\mathbf{u}^* = (\mathbf{k}'\mathbf{u} + l)^{-1}(A\mathbf{u} + \mathbf{a}). \tag{5.6}$$

But the same hypothesis applied to the frame K^* implies that t^*/τ depends only on u^{*2} as $\mathbf{u}$ varies. If this ratio is denoted by $g(u^{*2})$, it must thus satisfy

$$(\mathbf{k}'\mathbf{u} + l)f(u^2) = g(u^{*2}). \tag{5.7}$$

Now let $\mathbf{u}_1$, $\mathbf{u}_2$, $\mathbf{u}_3$ be three mutually orthogonal vectors in K with $\mathbf{u}_1$ parallel to $\mathbf{k}$ and with $\mathbf{u}_2$, $\mathbf{u}_3$ of equal magnitude. If the $\mathbf{u}$ of (5.4) is taken to be the function of θ defined by

$$\mathbf{u}(\theta) = \mathbf{u}_1 + \mathbf{u}_2 \cos\theta + \mathbf{u}_3 \sin\theta, \tag{5.8}$$

then the identity

$$(\mathbf{k}'\mathbf{u} + l)f(u^2) = (\mathbf{k}'\mathbf{u}_1 + l)f(u_1^2 + u_2^2) \tag{5.9}$$

shows that the left-hand side of (5.7) is independent of θ. But it follows from (5.6) that

$$u^{*2} = \tfrac{1}{2}(\mathbf{k}'\mathbf{u}_1 + l)^{-2}\{2(A\mathbf{u}_1 + \mathbf{a})^2 + (A\mathbf{u}_2)^2 + (A\mathbf{u}_3)^2 + 4(A\mathbf{u}_1 + \mathbf{a})'$$

$$\times (A\mathbf{u}_2 \cos\theta + A\mathbf{u}_3 \sin\theta) + [(A\mathbf{u}_2)^2 - (A\mathbf{u}_3)^2]\cos 2\theta$$

$$+ 2(A\mathbf{u}_2)'(A\mathbf{u}_3)\sin 2\theta\}. \tag{5.10}$$

Equation (5.7) can thus hold for all θ only if either g is constant or if u^{*2} is independent of θ. These two possibilities will be treated separately.

If g is constant, it follows from (5.7) that f is also constant and $\mathbf{k} = \mathbf{0}$. If both frames use identical clocks, and the clock whose motion is being studied is also identical with these, then $t = \tau$ when $\mathbf{u} = \mathbf{0}$ and $t^* = \tau$ when $\mathbf{u}^* = \mathbf{0}$. Constancy of f and g thus implies $f = g = l = 1$, and so $t = t^*$ from (5.3). Time is thus absolute, as in the Newtonian theory. Since $\mathbf{x} = \mathbf{0}$ now corresponds to $\mathbf{x}^* = \mathbf{a}t^*$, $\mathbf{a}$ is the velocity of the frame

K with respect to the frame K^*. If it could also be deduced that A must be orthogonal, the transformation would be Galilean and we would be back with the Newtonian situation. However, although the frames are now physically determinate according to the construction process given above, this orthogonality cannot be deduced without further knowledge of the behaviour of elastic solids in uniform motion. We return to this point in the next section. It is worth noting here, however, that this determinancy can now be confirmed mathematically, i.e. that an inertial frame is uniquely determined by the requirement that a single specified free particle be at rest. For the same particle can only be at rest in both the above frames if $\mathbf{a} = \mathbf{0}$. The rest states of both frames thus agree everywhere, and so in particular a measuring rod at rest in one frame is at rest in both. The transformation $\mathbf{x}^* = A\mathbf{x}$ must thus preserve distance, so that A is orthogonal, and (5.3) reduces to the transformation between two sets of natural coordinates for the same frame. Hence the two frames themselves are identical.

The other possibility is that u^{*2} is independent of θ. In this case the coefficients of $\cos\theta$, $\sin\theta$, $\cos 2\theta$ and $\sin 2\theta$ in (5.10) must vanish separately. This requires $A\mathbf{u}_1 + \mathbf{a}$, $A\mathbf{u}_2$ and $A\mathbf{u}_3$ to be mutually orthogonal, and $|A\mathbf{u}_2| = |A\mathbf{u}_3|$. Now $\mathbf{u}_1$ is constrained to be parallel to $\mathbf{k}$, but its length may be changed while $\mathbf{u}_2$ and $\mathbf{u}_3$ are kept fixed. As this change must preserve the orthogonality, $A\mathbf{u}_1$ must be parallel to $\mathbf{a}$.

With this additional information, the freedom to rotate or reflect the coordinate axes can now be used to its best advantage. If these transformations are given by $\mathbf{x}^* \to R\mathbf{x}^*$ and $\mathbf{x} \to S\mathbf{x}$, where R and S are (3×3) orthogonal matrices, the induced transformations of the coefficients occurring in (5.3) are

$$A \to RAS', \quad \mathbf{a} \to R\mathbf{a}, \quad \mathbf{k} \to S\mathbf{k}, \quad l \to l. \qquad (5.11)$$

The y and z components of both $\mathbf{a}$ and $\mathbf{k}$ may thus be made to vanish, while still leaving the freedom to make a further transformation in which the directions of the x-axes of both coordinate systems are kept fixed. If $\mathbf{u}_1$, $\mathbf{u}_2$ and $\mathbf{u}_3$ are taken to be the columns of the identity matrix, it now follows from the above conditions that A has the form

$$A = \begin{pmatrix} \alpha & 0 & 0 \\ 0 & \beta\cos\phi & \beta\sin\phi \\ 0 & -\epsilon\beta\sin\phi & \epsilon\beta\cos\phi \end{pmatrix}, \qquad (5.12)$$

where $\epsilon = \pm 1$, for some α, β and ϕ with $\beta > 0$. By a further rotation of either set of axes about its x-axis, together with a possible reversal of direction of the x- and/or z-axes, it is possible to make $\alpha > 0$, $\epsilon = 1$ and $\phi = 0$. The axes are then determined up to a simultaneous rotation of both sets through the same angle about their respective x-axes. If $\mathbf{a} = (a, 0, 0)'$ and $\mathbf{k} = (k, 0, 0)'$, (5.3) now takes the simple form

$$x^* = \alpha x + at, \quad y^* = \beta y, \quad z^* = \beta z, \quad t^* = kx + lt. \qquad (5.13)$$

Let us next return to equation (5.7). If $\mathbf{u}$ is taken as $(v, w, 0)'$, this gives

$$f(v^2 + w^2) = (kv + l)^{-1} g \left(\frac{(\alpha v + a)^2 + \beta^2 w^2}{(kv + l)^2} \right). \qquad (5.14)$$

Since the left-hand side is an even function of v, so also must be the right-hand side. Its partial derivative with respect to v thus vanishes at $v = 0$. If this is evaluated and the remaining variable in the result is changed from w to $x = (a^2 + \beta^2 w^2)/l^2$, we find that

$$2(A - x) \frac{dg}{dx} - g(x) = 0 \qquad (5.15)$$

for all $x \geqslant a^2/l^2$, where $A = a\alpha/kl$. This may be integrated to give

$$g(x) = B |A - x|^{-\frac{1}{2}} \qquad (5.16)$$

where B is a constant of integration. Due to the singularity which occurs at $x = A$, (5.16) holds for all $x \geqslant a^2/l^2$ if $A \leqslant a^2/l^2$, but only for $a^2/l^2 \leqslant x < A$ if $A > a^2/l^2$. In either case, there will be some open region of the (v, w) plane for which the argument of g in (5.14) lies in the range of validity of (5.16), and in this region the right-hand side of (5.14) must depend on v and w only in the combination $v^2 + w^2$. Substitution from (5.16) shows that this is so if and only if

$$A - a^2/l^2 = \beta^2 A/\alpha^2. \qquad (5.17)$$

The case $A \leqslant a^2/l^2$ is thus only possible if $A < 0$, when it necessarily does occur.

When $v = 0$ and (5.16) is valid, (5.14) reduces to

$$f(w^2) = B\beta^{-1} |C - w^2|^{-\frac{1}{2}} \operatorname{sgn} l, \qquad (5.18)$$

where

$$C = al/\alpha k = l^2 A/\alpha^2 \qquad (5.19)$$

and

$$\operatorname{sgn} x = \begin{cases} 1 & \text{if } x > 0 \\ 0 & \text{if } x = 0 \\ -1 & \text{if } x < 0. \end{cases} \qquad (5.20)$$

It follows from (5.19) that A and C have the same sign, which may be used together with the above ranges of validity for (5.16) to deduce that (5.18) holds for all w if $C < 0$, and for $w^2 < C$ if $C > 0$. An application of (5.14) to the special case $v = -a/\alpha$ now enables the range of validity of (5.16) to be extended. It can be seen that this equation must hold for all x when $A < 0$, and for $0 \leqslant x < A$ when $A > 0$. With this final deduction, all the information contained in (5.14) has been extracted. For that equation is identically satisfied with the above expressions for f and g, and both sides are defined for the same range of v and w, namely $v^2 + w^2 < C$ if $C > 0$, and for all v and w if $C < 0$.

So far, no connexion has been made between the time scales of the two frames. One can now be made if it is assumed, as in the previous case, that the moving clock and the standard clocks of both frames are all identical. Then $f(0) = g(0) = 1$, which gives

$$B = |A|^{\frac{1}{2}}, \quad \beta^2 = A/C, \quad l > 0. \tag{5.21}$$

As yet we are not in a position to be able to use theoretically an assumption that the standard measuring rods of the two frames are identical, so for the present the length scales of the frames will still be allowed to be independent.

As measured by K, the motion of the frame K^* is along the x-axis with speed $v = -a/\alpha$. This follows from (5.13) if x^* is set equal to zero. Let us now define v as this constant value. There should be no confusion with the variable v of (5.14), as that equation will not be referred to again. Then the above results may be combined together to give

$$\begin{aligned} f(u^2) &= (1 - u^2/C)^{-\frac{1}{2}} \quad (\text{with} \quad u^2 < C \quad \text{if} \quad C > 0), \\ g(u^2) &= (1 - u^2/A)^{-\frac{1}{2}} \quad (\text{with} \quad u^2 < A \quad \text{if} \quad A > 0), \\ A &= \beta^2 C, \end{aligned} \right\} \tag{5.22}$$

and
$$\left. \begin{aligned} x^* &= \beta(x - vt)(1 - v^2/C)^{-\frac{1}{2}}, \quad y^* = \beta y, \\ z^* &= \beta z, \quad t^* = (t - vx/C)(1 - v^2/C)^{-\frac{1}{2}}. \end{aligned} \right\} \tag{5.23}$$

Since f and g are defined in terms of the behaviour of moving clocks in the frames K and K^* respectively, it follows from (5.22) that the constants C and A may be determined by physical measurements made solely in the respective frames. If the particular frame K^* under consideration is changed while K is kept fixed, the constant C in the transformation (5.23) will thus remain unaltered. The value of v may be found by measurements in K of the motion of K^*, and is needed to specify the relationship between these two frames. Only the value of β thus remains unknown.

If the length scale in K^* is changed, C and v will remain constant but

β will change. It would thus be possible to define the length scales of the two frames to be equal when $\beta = 1$, in which case $A = C$ by (5.21). This seems a very natural choice, but it would leave open the question of whether a given measuring rod has the same length when at rest in each frame, when compared with the length units thus defined. The procedure used above has been to define length in terms of a standard rod. If this is to be retained, equality of the length units of the two frames must be defined by the use of identical standard rods in these frames. The value of β must then be determined by an investigation of the behaviour of moving rods.

6   The uniformity of space and time

This behaviour can be deduced from the *Principle of Uniformity*, which is a formalization of the hypothesis discussed above that space and time appear isotropic and homogeneous when viewed in an inertial reference frame. But the simplicity of this statement is misleading, and some clarification is needed before it can be used correctly. It essentially denies the existence of any features that distinguish one position in space and time from any other (homogeneity), or that distinguish, at any point, one direction in space from any other (isotropy). Its physical interpretation is that a given experiment will produce the same result wherever and whenever it is performed, and whatever the orientation of the apparatus, provided that the circumstances of the experiment are identical in all other respects.

If it is treated carefully, a great deal of information can be extracted from this extremely simple postulate. But one must be careful in deciding just what is meant by the 'circumstances of the experiment'. Two replications of an experiment can be considered as having the same circumstances if there exist two natural coordinate systems for the same inertial reference frame, one of which may be associated with each experiment in such a way that the initial conditions of the two are identical when each is referred to its associated coordinate system. Since natural coordinates are determined up to a translation and rotation of the axes, this gives the required homogeneity and isotropy. It is a rather cumbersome description, but any description that is sufficient to eliminate two replicas of an experiment which are in uniform rotation relative to each other, which clearly must be eliminated, must be equivalent to this and equally cumbersome. Thus, although the physical interpretation given above apparently makes no mention of any reference frame, the limitation to inertial frames made in the original form of the postulate is inherent in the 'circumstances of the experiment'.

Consider now the consequences of asserting this for all inertial frames. Suppose two replications of an experiment are given whose initial conditions are identical when referred to appropriate coordinate systems, as before. But this time suppose that they are natural coordinates for two different inertial frames, K_1 and K_2 say, which use identical standard rods and clocks. Now there is no necessity for the experimental apparatus to be at rest in the appropriate frame, as its velocity comprises one of the 'circumstances' referred to. The apparatus itself thus does not pick out any particular inertial frame that must be used in these considerations. The question thus arises of whether a third inertial frame K_3 can be found in which both experiments have identical circumstances.

Let us suppose

(i) that a frame K_3 can be found such that K_1 and K_2 have the same speed relative to K_3, and

(ii) that the transformation law between a given inertial frame K and any other inertial frame K^* which has the same standards of length and time as K can, by a suitable choice of natural coordinate systems in the two frames, be put in a form which depends only on the speed of K^* relative to K.

Then there exist two pairs of coordinate systems, (C_1, C_3) and (C_2, C_3'), with C_1 natural for K_1, C_2 natural for K_2, and both C_3 and C_3' natural for K_3, such that the transformation laws connecting each pair are identical. By an application of the Principle of Uniformity in K_1 and K_2, it may be seen that nothing is lost if it is assumed that C_1 and C_2 are the systems with respect to which the two experiments appear identical. Then, simply by a mathematical identity, they must also appear identical when referred to C_3 and C_3' respectively. It now follows from a third application of the principle, this time in K_3, that the results of both experiments must be the same.

Property (i) is ensured by simple considerations of continuity. If an inertial frame is considered whose velocity is slowly varied from that of K_1 to that of K_2, it must necessarily pass through a state in which K_1 and K_2 have the same speed, regardless of what law of composition of velocities operates. It will be shown below that property (ii) is itself a consequence of the Principle of Uniformity. We have thus deduced a result which is frequently taken as a hypothesis in the foundation of the special theory of relativity. This is the *Principle of Special Relativity*:

The laws of physics are identical in all inertial frames of reference.

Let us pause here to consider the nature of the assumptions that were needed to make this deduction. Essentially, it has been assumed that there is no qualitative difference between different inertial frames, and it has been deduced that there is no quantitative difference. To illustrate this from the preceding section, if it had been assumed that all inertial frames are quantitatively identical, the functions f and g of (5.4) and (5.7) could have been taken to be the same. But instead, the *a priori* possibility of their differing was accepted, and it was deduced from (5.22) that they are actually equal provided only that $\beta = 1$. If the Michelson–Morley experiment is accepted as evidence for the impossibility of an experimental determination of a state of absolute rest, then one is led to conclude that the laws of physics have the same qualitative form in all inertial frames. But at first sight it is possible, say, that the actual numerical constants that occur in these laws differ from frame to frame. The above deduction of the Principle of Special Relativity shows that in fact this is not possible; qualitative equality implies quantitative equality.

We are now in a position to complete the derivation of the transformation between inertial frames. There are still two cases to consider, according to whether f and g are, or are not, constant. The second of these, which only requires a determination of β in (5.23), will be treated first. Consider an elastic rod in its ground state which is parallel to the z-axis of the frame K^* and which has constant velocity $\mathbf{u}^* = (u^* \cos\theta, u^* \sin\theta, 0)$ in that frame. For all values of u^* and θ, the direction of motion is perpendicular to the length of the rod, and hence by the Principle of Uniformity the apparent length of the rod can only depend on its speed $|\mathbf{u}^*|$ and not on θ. Let this length be $G(u^{*2})$. The transformation (5.23) shows that when viewed in K the rod is again parallel to the z-axis, is of apparent length $\beta^{-1} G(u^{*2})$, and has a uniform velocity $\mathbf{u}(\theta)$ perpendicular to its direction of motion which is given by

$$\mathbf{u} = (\beta + u^* v \cos\theta/C)^{-1}(u^* \cos\theta + \beta v, u^* \sin\theta (1 - v^2/C)^{\frac{1}{2}}, 0). \quad (6.1)$$

But the Principle of Uniformity may be used in K to show that its length can depend only on $|\mathbf{u}|$, say $F(u^2)$. Hence

$$F(u^2) = \beta^{-1} G(u^{*2}), \quad (6.2)$$

identically for all θ. It is easily seen that this can only hold if F and G

are both constant, so that the apparent length of a rod does not change when it is given a uniform velocity perpendicular to its length. If both frames use identical standard rods, then by definition $F(0) = G(0)$; the constancy of F and G then gives $\beta = 1$ by (6.2). The transformation (5.23) is now complete, and it satisfies the property (ii) above, as required. It now follows from (5.22) that $A = C$. The inverse transformation to (5.23) is thus simply obtained by interchanging the starred and unstarred coordinates and replacing v by $-v$. The relationship between the frames K and K^* is thus completely symmetrical, and the constant C is a fundamental constant of nature, having the same value in all inertial frames.

Let us now return to the other possibility, namely that f and g are both constant. It has already been seen that time is absolute in this case. We shall now show that all the other Newtonian preconceptions concerning space and time can also be recovered. The transformation (5.3) has so far been reduced to the form

$$\mathbf{x}^* = A\mathbf{x} + \mathbf{a}t, \quad t^* = t. \tag{6.3}$$

Consider an elastic sphere in its ground state, at rest in the frame K, whose surface satisfies $\mathbf{x}'\mathbf{x} = 1$. At $t = 0$, this appears in K^* as an ellipsoid with surface $\mathbf{x}^{*\prime}(AA')^{-1}\mathbf{x}^* = 1$ which has uniform velocity $\mathbf{a}$. But the isotropy of space requires that this ellipsoid be invariant under rotation about its direction of motion. Hence the eigenvectors of AA' are $\lambda\mathbf{a}$ for any scalar $\lambda \neq 0$, and any vector orthogonal to $\mathbf{a}$, and all the latter eigenvectors correspond to the same eigenvalue.

There still remains the freedom expressed by (5.11) to rotate both sets of coordinate axes. If R is again chosen to make the y- and z-components of $\mathbf{a}$ vanish, the above discussion shows that AA' must take the form

$$AA' = \Delta^2, \quad \Delta = \begin{pmatrix} \alpha & 0 & 0 \\ 0 & \beta & 0 \\ 0 & 0 & \beta \end{pmatrix}$$

for suitable $\alpha > 0$ and $\beta > 0$. Since this shows that $\Delta^{-1}A$ is orthogonal, it is possible to take $S = \Delta^{-1}A$ in (5.11), in which case the transformed A is simply Δ. The transformation (6.3) then reduces to

$$x^* = \alpha x + at, \quad y^* = \beta y, \quad z^* = \beta z, \quad t^* = t. \tag{6.4}$$

Consider now a spherical body which is moving uniformly along the z-axis of K^* with speed u^*. By the above, this will appear in K^* as a spheroid with the coordinate z-axis as its axis. Its surface must thus satisfy

$$\frac{x^{*2}}{\lambda^2} + \frac{y^{*2}}{\lambda^2} + \frac{(z^* - u^* t^*)^2}{\mu^2} = 1 \qquad (6.5)$$

for suitable constants λ and μ. As seen in K, its surface will be given by

$$\frac{\alpha^2}{\lambda^2} (x + at/\alpha)^2 + \frac{\beta^2}{\lambda^2} y^2 + \frac{\beta^2}{\mu^2} (z - u^* t/\beta)^2 = 1. \qquad (6.6)$$

This is an ellipsoid with semi-axes λ/α, λ/β and μ/β directed along the three coordinate axes. But again, since it is a uniformly moving sphere also in K, it must appear as a spheroid whose axis is parallel to the velocity vector, namely $\mathbf{u} = (-a/\alpha, 0, u^*/\beta)$. This is consistent with the preceding result only if the surface (6.6) is actually a sphere, so that $\alpha = \beta$, $\lambda = \mu$. A repetition of the corresponding discussion for the case of non-constant f and g now shows that if both frames use identical rods as standards of length, then $\beta = 1$. Hence, with the special choice of coordinate axes made above,

$$x^* = x + at, \quad y^* = y, \quad z^* = z, \quad t^* = t. \qquad (6.7)$$

The coordinate transformations that were performed to achieve this simple form were the change of origin needed to obtain (5.3) from (5.2), and the rotation or reflection of the coordinate axes given by (5.11). If the inverse transformations are applied to (6.7), it can be seen that the original coordinates must have been related by a Galilean transformation (3.2) and that all such Galilean transformations are possible.

7 The invariant measure of interval

In rebuilding the concept of an inertial reference frame without the use of absolutely rigid rods, we have obtained two distinct possible transformation laws connecting different inertial frames. One of them, the Galilean transformation (6.7), implies that an absolute standard of simultaneity exists upon which all inertial observers agree, and furthermore that the distance between two simultaneous events is also the same for all such observers. This distance can thus be considered as having an absolute meaning independent of any observer, and may be thought of as a property of space itself. The time interval between two events similarly acquires an absolute meaning.

The other transformation law is (5.23) with $\beta = 1$, which preserves neither simultaneity nor distance. However, if δx, δy, δz, and δt are the coordinate differences between two events in both space and time, then it follows from (5.23), together with the transformation (5.11) and the changes of origin of the coordinate systems which were needed to reduce the general case to this special form, that

$$\Phi \equiv (\delta x)^2 + (\delta y)^2 + (\delta z)^2 - C(\delta t)^2 \qquad (7.1)$$

is the same for all inertial observers. The quantity $\delta s = |\Phi|^{\frac{1}{2}}$ is known as the *invariant interval* between the events, and again, its invariance gives it an absolute meaning independent of any observer. It is, however, a property associated with the united spacetime continuum, rather than with space or time separately. This is the first sign of the emergence of the spacetime continuum as a significant entity with a geometric structure of its own.

By a suitable choice of the unit of time, it is possible to make $|C| = 1$. If $C < 0$, (7.1) then reduces to

$$\delta s^2 = \delta x^2 + \delta y^2 + \delta z^2 + \delta t^2, \qquad (7.2)$$

which is the expression for the distance δs between two points in a four-dimensional Euclidean space. In such a space, all coordinates are completely equivalent and any axis is transformable into any other by a rotation. The rotations which connect the x-, y- and z-axes are simply the ordinary rotations of the spatial axes in a given inertial frame, while those which involve the t-axis are physically realized as the transformations between different inertial frames. To see this, we simply have to put $v = \tan \phi$ in (5.23). We obtain

$$x^* = x \cos \phi - t \sin \phi, \quad t^* = x \sin \phi + t \cos \phi, \qquad (7.3)$$

which is the standard form for a rotation in the (x, t) plane through an angle ϕ. This shows up a feature that makes this possibility virtually unacceptable as a model of the physical world. It requires an infinite velocity to be physically realizable. This is because a rotation through $\pi/2$, corresponding to $v = \infty$, can be built up as a finite succession of small rotations, each of which will correspond to a transformation between two inertial frames with a finite relative velocity. The first and last frames of a chain built up in this way will have infinite relative velocity and be related by the transformation

$$x^* = -t, \quad y^* = y, \quad z^* = z, \quad t^* = x. \qquad (7.4)$$

Space and time are thus essentially interchangeable at will. This is so much at variance with our experience that it can be rejected as unacceptable, although it is difficult to deny that it is logically possible.

This leaves the case $C > 0$ as the only acceptable alternative to the Newtonian theory. The resulting theory is the *special theory of relativity*, which was deduced by Einstein in 1905 on different grounds. Let us revert to an arbitrary unit of time and put $C = c^2$, $c > 0$. The invariant quadratic form (7.1) is now indefinite, so that event pairs may be classified according as whether $\Phi > 0$, $\Phi = 0$ or $\Phi < 0$. If δr is the spatial distance between the events as measured by some observer, these three possibilities correspond to $\delta r > c|\delta t|$, $\delta r = c|\delta t|$, and $\delta r < c|\delta t|$, so that all observers agree on which of these holds. Hence although two inertial observers do not necessarily agree on the simultaneity of two separated events, if one observer sees them as simultaneous then all observers see $|\delta t| < \delta r/c$. No inertial observer can thus perceive them as being coincident in space ($\delta r = 0$). Special relativity thus preserves a distinction between space and time. Two events are said to have spacelike separation if $\Phi > 0$, timelike separation if $\Phi < 0$ and null separation if $\Phi = 0$. For spacelike separation, δs is called the proper distance between the events, while for timelike separation it is known as the proper time interval between them. These are respectively the distance, and c times the time, between the events as measured by an inertial observer to whom they appear simultaneous, or spatially coincident, as appropriate. Note that proper time has the dimensions of length. It would be possible to define proper time as $\delta s/c$, but with the relativistic unification of space and time into a single geometric structure, it is artificial to keep independent units for their absolute measures.

For null separation $\delta s = 0$, so that $\delta r/\delta t = c$ to all inertial observers. A signal that propagates with speed c according to one such observer thus does so also according to all others. Since Maxwell's equations of electromagnetism predict that electromagnetic radiation has a constant speed c, these two constants c must be identical if special relativity is to be consistent with Maxwell's equations in all inertial frames. With this identification, Maxwell's equations become invariant under the transformation (5.23) provided that the electric and magnetic fields are suitably transformed. This will be proved in Chapter 5, and it completes the connexion with the usual development of special relativity.

The invariant quadratic form (7.1) can now be put in the form in which it is normally written:

$$\Phi \equiv \delta x^2 + \delta y^2 + \delta z^2 - c^2 \delta t^2 \tag{7.5}$$

where c is the speed of light. The special form (5.23) of the transformation between inertial frames can also be given its usual form:

$$x^* = \gamma(x - vt), \quad y^* = y, \quad z^* = z, \quad t^* = \gamma(t - vx/c^2),$$
$$\text{where} \qquad \gamma = (1 - v^2/c^2)^{-\frac{1}{2}} \qquad\qquad (7.6)$$

and v is the relative speed of the frames. The general transformation connecting an arbitrary pair of natural coordinate systems associated with two different inertial frames can be found from (7.6) either in the way that (3.2) was recovered from (6.7), or by application to both the starred and unstarred coordinates of (7.6) of the transformation (2.1) between natural coordinates of the same frame. The resultant transformations are usually known as orthochronous *Lorentz transformations*, although sometimes this name is reserved for the subclass which preserves the origin $x = y = z = t = 0$. When this is done, the inhomogeneous transformations are known as orthochronous *Poincaré transformations*.

The adjective 'orthochronous' indicates that the distinction between past and future is preserved. It will be shown below that the most general transformation which preserves the quadratic form Φ of (7.5) is either an orthochronous Lorentz transformation, or is obtained from such a transformation by following it with the time-reversal transformation

$$x \to x, \quad y \to y, \quad z \to z, \quad t \to -t. \qquad (7.7)$$

The unqualified term Lorentz (or Poincaré) transformation is used for a member of this extended class. However, as we shall seldom be concerned with nonorthochronous transformations, we shall normally use 'Lorentz transformation' to denote only the orthochronous transformations where this is not likely to cause ambiguity.

The particular transformation (7.6) is known as a *boost*. It is not the most general boost. This may be obtained from (7.6) by giving both the starred and unstarred spatial coordinate axes the *same* rotation or reflection. Two coordinate systems related by a boost are often considered as having parallel coordinate axes. However, this interpretation can be misleading as it suggests that a combination of two boosts performed in succession will also be a boost. This is not so unless the directions of the two boosts are parallel. A general Lorentz transformation can be decomposed into a boost followed by a transformation (2.1) and/or (7.7).

A spacetime coordinate system which is natural for some inertial frame is said to be *Minkowskian* in the relativistic theory and *Galilean*

in the Newtonian theory, so that the Lorentz (Galilean) transformations are the transformations which connect different Minkowskian
(Galilean) coordinate systems. It was seen in §3 that in the Newtonian
theory the Galilean coordinate systems are a dynamically significant
subclass of a wider class of coordinates which can be picked out by
geometrical considerations. These are the natural coordinates of the
frames which were there described as allowable. Since the invariant
form Φ has given a geometric structure to spacetime in the relativistic
theory, the question arises as to what extent the Minkowskian coordinate systems can be characterized geometrically. For a coordinate
system to be Minkowskian it is clearly necessary that it gives to Φ the
form (7.5). It will now be shown that this is also sufficient, so that in
contrast to the Newtonian situation the geometry contains enough
information to determine the inertial frames completely.

This will be proved in the equivalent form stated above, namely that
any coordinate transformation which preserves (7.5) must be a Lorentz
transformation. The proof is in two parts. The first step is to show that
any such transformation must be linear inhomogeneous, i.e. must have
the form (5.2). This will be deferred to §10, where techniques will be
developed that enable nonlinear transformations to be handled. Once
this has been shown, the remainder of the proof is straightforward.
By a change of origin of one of the coordinate systems, the transformation can be brought to the form (5.3). Preservation of (7.5) is then
equivalent to

$$\mathbf{x'x} - c^2t^2 = \mathbf{x^{*\prime}x^*} - c^2t^{*2}. \tag{7.8}$$

If (5.3) is substituted into this, it must become an identity in $\mathbf{x}$ and t.
It is thus necessary that

$$A'A - c^2\mathbf{kk'} = I, \tag{7.9}$$

$$\mathbf{a'a} = c^2(l^2 - 1) \tag{7.10}$$

and

$$A'\mathbf{a} = c^2l\mathbf{k}, \tag{7.11}$$

where I is the (3×3) unit matrix.

These are next simplified by a rotation of the two sets of coordinates
as given by (5.11). As in §5, this can be used to reduce $\mathbf{a}$ and $\mathbf{k}$ to the
forms $\mathbf{a} = a\mathbf{i}$, $\mathbf{k} = k\mathbf{i}$, where $\mathbf{i} = (1, 0, 0)'$. It then follows from (7.11)
that $\mathbf{i}$ is an eigenvector of A', so that $A_{12} = A_{13} = 0$. If this and $\mathbf{k} = k\mathbf{i}$
are substituted into (7.9), it is found that A_{21} and A_{31} also vanish, that
the (2×2) submatrix $B = (A_{ij})$, $i,j = 2, 3$ is orthogonal, and that

$$(A_{11})^2 - c^2k^2 = 1. \tag{7.12}$$

There still remains the freedom in (5.11) to make rotations or reflections of the coordinate systems that leave the directions of their

x-axes unchanged. This can be used to turn the submatrix B into the (2×2) unit matrix and to make $A_{11} > 0$. A time-reversal (7.7) can also be performed if necessary to ensure $l > 0$. If we define $v = -a/A_{11}$, the remaining equations are now easily solved for A_{11}, a, k and l in terms of v and c to give the transformation in the form (7.6). The original coordinate systems must thus have been related by a Lorentz transformation, as required.

8 Something for nothing?

The development of physics within the spacetime structure of special relativity will be followed up in Chapters 3 to 5. We shall leave it at this stage for the present to consider instead the effects of gravitation on this fundamental structure. But before we do so, there are a few points raised by the above treatment that call for further comment.

The existence of a fundamental constant c of the dimensions of a velocity has been deduced from physical principles which appear to contain insufficient information for this to be possible. There are three possible explanations:

(i) Its existence is a logical necessity and so really can be deduced by systematic reasoning from quite general considerations, or

(ii) Some equivalent assumption has been unintentionally introduced in addition to those postulates that have been stated explicitly, or

(iii) Despite appearances, its existence is implicit in the explicitly stated postulates.

Explanation (i) is in keeping with the spirit of the work of Sir Arthur Eddington, although he himself did not consider this particular problem. From a study of the foundations of relativity theory and quantum theory, he came to the conclusion that all physical laws, including the numerical values of any dimensionless constants of nature, could be arrived at by an analysis of the processes of measurement by which we study the physical world. The culmination of his endeavours in this direction was his posthumously published book *Fundamental Theory* (Eddington 1948), which includes calculations of the fine structure constant and the mass ratio of the proton and electron. However, this work has not received general acceptance. It now seems likely that even if he was right in his belief that such quantities can be theoretically calculated, the route to their calculation is probably considerably different from that followed by Eddington.

As care has been taken to avoid hidden assumptions, the most likely explanation is thus (iii). This becomes clear if one thinks about the nature of the elastic materials out of which our standard rods and clocks must be constructed. Their properties originate in the quantum nature of interatomic forces. The length scale is essentially determined by the spacing of a crystal lattice, and the time scale by the natural frequency of vibration of such a lattice. The latter may be used directly, as in a quartz crystal controlled clock, or it may manifest itself indirectly, say as the force constant of a spring, but its quantum origin is similar in either case. Thus, although indirectly, the chosen length standard is the fundamental atomic unit of length $\hbar^2/me^2$ and the time unit is the reciprocal of the fundamental unit of frequency $me^4/\hbar^3$. Here, $2\pi\hbar$ is Planck's constant while m and e are respectively the mass and charge of an electron. These standards inherently contain a natural velocity $e^2/\hbar$, and so it should not be surprising if it reveals itself in a study of their behaviour. Unfortunately the above treatment does not provide any information about the value of the dimensionless constant of proportionality between c and $e^2/\hbar$. This ratio $\alpha = e^2/\hbar c$ is known as the fine structure constant and its experimental value is given by $\alpha^{-1} \simeq 137\cdot036$.

Another point is that artificial difficulties may appear to have been created by the use of macroscopic standards based on elastic solids. The units of length and time that are now internationally adopted are based directly on atomic standards. The international standard metre is $1\,650\,763\cdot73$ wavelengths of the orange-red spectral line of the isotope ^{86}Kr of krypton, while the international standard second is $9\,192\,631\,770$ periods of the microwave transition between the two hyperfine levels of the ground state of the isotope ^{133}Cs of caesium. If the theoretical development had been based on these standards, all mention of elastic bodies and the conditions for their freedom from stress could surely have been avoided. But in fact this is not so. Since the frequencies of atomic transitions are affected by extraneous fields which perturb the energy levels, it is still necessary to insist that the atomic standards are used only under specified conditions. The macroscopic concept of stress is thus replaced by an equivalent microscopic concept. There is thus nothing to be gained from a development based on atomic standards, and it is interesting to see that the same conclusions can be reached with the use of purely mechanical macroscopic standards.

9 The Principle of Equivalence

Let us now consider the further upset that we know must come in the presence of gravitation. It was seen in §4 that the origin of the difficulties produced by gravitation is its universality. All particles, regardless of their constitution, fall with the same acceleration when acted upon by a gravitational field, and hence there can be no free particles with which to test a reference frame for inertiality. We shall show that it is this same property that forms the foundation on which to base the theory of gravitation.

At least two inequivalent precise interpretations can be placed on the above, somewhat vague, statement of this fundamental property. In any treatment of the motion of particles in a theory based on a spacetime continuum, it is convenient to think of the route followed by a moving particle as being represented by a curve in the four-dimensional spacetime. Such a curve is called the *world line* of the particle. Then the weaker interpretation is that the world line of a particle moving in a gravitational field depends only on the initial position and velocity of the particle and its instant of projection. It is independent of the material of which the particle is constructed. It is this that was being demonstrated by Galileo in the reputed experiment in which he simultaneously dropped balls of different masses from the Leaning Tower of Pisa and observed that they struck the ground also simultaneously. A more recent, although less direct, experiment to test this hypothesis was performed by Eötvös in 1889. Using improved forms of the Eötvös experiment, Roll, Krotkov & Dicke (1964) have found that aluminium and gold fall with accelerations that are equal to one part in 10^{11}, while Braginsky & Panov (1971) have found equality to one part in 10^{12} for aluminium and platinum.

This interpretation can thus be considered established to a very high accuracy. However, it says nothing about the behaviour of particles projected simultaneously from the same point but with different initial velocities. According to Newtonian gravitational theory, such particles also have the same initial acceleration when measured in an inertial frame. This common acceleration given to all particles, irrespective of their motion, is called the gravitational acceleration at that point of space and time. It is this stronger interpretation that is needed in the foundation of the relativistic theory of gravitation, but in its present form it relies on Newtonian concepts. To avoid this, it must first be reformulated in such a way that it defines its own frames of validity, as does the principle of inertia. In the foundation of the special theory of relativity, it was found necessary to extend the range of phenomena considered in the Newtonian principle of inertia, to form the extended principle. This was because the principle was given a larger task to perform, in that the inertial frames had to be selected from a more general class of reference frames than was the case in the Newtonian theory. Similarly, the scope of this fundamental gravitational principle must be extended when it is given the additional task of selecting its own frames of validity.

To see what this involves, let us first consider the situation in a static uniform gravitational field. In reality all gravitational fields are more or less inhomogeneous, but in a small enough region of space, for example in a terrestrial laboratory, it is a reasonable approximation to neglect the nonuniformity. In Newtonian terms, the gravitational acceleration will then be independent of both position and time. Suppose a reference frame is now used which has this same acceleration. Such a frame is said to be freely falling. Then all particles which are acted upon solely by gravitation will appear to be either at rest or moving uniformly in a straight line. Their behaviour is thus identical to that of free particles in an inertial frame. Indeed, according to Newtonian theory, no dynamical experiment can distinguish between a freely falling frame in a static uniform gravitational field and an inertial frame in the absence of gravitation. The extended principle of inertia thus also holds in such a frame if 'free from external forces' is replaced by 'free from all external forces except gravitation'. The same replacement must also be made in the characterization of the ground state of an elastic body. Since the extended principle of inertia is free from Newtonian preconceptions, this thus provides the necessary modification of our basic gravitational principle for uniform fields.

Before nonuniform fields are considered, it must be remarked that there is an alternative form for the case of a uniform field which is less precise but is of historical significance. In the above discussion two situations were compared. One used a freely falling frame in a uniform field, while the other used an inertial frame in the absence of gravitation. Suppose now that the first of these situations is referred to a Newtonian inertial frame. If the mathematical transformation that this involves is also applied to the second situation, the frame that results is uniformly accelerated. The results of all dynamical experiments will necessarily still be the same when performed in either of the new frames. Thus, for dynamical experiments, placing a reference frame in a uniform gravitational field is equivalent to giving it a uniform acceleration. It is from this version that the principle that we are being led towards takes its name – the *Principle of Dynamical Equivalence*. Its lack of precision in comparison with the previous version arises from its retention of the Newtonian concept of an inertial frame.

There is another crucial step in the historical development of relativity theory that is relevant here. The successful development of special relativity had taught the dangers of believing that situations which are equivalent for dynamics could be distinguished nevertheless by more far-ranging physical experiments. Einstein thus put forward the more general hypothesis that the above equivalence holds for all physical phenomena, this being called the (unqualified) *Principle of Equivalence*. However, it must be treated with care because of its essentially approximate nature in dealing only with uniform fields. Einstein's first application of it was in 1911 to predict the deflection of light by a gravitational field, but although the effect has since been observationally verified, its magnitude disagrees by a factor of two with that given by a naive application of the principle.

The following study of the nature of space and time in a gravitational field will be based purely on dynamical equivalence, but this does not remove the need for care. The first step must be to make a precise statement of the equivalence principle that is meaningful even in inhomogeneous fields. An inhomogeneous field can be considered as uniform to better and better approximation as the size of the region of spacetime under consideration is reduced. The freely falling frames of the principle can thus be defined more and more accurately as their domain of definition is shrunk. This suggests that the only thing capable of precise definition is a 'freely falling frame at *P*', where *P* is a point in spacetime. Such a frame will extend throughout a region of spacetime, but the implication is that the free fall condition will only be exactly satisfied at one point. This is our first indication that gravitation theory needs the full generality of the reference frame concept as defined in §2. To be meaningful, a reference frame must extend over a finite region of spacetime, and yet in a realistic gravitational field a frame can only satisfy the 'preferred' free fall condition at one point. Elsewhere, the frame will be arbitrary. Any theory built on this basis will thus have no preferred class of observers unless their observations are restricted to their immediate vicinity. This goes a long way towards Einstein's hope that all observers should be on an equal footing.

The essence of the principle of equivalence is that there is no clear-cut separation between inertia and gravitation. In keeping with this idea, a particle which is acted upon solely by gravitation should be considered as 'freely moving'. Similarly, a frame that is falling freely and without rotation in a gravitational field should be considered as inertial. From now on these terms will be used with these generalized meanings. This cannot cause confusion if it is used consistently, but it should be borne in mind that such a frame is quite distinct from the inertial frames customarily defined in the Newtonian theory of gravitation.

The precise form of the principle of dynamical equivalence valid in an inhomogeneous gravitational field is now obtained by rewriting the extended principle of inertia in such a way that all properties are required to hold only instantaneously, at some fixed but arbitrary point P of spacetime. The reference frames so picked out are said to be *inertial at P*. There is no difficulty in doing this – for example, 'uniform motion' becomes 'instantaneous zero acceleration at P' – but it is rather laborious. It will not be given here in full, as nothing is really gained by doing so, and any points of apparent difficulty should be cleared up in the next section.

In the absence of gravitation, definite theoretical results were only achieved when the principle of uniformity was additionally adopted. A suitably localized form of this principle will similarly be needed in the gravitational case. It may seem less natural in a gravitational field, where one intuitively thinks of well-defined directions as 'up' and 'down', which apparently violates isotropy. But it must be remembered that it is being asserted in a freely falling frame, and only to the extent to which dynamical experiments are unable to distinguish the inhomogeneity of the field.

10   Gravitation as curvature

Before we turn to the mathematical implications of the above hypotheses, it is convenient to introduce the *kernel–index convention* which will be used throughout the book. This convention has been developed in great detail by Schouten (1954) to enable formulae involving co-ordinate-dependent indexed variables to be handled concisely and unambiguously.

Arrays of variables, for example matrix elements or the coordinates of a point of spacetime, are usually denoted by indexed symbols such as a_{ij}, x^α, whose indices run over a stated range of numerical values. In the case of an element a_{ij} of a specific matrix, its value is determined when the matrix is known and when the numerical values of the indices are specified. For the coordinates x^α of a point x of spacetime, there is however an additional dependence. When the numerical value of the index is given, it is still necessary to know both the point and the coordinate system before its value becomes determinate. The kernel–index convention gives a way to distinguish notationally between a change of point and a change of coordinate system. The kernel symbol x is taken as labelling the point, while the coordinate system is represented by the particular alphabet used for the index. Thus x^α and y^α denote the coordinates of two different points x and y in the same coordinate system, while x^α and x^a denote the coordinates of the same point x in two different coordinate systems. A coordinate system may be referred to by a typical letter of the alphabet that is being used for it. Both the kernel symbol and the alphabet can be modified by primes or other similar affixes, so that with respect to x^α, $\bar{x}^\alpha$ denotes a change of point from x to $\bar{x}$ and $x^{\alpha'}$ a change of coordinate system from (α) to (α'). When primes or asterisks are used to modify the kernel symbol, they are generally written as prefixes to separate them from suffixed indices, e.g. $*x^a$. When a specific numerical value is given to an index, some method must be used to distinguish the alphabet that was used for the index, as otherwise the labelling of the coordinate system would be lost. If an alphabet is modified by some affix, then the numerical values should be similarly affixed, e.g. $\alpha' = 1', 2', 3', 4'$. If a distinction has to be drawn between the numerical values taken by indices of two unmodified alphabets such as upper and lower case latin indices, then an *ad hoc* distinction must be introduced each time.

These rules apply to all coordinate-dependent indexed variables. They even allow for the possibility that different indices of a variable with several indices may be associated with different coordinate systems. Such variables will indeed be introduced below. For reasons that will become clear later, both superscript and subscript indices are used. The coordinates of a point will always be labelled by a superscript index which for spacetime will run from 1 to 4. For Minkowskian and Galilean coordinate systems the fourth coordinate will always be taken as the time, t.

A few general points concerning equations between indexed variables also need mention here, although they are not specifically concerned with coordinate dependence. Indices which have been given specific values are known as *fixed* indices, while those which take a range of values are called *running* indices. The running indices further subdivide into *free* indices, which can be given any particular value from their range, and *dummy* indices, which occur in some operation in the equation such as a summation or repeated product. For example, in

$$w = \sum_{\alpha=1}^{n} u^{\alpha\beta} v^{\alpha 3} \tag{10.1}$$

α is a dummy index, β is a free index and 3 is a fixed index. Unless the contrary is explicitly stated, equations must hold for the whole range of allowed values of their free indices, with each such index running independently over its range. The right-hand side of (10.1) must thus have the same value for each value of β, as the left-hand side does not involve β. The meaning of an equation is unchanged if its dummy indices are relabelled, provided that if the kernel–index convention is being used, the relabelling should be kept within the same alphabet.

We can now return to the problem at hand. If (α) and (α') are any two coordinate systems in the same region of spacetime, each of the variables $x^{\alpha'}$ can be considered as a function of the four coordinates x^{α}. Their partial derivatives

$$A_{\alpha}^{\alpha'}(x) \equiv \partial x^{\alpha'} / \partial x^{\alpha} \tag{10.2}$$

are an example of an array which has indices associated with two different coordinate systems. Now suppose that (α) and (α') are natural coordinates of reference frames K and K' respectively, each of which is inertial at the same fixed point z. Let the fourth coordinate in each case be the time coordinate, and let $t = x^4$, $t' = x^{4'}$. Then the world line L of a freely moving clock which passes through z may be parametrized by t, or by t', or by the reading τ of the clock itself. Let us first see what can be deduced about these parametrizations due to the frames being inertial at z.

Since the clock is freely moving, it has zero acceleration at z and hence $d^2x^{\alpha}/dt^2 = 0$ there. Further, the rate at which the moving clock runs in comparison with the time coordinate t must be stationary at z by the latter part of the extended principle of inertia. This gives $d^2t/d\tau^2 = 0$ at z, which in combination with the preceding result implies

$$d^2x^{\alpha}/d\tau^2 = 0 \quad \text{at} \quad z. \tag{10.3}$$

But in addition to being stationary, this rate must also depend only on the magnitude of the instantaneous velocity of the clock. As in §5, let $\mathbf{x} = (x^1, x^2, x^3)'$ and put $\mathbf{u} = d\mathbf{x}/dt$, $u^2 = \mathbf{u}'\mathbf{u}$, where a prime denotes the matrix transpose rather than being a part of the kernel–index convention. Then there exists a function f such that

$$dt/d\tau = f(u^2).\qquad(10.4)$$

Since the same considerations also hold in the frame K', the situation is almost identical to that in §5, with (10.2) and (10.4) replacing (5.3) and (5.4) respectively. If we now put

$$A = \begin{pmatrix} A_1^{1'} & A_2^{1'} & A_3^{1'} \\ A_1^{2'} & A_2^{2'} & A_3^{2'} \\ A_1^{3'} & A_2^{3'} & A_3^{3'} \end{pmatrix}, \quad \mathbf{a} = \begin{pmatrix} A_4^{1'} \\ A_4^{2'} \\ A_4^{3'} \end{pmatrix}, \quad \mathbf{k} = \begin{pmatrix} A_1^{4'} \\ A_2^{4'} \\ A_3^{4'} \end{pmatrix}, \quad l = A_4^{4'}, \qquad(10.5)$$

where the $A_\alpha^{\alpha'}$ are all evaluated at z, then the calculations of §§5 and 6 can be repeated with these new definitions virtually without alteration. The matrices (10.5) are found to have precisely the same values as those denoted by the corresponding symbols in those sections. We shall consider only the case in which f and g are not constant, as that is the one which in the absence of gravitation leads to special relativity. The case of constant f and g is also of interest as it sheds light on the relationship between the Newtonian and relativistic theories of gravitation, but we shall not pursue it further here.

One of the cornerstones of special relativity is the invariance of the quadratic form Φ of (7.5). If we define a matrix η by

$$\eta = \operatorname{diag}(1,1,1,-c^2)\qquad(10.6)$$

and a coordinate system (α) by

$$x^1 = x, \quad x^2 = y, \quad x^3 = z, \quad x^4 = t,$$

then (7.5) can be written as

$$\Phi = \sum_{\alpha,\,\beta=1}^{4} \eta_{\alpha\beta}\,\delta x^\alpha \delta x^\beta.\qquad(10.7)$$

The interpretation of (10.6) is that η is a diagonal matrix whose leading diagonal has the given elements in the given order. To state the result which corresponds in the gravitational case to the invariance of Φ, let $x(\lambda)$ be any spacetime curve through z. Since

$$\frac{dx^{\alpha'}}{d\lambda} = \sum_{\alpha=1}^{4} A_\alpha^{\alpha'}(x)\,\frac{dx^\alpha}{d\lambda}\qquad(10.8)$$

from (10.2), the derivatives $dx^\alpha/d\lambda$ transform under a change of coordinates in exactly the same way as the coordinate differences δx^α of §7. It follows that the quantity Φ defined at z by

$$\Phi \equiv \sum_{\alpha,\,\beta=1}^{4} \eta_{\alpha\beta}\,\frac{dx^\alpha}{d\lambda}\frac{dx^\beta}{d\lambda}\qquad(10.9)$$

has the same value in all natural coordinate systems of all reference frames that are inertial at z and which use there a given standard of length. It thus has an absolute value which depends on the curve, its parametrization, and this unit of length, but which is free from any choice of reference frame. As in §7, this enables Φ to be considered as a geometric property of spacetime itself.

This appears to differ from the situation in the special theory in that the invariant Φ is only defined at one point in spacetime. However, that point is arbitrary, so that Φ can be considered as a function $\Phi(x)$ along the curve if it is evaluated at each x in a frame that is inertial at x. The real difference is that the expression (10.9) cannot be used in a single frame to evaluate Φ everywhere. It will now be shown that there is a generalization of (10.9) which does enable a single frame to be used at all points.

Let (a) be an arbitrary coordinate system. Since the definition (10.2) was made for any two coordinate systems, it follows from the kernel-index convention that A_a^α is already defined and is given by

$$A_a^\alpha = \partial x^\alpha / \partial x^a.$$

If the identity
$$\frac{dx^\alpha}{d\lambda} = \sum_a A_a^\alpha \frac{dx^a}{d\lambda} \tag{10.10}$$

is substituted into (10.9), it is found that

$$\Phi(z) = \sum_{a,b} g_{ab}(z) \frac{dx^a}{d\lambda} \frac{dx^b}{d\lambda}\bigg|_{x=z}, \tag{10.11}$$

where
$$g_{ab}(z) = \sum_{\alpha,\beta} \eta_{\alpha\beta} A_a^\alpha(z) A_b^\beta(z). \tag{10.12}$$

It is easily seen that $g_{ab}(z)$ is independent both of the choice of inertial frame at z, with its corresponding natural coordinate system (α), and of the choice of curve $x(\lambda)$. Since z could have been chosen arbitrarily, $g_{ab}(x)$ is thus a matrix function of position defined throughout the domain of the coordinate system and dependent on nothing but this coordinate system. Its elements are known as the components of the *metric tensor* in this coordinate system. The expression (10.11) is the required generalization of (10.9) in that once the functions $g_{ab}(x)$ are known, it can be used to evaluate Φ everywhere and for every curve $x(\lambda)$, using only the given coordinate system (a).

We are now in a position to prove that any coordinate transformation which preserves (7.5) is necessarily linear inhomogeneous. This was used without proof in §7. In any coordinate system in which Φ

takes the form (7.5), the metric tensor will everywhere take the constant value (10.6). If this is true of both the coordinate systems (a) and (α) of (10.12), that equation will hold with $g_{ab} = \eta_{ab}$ at all points x and not merely at the fixed point z. Differentiation of (10.12) with respect to x^c then yields

$$\sum_{\alpha,\beta} \eta_{\alpha\beta}(A_a^\alpha \, \partial_c A_b^\beta + A_b^\beta \, \partial_c A_a^\alpha) = 0, \tag{10.13}$$

where

$$\partial_c \equiv \partial/\partial x^c. \tag{10.14}$$

But a relabelling of the indices gives us also

$$\sum_{\alpha,\beta} \eta_{\alpha\beta}(A_b^\beta \, \partial_a A_c^\alpha + A_c^\alpha \, \partial_a A_b^\beta) = 0 \tag{10.15}$$

and

$$\sum_{\alpha,\beta} \eta_{\alpha\beta}(A_c^\beta \, \partial_b A_a^\alpha + A_a^\alpha \, \partial_b A_c^\beta) = 0, \tag{10.16}$$

since the matrix η is symmetric. Now subtract (10.16) from the sum of the equations (10.13) and (10.15). Some cancellation occurs since, for example,

$$\partial_c A_b^\beta = \partial_b A_c^\beta = \frac{\partial^2 x^\beta}{\partial x^b \, \partial x^c}, \tag{10.17}$$

which leaves simply

$$2\sum_{\alpha,\beta} \eta_{\alpha\beta} A_b^\beta \partial_a A_c^\alpha = 0. \tag{10.18}$$

But the matrices (A_b^β) and $(\eta_{\alpha\beta})$ are nonsingular. This can thus hold only if

$$\frac{\partial^2 x^\alpha}{\partial x^a \partial x^b} = 0, \tag{10.19}$$

which on integration gives the desired linearity.

The matrix g_{ab} given by (10.12) possesses certain algebraic properties purely as a result of its method of construction. It is clearly symmetric and nonsingular, but this is not all that can be deduced. To show this, let matrices G, H and A be defined by

$$G = (g_{ab}), \quad H = (\eta_{\alpha\beta}), \quad A = (A_a^\alpha), \tag{10.20}$$

in the latter of which α labels the rows and a the columns. Then (10.12) can be written as $G = A'HA$, so that

$$H = B'GB \tag{10.21}$$

where $B = A^{-1}$. Now it is known from matrix theory that any real symmetric $(k \times k)$ matrix G is reducible by a transformation of the form

$$G \to B'GB, \tag{10.22}$$

where B is nonsingular, to a diagonal form in which all the diagonal elements are either 1, 0 or -1. Moreover, the numbers p of 1s, z of zeroes and n of -1s that occur in the diagonal are invariants of the matrix, in that they are independent of the particular transformation used to achieve this form. The numbers $p+n$ and $p-n$ are known respectively as the *rank* and *signature* of the matrix. When G is non-singular, its rank is necessarily k and hence its signature is the only independent invariant under this transformation. It follows from (10.21) and (10.6) that the matrix (g_{ab}) of metric tensor components must have signature $+2$.

We can now conclude that when gravitation is taken into consideration, the geometry of spacetime is described by a metric tensor. This associates with every coordinate system a position-dependent matrix (g_{ab}) of components which is symmetric, nonsingular and has signature $+2$. If there exists a coordinate system in which the g_{ab} are independent of position, then a linear coordinate transformation can make $g_{ab} = \eta_{ab}$ throughout its domain, in the notation of (10.6). This is the case of special relativity, characterized physically by the absence of gravitation. The corresponding geometry is said to be *flat*. When it is not possible to make the g_{ab} constant by a suitable choice of coordinates, the geometry is said to be *curved*. The terminology arises from the geometric theory of surfaces in ordinary three-dimensional Euclidean space, where a closely analogous situation occurs. In this language we can say that *spacetime is curved by a gravitational field*. We shall leave the theory of gravitation at this point. When these ideas are followed up in more detail, they lead to the general theory of relativity, which is beyond the scope of the present book.

The discussion in §9 showed from a physical viewpoint what approximations are involved in the use of the laws of special relativity within a gravitational field. We can now see the same approximations also from a mathematical viewpoint. To use special relativity in such conditions a coordinate system must first be chosen, in the spacetime region of interest, in which the metric tensor components g_{ab} approximate to the constant matrix η_{ab} of (10.6). The reference frame for which these coordinates are natural, in the sense of §2, can be treated as an inertial frame of special relativity to the extent to which the departure of g_{ab} from η_{ab} may be neglected.

2
Affine spaces in mathematics and physics

1 Introduction

In theoretical physics, coordinate systems are a necessary evil. Space-time without a coordinate system is like the Earth without place names or a latitude and longitude grid. It is impossible to say where anything is, or when anything happens, or to describe the arrangement and outcome of any experiment. But although necessary, they are evil because they are artificial constructs. The outcome of an experiment is in some sense independent of the coordinate system used to describe it, even though the numerical content of the results may be affected by a change in the coordinate system used. A task that must be faced before the full implications of any law of physics can be understood is thus to unravel its true physical content from any dependence on the coordinate system in which it is stated.

This task is made harder, rather than easier, by the existence of the preferred coordinate systems discussed in the preceding chapter. Preferred coordinate systems certainly reduce the arbitrariness in any physical description, but they obscure the physical phenomena which are responsible for their existence. This is of particular significance in a theory of gravitation, as we have already seen that gravitation has an effect on the physical construction of inertial reference frames that is unique among all forces in physics. What is needed is a method of analysis which enables statements to be made about coordinate-dependent variables in a manner which ensures that their validity does not depend on the particular coordinate system used. This would be useful if it only ensured independence from the choice of coordinate system within some preferred family. If it truly ensured independence from a completely free choice of coordinates, it would also force into the open those variables which give the preferred coordinates their specially advantageous position.

General tensor analysis does precisely this, and it is indispensable in general relativity for the reasons just given. However, special relativity neither requires nor benefits from the full power of general tensor analysis. Instead it is more appropriate to use *affine tensor analysis,* which will be developed in the present chapter. In this, there is still a preferred family of coordinate systems, but it is a large one. The transformations which connect the preferred systems are all those of the form

$$x^a = \sum_\alpha A^a_\alpha x^\alpha + k^a, \qquad (1.1)$$

where the matrix (A^a_α) is nonsingular and the range of the indices is from 1 to n. These are known as *affine transformations.* When $n = 4$ they include the Galilean and Lorentz transformations as special cases, so that there is enough generality to cope with the distinctive features of both Newtonian and special relativistic physics.

A feature of §1-10 that was much in evidence was the large number of summation signs in the equations. Such summations occur frequently throughout the whole of tensor analysis. However, they do so with sufficient regularity that the summation signs may be omitted, leaving the summations implicit in the structure of the equations. This is a considerable simplification of the notation. If the equations of §1-10 are examined, they are seen to have the following properties in common:

(i) No index occurs in the same term more than twice.

(ii) Those indices that occur only once in a term are free.

(iii) With the exception of zero, every term of an equation has the same free indices, which occur consistently either as superscripts or subscripts.

(iv) When an index occurs twice in a term, one occurrence is as a superscript and the other as a subscript.

(v) Such repeated indices are dummy indices which are summed over their entire range of values.

In this, a 'term' is to be interpreted as any summand of an equation, even if it consists of a product of several separate quantities. Any brackets present, such as in (1-10.13), are to be considered as expanded. It will be seen below that the notation can be developed in such a way that almost all useful equations naturally obey these rules. Rules (iv) and (v) together are then sufficient to indicate summations, and all summation signs may be omitted. This assumption of summation over any index that occurs in the same term once as a superscript and once as a subscript is known as the *summation convention.* From now on it will be adopted in every equation that satisfies all the five rules, unless the contrary is stated explicitly. It should be noted that equations which violate these rules can still be mathematically meaningful, as in the example (1-10.1). On the occcasions when such equations must be used, the summation convention should be suspended and all necessary summation signs written in explicitly.

2 Basic properties of affine spaces

2a Algebraic properties

An n-dimensional *affine space*, which for conciseness will be called† an E_n, is an n-dimensional space which possesses a family of preferred coordinate systems satisfying the following three conditions:

(i) the coordinates are real-valued and have an unrestricted range,

(ii) any two systems of the preferred family are related by a transformation of the form (1.1), and

(iii) the result of transforming any preferred system by a transformation of that form is again a preferred system.

Coordinate systems of the preferred family are said to be *rectilinear*. Any space which possesses a preferred family of coordinate systems satisfying only conditions (i) and (ii) may be considered as an affine space if the family is enlarged to comply also with (iii). The spacetimes of both special relativity and Newtonian physics are thus E_4s, while the space of (three-dimensional) Euclidean geometry is an E_3.

The basic mathematical model of an E_n is constructed from an n-dimensional real vector space V. Since the components of any vector in V transform under a change of basis according to (1.1) with $k^a = 0$, the bases of V provide a family of coordinate systems for V which satisfy (i) and (ii). This may thus be enlarged to a full rectilinear family in the manner just described. The construction is reversible, for if a fixed point O is arbitrarily chosen in an E_n and a restriction is made to those rectilinear coordinate systems which have O as origin, then the E_n acquires a natural structure as a real vector space.

These examples of an E_n have one important feature in common. In none of them are all rectilinear coordinate systems in any sense mutually equivalent, since by construction the full rectilinear family has been obtained as an extension of an even more restrictive subfamily. In §3 we shall see how to characterize these original subfamilies, thus 'forcing into the open' some hidden variables as discussed in §1.

Before we develop the algebraic consequences of (1.1), one point

---

† The description of different types of space by a characteristic letter with the dimension as a subscript is taken from Schouten (1954).

needs to be made concerning its notation. Since it connects two coordinate systems, the transformation coefficients A_α^a and k^a should be labelled to show their association with both systems. Although A_α^a satisfies this by the rules of the kernel–index convention, k^a does not. To make the notation precise, a reference to the (α) coordinates should be added. A refinement of the convention which can be used in such circumstances is to add this in brackets as a prefix to the kernel, thus $^{(\alpha)}k^a$, where the α is a generic letter rather than an index which could be given a specific value. We thus write

$$x^a = A_\alpha^a x^\alpha + {}^{(\alpha)}k^a. \tag{2.1}$$

The symbols A_α^a and $^{(\alpha)}k^a$ will from now on be reserved exclusively for the coefficients defined by (2.1), although extensive use will be made only of A_α^a. This notation enables any number of coordinate systems to be handled simultaneously and without ambiguity. If a third system (A) is introduced, no further definitions are needed before the transformations

$$x^A = A_a^A x^a + {}^{(a)}k^A \tag{2.2}$$

and

$$x^A = A_\alpha^A x^\alpha + {}^{(\alpha)}k^A \tag{2.3}$$

can be written down, from which it follows that

$$A_\alpha^A = A_a^A A_\alpha^a \tag{2.4}$$

and

$$^{(\alpha)}k^A = A_a^A {}^{(\alpha)}k^a + {}^{(a)}k^A. \tag{2.5}$$

These three coordinate systems need not all be different. When (A) and (α) are the same, it can be seen from (2.3) that $^{(\alpha)}k^A = 0$ and that A_α^A is the unit matrix, i.e.

$$A_\beta^\alpha = \begin{cases} 1 & \text{if} \quad \alpha = \beta \\ 0 & \text{if} \quad \alpha \neq \beta. \end{cases} \tag{2.6}$$

As special cases of (2.4) and (2.5) we thus have

$$A_a^\beta A_\alpha^a = A_\alpha^\beta \tag{2.7}$$

and

$$^{(\alpha)}k^\alpha = -A_a^\alpha {}^{(\alpha)}k^a, \tag{2.8}$$

so that the inverse of (2.1) can be written as

$$x^\alpha = A_a^\alpha (x^a - {}^{(\alpha)}k^a). \tag{2.9}$$

The coordinate differences†

$$\delta x^\alpha = x^\alpha - y^\alpha \tag{2.10}$$

† In δx^α, δx is to be considered as a composite kernel symbol with index α, rather than as an operator δ acting on the coordinates x^α.

between two points x and y have from (2.1) the transformation law

$$\delta x^a = A^a_\alpha \, \delta x^\alpha. \tag{2.11}$$

With this as a prototype, a *contravariant affine vector* is defined as an association with every rectilinear coordinate system (α) of a set of n numbers u^α, say, labelled by a superscript index, whose values in any two coordinate systems are related to one another by

$$u^a = A^a_\alpha u^\alpha. \tag{2.12}$$

When no ambiguity can arise, the descriptions 'contravariant' and 'affine' may be omitted. 'Affine' refers to the group of coordinate transformations involved, which is usually clear, and as will be seen below, the contravariant character is indicated by the index position. This can thus almost always be done. Each of the numbers u^α is said to be a *component* of the vector in the coordinate system (α). The prototype δx^α is known as the *position vector* of x relative to y.

An affine scalar is naturally defined as a number whose value is independent of the choice of coordinate system. A scalar or vector field is an assignment of a scalar or vector respectively to each point of some subset of the space. If C is a curve described by a scalar parameter λ, it follows from (2.1) and (2.12) that $dx^\alpha/d\lambda$ is a contravariant vector field along C. Its value at any point z is called the *tangent vector* to C at z.

If $\phi(x)$ is a scalar field defined on an open set, it similarly follows from (2.9) that the partial derivatives $\partial\phi/\partial x^\alpha$ satisfy

$$\frac{\partial \phi}{\partial x^a} = \frac{\partial x^\alpha}{\partial x^a} \frac{\partial \phi}{\partial x^\alpha} = A^\alpha_a \frac{\partial \phi}{\partial x^\alpha}. \tag{2.13}$$

This differs from (2.12) in the position of the indices on A, and for consistency with the rules of §1, the index on the partial derivative must be considered as a subscript. It motivates the definition of a *covariant affine vector* as an association with every rectilinear coordinate system (α) of a set of n numbers v_α, say, distinctively labelled by a subscript index, whose values in any two coordinate systems satisfy

$$v_a = A^\alpha_a v_\alpha. \tag{2.14}$$

The prototype $\partial\phi/\partial x^\alpha$ is known as the *gradient* of the scalar field ϕ.

A vector is uniquely determined when its components are known in only one coordinate system, as they may be found in every other system from the appropriate transformation law (2.12) or (2.14). In any given coordinate system, the vectors which have unity for one component and zeroes for all other components are called the *basis vectors* of that coordinate system. In an E_n there are n contravariant basis vectors and n covariant basis vectors, either set of which has a natural ordering according to the position of the nonzero component.

It follows from (2.12) and (2.14) that

$$u^a v_b = A^a_\alpha A^\beta_b u^\alpha v_\beta, \qquad (2.15)$$

and with the additional use of (2.6) and (2.7) also that

$$u^a v_a = A^a_\alpha A^\beta_a u^\alpha v_\beta = A^\beta_\alpha u^\alpha v_\beta = u^\alpha v_\alpha. \qquad (2.16)$$

The final step in (2.16) illustrates the action of A^β_α as a 'substitution operator', so called as it has the effect in such a term as $A^\beta_\alpha v_\beta$ of substituting the free index α for the dummy index β to give v_α. This follows immediately from (2.6) and the summation convention. The fact that in (2.16) the α is also a dummy index is irrelevant. We could have simplified $A^\beta_\alpha u^\alpha$ to u^β by the same process, to give $u^\beta v_\beta$ as the final result. But since β is a dummy index and α and β belong to the same alphabet, $u^\alpha v_\alpha$ and $u^\beta v_\beta$ have identical meanings.

Since a and α belong to different alphabets, $u^a v_a$ and $u^\alpha v_\alpha$ represent the same single-component quantity evaluated in two different coordinate systems. Equation (2.16) is thus a proof that this quantity is a scalar. It is the *inner product* of the vectors u^α and v_α. On the other hand, (2.15) is a transformation law for the array of n^2 elements which consists of all possible products of a pair of components, one from each of the two vectors. If one starts with m different vectors, one can generalize this to an array of n^m elements whose transformation law involves a product of m of the matrices (A^a_α) or (A^α_a). Such an array is the prototype of a tensor, which leads to the following definition. An *affine tensor of valence* (r, s) is an association with every rectilinear coordinate system (α) of a set of n^{r+s} numbers $t^{\alpha \dots \beta}{}_{\gamma \dots \delta}$ say, labelled by r superscript and s subscript indices each of which ranges independently from 1 to n, such that

$$t^{a \dots b}{}_{c \dots d} = A^a_\alpha \dots A^b_\beta A^\gamma_c \dots A^\delta_d t^{\alpha \dots \beta}{}_{\gamma \dots \delta}. \qquad (2.17)$$

As before, each number is said to be a component of the tensor in the appropriate coordinate system. An abbreviation of the notation of (2.17) which is often useful is to write the kernel symbol A only once in the product of transformation coefficients. We thus write

$$A^{a...by...\delta}_{\alpha...\beta c...d} \equiv A^a_\alpha...A^b_\beta A^\gamma_c...A^\delta_d. \tag{2.18}$$

A tensor of valence (r, s) is also said to have contravariant valence r, covariant valence s and total valence $r + s$.

Apart from a few special cases, the indices of a tensor are always spaced so that no subscript index is vertically below a superscript, although it is not necessary for the superscripts to precede the subscripts in the manner given in (2.17). This spacing is to allow indices to be raised and lowered without ambiguity, as certain general operations on tensors will be introduced in the next section which are conveniently denoted simply by such a change in the positions of the indices. To remove any ambiguity as to the vertical alignment, vacant spaces may be marked by dots, e.g. $t_{\alpha \cdot \gamma}^{\cdot \beta}$. When the index α has been raised and β lowered, the result is written as $t^\alpha_{\cdot \beta \gamma}$.

Due to the substitution action of A^β_α, (2.7) can be written in the form

$$A^a_\alpha A^\beta_b A^b_a = A^\beta_\alpha. \tag{2.19}$$

If this is compared with (2.17), it is seen that A^β_α is a tensor. It is known as the *unit tensor*, and is one of the exceptions mentioned above whose indices are written in the same vertical line. We see from (2.6) that it has the unusual property that its components have the same numerical values in every coordinate system. Such a tensor is said to be *isotropic*.

Consider now the equation (2.4). If the index α is given a fixed numerical value and A^a_α is consequently considered as a variable with a single free index a, it shows that A^a_α behaves as a contravariant vector under a change of the (a) coordinates. The array A^a_α may thus be obtained from the unit tensor A^β_α by transformation solely of the superscript index. This incidentally shows that A^a_α is, for fixed α, a basis vector of the (α) coordinates expressed in terms of the (a) coordinates. Such *mixed components* may be similarly defined for any tensor, e.g.

$$t_{\alpha \cdot c}^{\cdot \beta} = A^\gamma_c t_{\alpha \cdot \gamma}^{\cdot \beta}, \tag{2.20}$$

although only a few special cases such as A^a_α are much used. They have been introduced here for completeness and to show that this dual use of A as the kernel letter in both the unit tensor and the transformation coefficients is part of a general scheme. The definition of an isotropic tensor assumes that all indices are referred to the same coordinate system.

The tensor of valence (1, 1) given by $u^{\alpha}v_{\beta}$ is known as the *outer product* of the two vectors. It has an obvious extension to any two tensors, the outer product of a tensor of valence (r, s) with one of valence (r', s') being a tensor of valence $(r + r', s + s')$. This is one of the three algebraic processes by which new tensors can be formed from given ones. Another is that since the transformation law (2.17) is linear homogeneous, any linear combination with scalar coefficients of two tensors of the same valence and with the same free superscript and subscript indices is also a tensor of that valence. The tensors of a given valence thus form a vector space. The third process is applicable to a single tensor of valence (r, s) with $r \geqslant 1$, $s \geqslant 1$, and is the *contraction* of a free superscript index with a free subscript index. The two indices concerned are simply given the same letter. By the rules of the summation convention they then become dummy indices and there is an implicit summation from 1 to n. An example is the construction of the scalar $u^{\alpha}v_{\alpha}$ from the tensor $u^{\alpha}v_{\beta}$ of valence (1,1). In the general case, contraction of a tensor of valence (r, s) yields a tensor of valence $(r - 1, s - 1)$, as may be proved by the method of equation (2.16).

A fourth process is available when one starts from a tensor field. The partial derivatives of the components of a tensor field of valence (r, s) form a field of tensors of valence $(r, s + 1)$, known as the *gradient* of the original field. This follows immediately from (2.17) and generalizes the construction (2.13) of the prototype covariant vector. As it is useful to have an abbreviated notation for this often used process, we write

$$\partial_{\alpha} \equiv \partial/\partial x^{\alpha}. \tag{2.21}$$

If the derivative index is contracted with a superscript index of the tensor, as in $\partial_{\alpha} t^{\alpha\beta}$, the result is said to be a *divergence*. A given tensor field may have several different divergences, as the contraction may be performed with any of the superscript indices and in general the resulting tensors will be unequal. All the divergences of a tensor field of valence (r, s) are of valence $(r - 1, s)$.

A partial converse of the process of contraction exists, known as the *quotient rule* for the recognition of tensor character. It is best illustrated by an example. Suppose that for every coordinate system (α) there is given an indexed array $u^{\alpha\beta}_{\cdot\cdot\gamma}$ of n^3 numbers. Suppose further that for every tensor $v^{\alpha}_{\cdot\beta}$ of valence (1, 1), the array w^{α} defined by

$$w^{\alpha} = u^{\alpha\beta}_{\cdot\cdot\gamma} v^{\gamma}_{\cdot\beta} \tag{2.22}$$

satisfies the contravariant vector transformation law (2.12). Then we can 'uncontract' $v^\gamma_{.\beta}$ to deduce that $u^{\alpha\beta}_{..\gamma}$ is a tensor of valence (2, 1). To prove this, we note that

$$w^a = u^{ab}_{..c} v^c_{.b} \tag{2.23}$$

by the definition (2.22), that

$$v^c_{.b} = A^c_\gamma A^\beta_b v^\gamma_{.\beta} \tag{2.24}$$

since $v^\gamma_{.\beta}$ is a tensor, and that

$$w^a = A^a_\alpha w^\alpha \tag{2.25}$$

by hypothesis. If these are combined with (2.22), we can deduce that

$$(A^a_\alpha u^{\alpha\beta}_{..\gamma} - A^c_\gamma A^\beta_b v^{ab}_{..c}) v^\gamma_{.\beta} = 0. \tag{2.26}$$

But by hypothesis $v^\gamma_{.\beta}$ is arbitrary. We thus have a linear combination (from the implied summations over β and γ) of n^2 arbitrary variables which must vanish identically. This can only happen if the coefficients in the sum all vanish separately, which gives

$$A^a_\alpha u^{\alpha\beta}_{..\gamma} = A^c_\gamma A^\beta_b u^{ab}_{..c}. \tag{2.27}$$

If both sides are multiplified by A^δ_a, we obtain

$$u^{\delta\beta}_{..\gamma} = A^\delta_a A^\beta_b A^c_\gamma u^{ab}_{..c}. \tag{2.28}$$

As this is the tensor transformation law, the proof is complete. In the general case the array $u^{\cdots}_{\ \ \cdots}$ can have any number of indices and any number of them can be formally contracted with indices of an arbitrary tensor $v^{\cdots}_{\ \ \cdots}$. The important features are that $v^{\cdots}_{\ \ \cdots}$ must be arbitrary and that the result of the formal contraction must transform as a tensor. The contractions are described as formal since the operation of contraction is strictly only defined on tensors, and at the outset it is not known that $u^{\cdots}_{\ \ \cdots}$ is a tensor.

The homogeneity of the transformation law (2.17) shows that there is a tensor of any given valence whose components all vanish in every rectilinear coordinate system. This is naturally called the *zero tensor* of that valence and is written as 0 with the indices omitted. Suppose now that an equation is given which involves only scalars and tensors, and which is known to be valid in a particular rectilinear coordinate system. Suppose further that it is constructed according to the five rules given in §1. It then follows from the three algebraic operations just given that every term of the equation is a tensor of the same valence. If all the terms of the equation are brought on to its left-hand side, it then equates some particular tensor to the zero tensor. Such an equation is invariant under affine coordinate transformations, and hence the original equation must hold in *every* rectilinear coordinate system. Within the structure of an affine space it thus has a meaning which is coordinate independent. This shows the significance of the five rules for the theory of affine tensors.

A simple but important illustration of these remarks is provided by the relationship between a coordinate system and its basis vectors. The coordinates of a point are the coefficients in the expression of the position vector of that point with respect to the origin as a linear combination of the contravariant basis vectors. They are also given by the inner products of that position vector with the covariant basis vectors. Both of these results are trivially verified in the particular coordinate system concerned. They show that the coordinate system is determined completely by its origin and either set of basis vectors. Moreover, if a point z and an ordered set of n linearly independent vectors of the same valence are chosen arbitrarily, there exists a unique coordinate system which has origin at z and these vectors as an ordered set of basis vectors. The above results show how the coordinates of any other point x may be evaluated from this data. For this reason, any such ordered set will be said to form a *basis* of the E_n.

Another consequence of these remarks concerns symmetry properties of tensors. If $t_{\alpha\beta}$ is a tensor of valence $(0, 2)$, the equation

$$t_{\alpha\beta} = t_{\beta\alpha} \tag{2.29}$$

must hold in all rectilinear coordinate systems if it holds in any one such system. When it holds, the tensor is said to be *symmetric*, while if

$$t_{\alpha\beta} = -t_{\beta\alpha} \tag{2.30}$$

it is said to be *antisymmetric*. The terminology is taken from matrix theory, and corresponding definitions are made for tensors of valence $(2, 0)$. In contrast, however, a tensor $t^{\alpha}{}_{\beta}$ of valence $(1, 1)$ may be symmetric when considered as a matrix in one coordinate system without this being the case in all such systems. This is because the equation

$$t^{\alpha}{}_{.\beta} = t^{\beta}{}_{.\alpha}, \tag{2.31}$$

which expresses this symmetry, violates rule (iii). Symmetries of tensors of higher valence will be considered in §5.

2b Geometric properties

This completes our development of basic algebraic techniques. We now turn to the fundamental concepts of geometry in an E_n. Choose a fixed point y and let $a^\alpha, b^\alpha, \ldots, c^\alpha$ be r linearly independent vectors, $1 \leqslant r \leqslant (n-1)$. Then the r-dimensional surface Π described by r real parameters $\lambda, \mu, \ldots, \nu$ and given by

$$x^\alpha(\lambda, \mu, \ldots, \nu) = y^\alpha + \lambda a^\alpha + \mu b^\alpha + \ldots + \nu c^\alpha \qquad (2.32)$$

is called an *r-plane*. The parameters $\lambda, \mu, \ldots, \nu$ form a coordinate system on Π, and if the choice of y and the generating vectors is changed in such a way that the r-plane formed remains fixed, then this coordinate system undergoes an affine transformation. There is thus a natural structure for Π as an E_r, so that it is also called an affine subspace of E_n. If E_n and this E_r are both given the natural vector space structures that arise when the rectilinear coordinates are restricted to those with origin at y, the E_r becomes an r-dimensional vector subspace of E_n. This can be used as an alternative definition of an r-plane.

A 1-plane and an $(n-1)$-plane are alternatively known as a *straight line* and a *hyperplane* respectively. It follows from the elementary properties of linear equations that a hyperplane may also be specified by a single linear equation

$$u_\alpha(x^\alpha - y^\alpha) = 0, \qquad (2.33)$$

where u_α is a covariant vector. Similarly, if $u_\alpha, \ldots, w_\alpha$ are r linearly independent covariant vectors, then the r hyperplanes

$$u_\alpha(x^\alpha - y^\alpha) = 0, \quad \ldots, \quad w_\alpha(x^\alpha - y^\alpha) = 0 \qquad (2.34)$$

intersect in an $(n-r)$-plane. The coordinate invariance of (2.32) to (2.34) follows since $(x^\alpha - y^\alpha)$ is a vector even though x^α and y^α separately are not. In a given coordinate system the straight lines through the origin along which only one coordinate is nonzero are called the *coordinate axes*, while the hyperplanes on which precisely one coordinate vanishes are known as the *coordinate hyperplanes*.

If a^α is a fixed vector, the transformation of a geometrical figure in which every point is moved according to the rule $x \to y$, where

$$y^\alpha = x^\alpha + a^\alpha,$$

is known as a *translation*. Two r-planes which can be brought into coincidence by a translation are said to be *parallel*. If $1 \leqslant s < r < n$, an s-plane is said to be parallel to a given r-plane Π if it can be brought by a translation to lie entirely within Π.

These constructions enable both covariant and contravariant vectors to be given a geometric representation. Although this is largely just an inversion of the definitions of the geometric concepts, it can help one to visualize these otherwise rather abstract concepts, since the geometric objects are generalizations of familiar objects of Euclidean geometry. A contravariant vector a^α has a simple representation by a directed straight line segment from any point y to the point with coordinates $y^\alpha + a^\alpha$. When considered in this manner, it is often referred to as a 'vector at y', and the line segment is identified with the vector. If a^α is an isolated vector, y will usually be taken as some fixed origin, while if $a^\alpha(x)$ is a vector field, it is most naturally taken as the field point x. The hyperplane (2.33) determines the covariant vector u_α up to a scalar multiple. This multiple is fixed if the parallel hyperplane

$$u_\alpha(x^\alpha - y^\alpha) = 1 \qquad (2.35)$$

is known as well. A covariant vector is thus representable up to a scalar multiple by a hyperplane through a fixed point y, and uniquely by an ordered pair of parallel hyperplanes, the first of which passes through the fixed point. Two vectors of the same type may be called parallel if the lines or hyperplanes which represent them are parallel. This is so if and only if one vector is a nonzero scalar multiple of the other. Sometimes this is refined by calling them parallel or antiparallel according as this proportionality factor is positive or negative. It is generally clear from the context which meaning for parallelism is intended.

The basis vectors of a coordinate system have a particularly simple representation. The contravariant ones correspond to the line segments which join the origin to the unit points on the coordinate axes. The covariant ones correspond to the pairs of hyperplanes $x^\alpha = 0$, $x^\alpha = 1$, for $\alpha = 1, 2, ..., n$. Both together, they may be represented by the edges and faces of the n-dimensional parallelepiped formed by the intersection of these $2n$ hyperplanes.

We often need to use smooth surfaces in an E_n which are more general than the r-planes defined above, so let us now make this concept more precise. This needs a formalization and generalization of intuitive notions based on curves and surfaces in three dimensions, and is most easily given in terms of a specific parametrization. So let (α) be a fixed coordinate system in E_n and let $x^\alpha = x^\alpha(\lambda_1, ..., \lambda_r)$ be n differentiable functions of r parameters. If the generated r-surface is to be in one piece and truly r-dimensional, the range of the parameter set must be a connected open subset U of the space R^r of ordered real r-tuples with its natural topology. To prevent the surface from intersecting itself, the correspondence between U and E_n has to be one-to-one. Finally, to ensure smoothness, at every point of U the r vectors $\partial x^\alpha/\partial\lambda_i$, $i = 1, 2, \ldots, r$ are required to be nonzero and linearly independent. These three conditions are clearly invariant under a change of rectilinear coordinate system in E_n. When they are satisfied, the image of U in E_n will be called a *parametrizable r-surface*, and the λ_i will be said to give it an *admissible parametrization*.

Parametrizable r-surfaces are certainly smooth, but as a class they are too restrictive to be considered as the only smooth r-surfaces. Some very simple surfaces, such as the surface of a sphere, cannot be given a parametrization which satisfies these conditions everywhere. For example, the usual polar angles (θ, ϕ) are degenerate on the axis $\theta = 0$. A nondegenerate description of a sphere requires two parametrizations, neither of which covers the whole sphere on its own but which overlap and cover it between them. A *smooth r-surface* will thus be defined as one which can be constructed from a finite number of overlapping parametrizable r-surfaces.

A 1-surface and an $(n-1)$-surface in an E_n are known also as a *curve* and a *hypersurface* respectively. An n-surface is also meaningful. It is an open set in E_n, and as such has a natural parametrization by any rectilinear coordinate system. To give it a more appropriate name, an n-surface will be called a *region* of the E_n. A hypersurface is often described more conveniently by a single equation than by $(n-1)$ parameters. Let ϕ be a scalar function, and S be the set of points at which ϕ is zero. Suppose further that $\partial_\alpha\phi \neq 0$† at all points of S. Then if $y \in S$, it follows from the implicit function theorem of calculus that the equation

$$\phi(x^1, \ldots, x^n) = 0 \qquad (2.36)$$

can be solved in the neighbourhood of y for one of the coordinates, x^1 say, as a differentiable function of the others. The coordinates $x^2, \ldots, x^n$ thus serve as a parametrization of S around y which satisfies the conditions given above. This shows that the portion of S near y is a parametrizable hypersurface, and since y was chosen arbitrarily, S itself is formed from overlapping parametrizable hypersurfaces. S thus consists either of a single smooth hypersurface or of a number of disconnected smooth hypersurfaces. Note that the single equation (2.36) may describe a hypersurface that is not expressible by a single

---

† This means that $\partial_\alpha\phi$ is not the zero vector. If it were taken as implying that $\partial_\alpha\phi$ is nonzero for each α, there could be situations in which neither of $\partial_\alpha\phi = 0$ and $\partial_\alpha\phi \neq 0$ hold. Tensor inequalities are always taken as the negation of the corresponding equality.

admissible parametrization. This, of course, is the situation for a sphere.

If y is a point of an r-surface S and $x^\alpha(\lambda_1, \ldots, \lambda_r)$ is a parametrization of S which is admissible around y, the r-plane through y determined by the r vectors $[\partial x^\alpha/\partial \lambda_i]_y$ is known as the *tangent r-plane* to S at y. It is easily seen to be independent of the particular parametrization chosen. A vector at y is said to be tangent to S at y if it lies in the tangent r-plane at that point. If S is the hypersurface given by (2.36), its tangent hyperplane at y is also given by

$$(x^\alpha - y^\alpha)\,[\partial_\alpha \phi]_y = 0. \qquad (2.37)$$

3 Fundamental tensors

The tensor notation of the preceding section has been devised to enable all rectilinear coordinate systems to be treated equally. The fact must now be faced that in the physical and geometrical theories with which we are concerned, they are not all equal. There exists a preferred subclass whose members are mutually equivalent, but which is distinguished by some physical or geometrical property from other rectilinear systems. Our task is now to translate this distinctive property into a mathematical form.

The key to this is provided by a type of isotropic tensor. We have seen that the unit tensor A^α_β is fully isotropic, i.e. its components have the same numerical values in *all* coordinate systems. The tensors that take this key role have a more restricted type of isotropy. Their components take fixed values in all coordinate systems of the preferred subclass, *but in no others*. They thus characterize that subclass completely, and will be called the *fundamental tensors* of the theory concerned. There may be more than one for a given theory, all of which have to take given values simultaneously in the preferred coordinates.

This association of tensors with preferred coordinate systems can be taken in either direction. Although not all preferred subclasses can be represented in this way by suitable tensors, all tensors can be used to define a preferred subclass. If an arbitrary tensor is chosen and arbitrary values are specified for its components, then the class of all coordinate systems in which the components of that tensor take those values may be considered as preferred. In general it will not be an important subclass, but it shows that there is nothing distinctive about the tensors that can be used in this way. If different component values are specified for the same tensor then a different preferred subclass is obtained.

Let us now consider the significance of these remarks for physics. The existence of the tensors which determine the preferred coordinates of physics is obscured when attention is restricted to such coordinate systems, as its components are then fixed numbers which are indistinguishable from scalars. Nevertheless, these tensors are objects of fundamental physical importance, as they have an existence independent of any coordinate system. In contrast, the numerical values of their components in the preferred systems have only a limited significance. They are characteristic of the fundamental tensor only to the extent that not every tensor of the same valence may be given these component values by a suitable choice of coordinate system. Similarly, the preferred systems themselves have no unique part to play. As remarked above, the numerical values could be changed and the preferred coordinate systems would change accordingly, but nothing of physical significance would be altered. This is analogous to the situation discussed in §1-2, where a change from Cartesian to polar coordinates was seen to alter the mathematical description of phenomena without it changing the frame of reference in any real sense.

In conclusion, the important quantities to extract from the preferred frames are the fundamental tensors. Each of these must be specified by its valence and such additional information as is needed to ensure that there exist coordinate systems in which its components take the appropriate values. This is all that has any absolute existence. Everything else, including these specific values, is conventional.

This discussion has been aimed at the situation in physics, but similar remarks are valid also for Euclidean geometry. The Cartesian coordinate systems are picked out by the specially simple expression that they give for the distance between two points, but this is purely by convention. Apart from the distance function itself, the quantity of absolute significance is the fundamental tensor needed to specify the Cartesian systems and which enables the distance function to be evaluated in any rectilinear coordinate system.

As the simplest example, this case forms a good starting point for the mathematical development that must now follow. In any Cartesian coordinate system, the distance s between two points x and y is given by

$$s^2 = (x^1 - y^1)^2 + (x^2 - y^2)^2 + (x^3 - y^3)^2, \qquad (3.1)$$

and conversely any coordinate system in which this holds is Cartesian. If indices run from 1 to 3 and we define

$$\delta_{\alpha\beta} = \begin{cases} 1 & \text{if } \alpha = \beta \\ 0 & \text{if } \alpha \neq \beta \end{cases} \tag{3.2}$$

then (3.1) can be written with the notation (2.10) as

$$s^2 = \delta_{\alpha\beta}\,\delta x^{\alpha}\delta x^{\beta}. \tag{3.3}$$

It is easily seen that the transformation (2.11) preserves this form if and only if

$$\delta_{\alpha\beta} = A^a_{\alpha}A^b_{\beta}\,\delta_{ab}. \tag{3.4}$$

The equation (3.4) has two roles. First, it characterizes those affine transformations which connect two Cartesian coordinate systems. This can be expressed in more familiar language if we write A for the matrix (A^a_{α}), where a and α are the row and column labels respectively. Then since $I = (\delta_{\alpha\beta})$ is the unit matrix, (3.4) becomes

$$A'A = I, \tag{3.5}$$

which is the condition for A to be orthogonal. Secondly, it shows that $\delta_{\alpha\beta}$ behaves as an isotropic tensor of valence $(0, 2)$ if the coordinate systems are restricted to be Cartesian. But a tensor is uniquely determined if its components are known in only one coordinate system. There thus exists a unique affine tensor ${}^{(e)}g_{\alpha\beta}$ of valence $(0, 2)$ such that

$${}^{(e)}g_{\alpha\beta} = \delta_{\alpha\beta} \tag{3.6}$$

in every Cartesian coordinate system. With its aid, (3.3) can be written as

$$s^2 = {}^{(e)}g_{\alpha\beta}\,\delta x^{\alpha}\delta x^{\beta}, \tag{3.7}$$

which now holds in all coordinate systems. The Cartesian systems are picked out as the only ones in which ${}^{(e)}g_{\alpha\beta}$ satisfies (3.6). The prefix (e) is to distinguish this Euclidean case from other tensors which will be introduced below and also given a kernel letter g.

The distinction between the notations ${}^{(e)}g_{\alpha\beta}$ and $\delta_{\alpha\beta}$ should be noted carefully. The former is a tensor and complies with the kernel–index convention. The latter is a pure numerical array which lies outside this convention. Although this perhaps seems a subtle distinction, it is quite essential if (3.6) is to be anything other than a trivial identity. It is useful to extend the δ-notation of (3.2) to the n-dimensional case and to write its indices in any convenient position, so that

$$\delta_{\alpha\beta} = \delta^{\alpha}_{\beta} = \delta^{\alpha\beta} = \begin{cases} 1 & \text{if } \alpha = \beta \\ 0 & \text{if } \alpha \neq \beta. \end{cases} \tag{3.8}$$

It is known as the *Kronecker symbol*. Its relation to the unit tensor is that

$$A_\beta^\alpha = \delta_\beta^\alpha \qquad (3.9)$$

for any single rectilinear coordinate system (α), but that in general

$$A_\alpha^a \ne \delta_\alpha^a \qquad (3.10)$$

when the two systems (α) and (a) are distinct.

The fundamental tensor of special relativity has a similar derivation. Let indices now run from 1 to 4 and define

$$(\eta_{\alpha\beta}) = \text{diag}\,(1, 1, 1, -c^2) \qquad (3.11)$$

as in (1-10.6). This, like $\delta_{\alpha\beta}$, is simply a numerical array. It follows from the results of §§1-7 and 1-10 that the Minkowskian coordinate systems are precisely those for which the quadratic invariant Φ associated with any pair of points takes the form

$$\Phi = \eta_{\alpha\beta}\,\delta x^\alpha \delta x^\beta \qquad (3.12)$$

given by (1-10.7). The identical argument which led from (3.3) to (3.6) now shows that there exists a unique affine tensor $^{(r)}g_{\alpha\beta}$ of valence $(0, 2)$ such that

$$^{(r)}g_{\alpha\beta} = \eta_{\alpha\beta} \qquad (3.13)$$

in every Minkowskian coordinate system but in no others. The general form for Φ is thus

$$\Phi = {}^{(r)}g_{\alpha\beta}\,\delta x^\alpha \delta x^\beta. \qquad (3.14)$$

The situation in the Newtonian theory is a little different. Here, we start from the explicit form for a general Galilean transformation given by (1-3.2). With a slight change of notation, this shows that the corresponding matrix (A_α^a) of (2.1) has the $(3+1) \times (3+1)$ block form

$$(A_\alpha^a) = \begin{pmatrix} L & \mathbf{v} \\ 0 & 1 \end{pmatrix}, \qquad (3.15)$$

where L is a (3×3) orthogonal matrix and the (3×1) column vector $\mathbf{v}$ is arbitrary. Since (3.15) satisfies

$$A_\alpha^a \delta_a^4 = \delta_\alpha^4, \qquad (3.16)$$

there exists an affine vector t_α such that

$$t_\alpha = \delta_\alpha^4 \qquad (3.17)$$

in every Galilean coordinate system. But (3.17) alone is not sufficient to characterize the Galilean systems, as although (3.16) implies that A_α^a has the block form (3.15), it does not imply that L is orthogonal.

This orthogonality may be expressed as a condition on A_{α}^{a} with the aid of the matrix identity

$$\begin{pmatrix} L & \mathbf{v} \\ 0 & 1 \end{pmatrix}\begin{pmatrix} I & 0 \\ 0 & 0 \end{pmatrix}\begin{pmatrix} L' & 0 \\ \mathbf{v}' & 1 \end{pmatrix} = \begin{pmatrix} LL' & 0 \\ 0 & 0 \end{pmatrix}, \tag{3.18}$$

where I is the (3×3) unit matrix. The required condition $LL' = I$ is thus equivalent to

$$A_{\alpha}^{a}\,\zeta^{\alpha\beta}A_{\beta}^{b} = \zeta^{ab}, \tag{3.19}$$

where

$$(\zeta^{ab}) = \mathrm{diag}\,(1,1,1,0). \tag{3.20}$$

Once again, ζ^{ab} is to be considered simply as a numerical array. It follows from (3.19) that there exists an affine tensor $^{(n)}g^{\alpha\beta}$ such that

$$^{(n)}g^{\alpha\beta} = \zeta^{\alpha\beta} \tag{3.21}$$

in every Galilean coordinate system. These systems are now completely characterized by both (3.17) and (3.21) together. The Newtonian theory thus has two fundamental tensors.

The valences of the fundamental tensors have been found for the three cases of particular interest, but according to the general discussion we also require 'such additional information as is needed to ensure that there exist coordinate systems in which their components take the appropriate values'. It is obvious from the values (3.2), (3.11) and (3.20) that the three tensors $^{(e)}g_{\alpha\beta}$, $^{(r)}g_{\alpha\beta}$ and $^{(n)}g^{\alpha\beta}$ must all be symmetric, but this alone is not enough. Now under a coordinate transformation the matrix G of components of a symmetric tensor of valence $(0, 2)$ or $(2, 0)$ undergoes a transformation of the form

$$G \to A'GA \tag{3.22}$$

where A is a nonsingular matrix. Such transformations were discussed in §1-10 in respect of equation (1-10.22), where it was seen that by a suitable choice of A, G can be reduced to a diagonal form in which the first p diagonal elements are $+1$, the next z are 0 and the final n are -1. These values of p, z and n thus completely characterize the component values obtainable for that tensor. In matrix theory, the transformations (3.22) arise in the treatment of quadratic forms. The terminology of that theory is taken over into tensor theory so that a symmetric tensor of total valence 2 is said to be

nonsingular	if $z = 0$
singular	if $z > 0$
positive semi-definite	if $n = 0$
negative semi-definite	if $p = 0$
positive definite	if $z = n = 0$
negative definite	if $z = p = 0$
indefinite	if $p > 0$ and $n > 0$.

If $p = n = 0$ it is, of course, the zero tensor. The numbers $p + n$ and $p - n$ are known as the *rank* and *signature* of the tensor, while $p + n + z$ is necessarily the dimension of the space on which it is defined. Hence

$^{(e)}g_{\alpha\beta}$ is positive definite on an E_3,

$^{(r)}g_{\alpha\beta}$ is nonsingular and of signature $+2$ on an E_4,

$^{(n)}g^{\alpha\beta}$ is positive semi-definite and of rank 3 on an E_4.

Only the Newtonian theory requires further examination, as nothing has yet been said about the vector t_α. It follows from (3.17) and (3.20) that t_α satisfies

$$^{(n)}g^{\alpha\beta}t_\beta = 0. \tag{3.23}$$

To see that this is a sufficient characterization of t_α, it is only necessary to observe that if (3.23) and (3.21) both hold, t_α must have the form

$$t_\alpha = K\delta_\alpha^4 \tag{3.24}$$

for some real K. A coordinate transformation $(\alpha) \to (a)$ with

$$(A_a^\alpha) = \mathrm{diag}\,(1, 1, 1, K^{-1}) \tag{3.25}$$

now leaves $^{(n)}g^{\alpha\beta}$ unchanged but makes $t_a = \delta_a^4$ as required.

An E_n on which is given a symmetric nonsingular fundamental tensor of valence $(0, 2)$ is said to have a *metric*, and is called an R_n. The fundamental tensor is known as the *metric tensor* of the R_n, and is generally denoted by $g_{\alpha\beta}$. If it is positive definite, the R_n is said to be *Euclidean*. The signature of $g_{\alpha\beta}$ is known also as the signature of the R_n. The space of Euclidean geometry is thus a Euclidean R_3. It is the relationship (3.7) between $^{(e)}g_{\alpha\beta}$ and distance which is the origin of the term 'metric'. The spacetime of special relativity is an R_4, but that of Newtonian physics is not.

The components of an arbitrary metric tensor $g_{\alpha\beta}$ transform under a coordinate transformation $(\alpha) \to (a)$ according to (3.22) if

$$G = (g_{\alpha\beta}) \quad \text{and} \quad A = (A_a^\alpha), \tag{3.26}$$

where the index labelling the rows is either the first index or the superscript index as appropriate. With this convention it follows from (2.6) and (2.7) that

$$A^{-1} = (A_\alpha^a), \tag{3.27}$$

so that if $G^{-1} = (g^{\alpha\beta})$ then the transformation

$$G^{-1} \to A^{-1}G^{-1}(A^{-1})'$$

induced by (3.22) has the index form

$$g^{\alpha\beta} \to g^{ab} = A_\alpha^a A_\beta^b g^{\alpha\beta}. \tag{3.28}$$

The variables $g^{\alpha\beta}$ thus form a tensor of valence $(2,0)$, known as the *contravariant metric tensor*. Since $g_{\alpha\beta}$ is symmetric, so also is $g^{\alpha\beta}$. Its relationship to $g_{\alpha\beta}$ can be written in the tensorial form

$$g^{\alpha\gamma}g_{\gamma\beta} = A^{\alpha}_{\beta}. \tag{3.29}$$

In an R_n, the distinction between covariant and contravariant vectors loses much of its significance. From any contravariant vector u^{α} we can construct a covariant vector $g_{\alpha\beta}u^{\beta}$. This will be denoted by u_{α}, which retains the same kernel letter and simply has the index moved to the appropriate position. Similarly, a contravariant vector

$$v^{\alpha} \equiv g^{\alpha\beta}v_{\beta} \tag{3.30}$$

can be associated with any covariant vector v_{α}. The two processes are consistent since it follows from (3.29) that

$$u_{\alpha} = g_{\alpha\beta}u^{\beta} \quad \text{implies} \quad u^{\alpha} = g^{\alpha\beta}u_{\beta}.$$

There is a natural extension to tensors of any valence, in which each index is moved independently. The same kernel letter is used for all the resulting tensors and the indices are moved vertically. This is the reason why superscript and subscript indices are not usually written on a tensor in the same vertical line. As an example, if we start from a tensor $T^{\alpha\beta}$ of valence $(2,0)$ we can produce three more tensors:

$$T^{\alpha}_{.\beta} = g_{\beta\gamma}T^{\alpha\gamma}, \quad T^{.\beta}_{\alpha} = g_{\alpha\gamma}T^{\gamma\beta}, \quad T_{\alpha\beta} = g_{\alpha\gamma}g_{\beta\delta}T^{\gamma\delta}, \tag{3.31}$$

with such interrelations as

$$T^{.\beta}_{\alpha} = g_{\alpha\gamma}g^{\beta\delta}T^{\gamma}_{.\delta}. \tag{3.32}$$

These processes are known simply as *raising and lowering indices*.

Although the indices on the unit tensor could be raised and lowered, we would find that

$$A^{\alpha\beta} = g^{\alpha\beta} \quad \text{and} \quad A_{\alpha\beta} = g_{\alpha\beta}. \tag{3.33}$$

Similarly, if one index of $g_{\alpha\beta}$ were raised, we would obtain

$$g^{.\beta}_{\alpha} = A^{\beta}_{\alpha}. \tag{3.34}$$

Such a notation would be redundant, and it is thus conventional not to apply these processes to A^{β}_{α} and $g_{\alpha\beta}$. There is thus no need to space out the indices on A^{β}_{α}. It should be noted, however, that (3.29) implies

$$g^{\gamma\alpha}g^{\delta\beta}g_{\alpha\beta} = g^{\gamma\delta}, \tag{3.35}$$

so that the relationship between $g_{\alpha\beta}$ and $g^{\alpha\beta}$ is consistent with the raising and lowering operations.

In a Newtonian spacetime it is possible to use $^{(n)}g^{\alpha\beta}$ in (3.30) to define a raising operation, but this differs in nature from the corresponding operation in an R_n since $^{(n)}g^{\alpha\beta}$ is singular. There is thus no inverse process of lowering an index, and the vector u_α and its raised version u^α are inequivalent. This is well illustrated from (3.23), which implies that $t^\alpha = 0$. Despite these drawbacks it remains a useful convention and we shall adopt it. However, the distinction between covariant and contravariant tensor indices remains of much greater significance in Newtonian physics than in the metric spacetime of special relativity.

4 Specializations and limits

4a Cartesian tensors

In the preceding section it was seen that the use of a preferred subclass of the rectilinear coordinate systems can be avoided if the fundamental tensors which are responsible for its existence are introduced explicitly. This is important for general theoretical considerations, as it shows that nothing has been hidden in the choice of coordinates. However, when a specific physical or geometrical situation is being treated, it is often more convenient to reintroduce a special coordinate system. This is normally one which is in some way linked to the situation under consideration. Let us now consider the mathematical simplifications that this can produce.

A restriction to a Galilean or Minkowskian system in spacetime enables the fundamental tensors to be replaced by their constant numerical values, but it does little more than this. The corresponding choice in a Euclidean R_n, however, does produce a more significant simplification. The natural choice of preferred coordinates here is that given by

$$g_{\alpha\beta} = \delta_{\alpha\beta}, \quad g^{\alpha\beta} = \delta^{\alpha\beta}. \tag{4.1}$$

As a generalization of the usual $n = 3$ situation, such coordinate systems are said to be Cartesian. In such systems, the raising and lowering operations such as (3.31) do not alter the numerical values of the components of any tensor. The distinction between covariant and contravariant indices thus disappears completely. This enables us to write all indices as subscripts, provided that the five rules of §1 are amended accordingly. A tensor which is so written, with the corresponding assumption that only Cartesian coordinate systems are to be considered, is known as a *Cartesian tensor*. As only its total valence is of significance, the separate contravariant and covariant valences are not defined. Nothing is lost by the use of this simplified notation, as validity in a general rectilinear coordinate system may be restored simply by raising some indices so as to comply with the original form of the five rules. It should be noted that no corresponding simplification is possible in an R_n whose metric tensor is not positive definite. For although coordinate systems exist in which $g_{\alpha\beta}$ is diagonal and in which all the diagonal elements are ± 1, the raising or lowering of an index in such a system will still change the sign of some of the components. A distinction, even if only a slight one, thus still exists between the covariant and contravariant forms.

Let us now consider the situation in physics under restriction to the natural coordinate systems of a fixed inertial frame. We recall that the remaining coordinate freedom, expressed by (1-2.1), corresponds to a change of the zero of time together with a change of Cartesian coordinate system in the Euclidean geometry of space. Of these changes, only the latter affects the components of a tensor. Its effect can be found by substituting into (2.17) the appropriate value for (A^a_α), which from (1-2.1) has the $(3+1) \times (3+1)$ block form

$$(A^a_\alpha) = \begin{pmatrix} L & 0 \\ 0 & 1 \end{pmatrix}, \tag{4.2}$$

where L is orthogonal as in (3.15). The four index values are no longer all intermingled under (4.2). Instead, the values 1, 2, 3 transform among themselves like the index of a Cartesian tensor under the change of axes described by L, while the value 4 acquires an invariant character. Consequently, a four-dimensional affine tensor of total valence m separates into 2^m three-dimensional Cartesian tensors, ^mC_r of which have valence r.

To illustrate this, let α, β range from 1 to 4 and κ, λ range from 1 to 3. Note that both pairs of letters represent the same coordinate system as they come from the same alphabet. Then the affine tensor $T^\alpha_{.\beta}$ splits into

$$\left.\begin{array}{l} \text{a Cartesian scalar, } t = T^4_{.4}; \\ \text{two Cartesian vectors, } u_\kappa = T^\kappa_{.4}, \; v_\kappa = T^4_{.\kappa}; \\ \text{a Cartesian tensor of valence 2,} \; W_{\kappa\lambda} = T^\kappa_{.\lambda}. \end{array}\right\} \tag{4.3}$$

In accordance with the above discussion, all Cartesian tensor indices are written as subscripts. The decomposition (4.3) will be written for conciseness in the hybrid matrix notations

$$T^{\alpha}_{.\beta} = \left(\begin{array}{c|c} W_{\kappa\lambda} & u_{\kappa} \\ \hline v_{\kappa} & t \end{array}\right) \quad \text{or} \quad \left(\begin{array}{c|c} \mathbf{W} & \mathbf{u} \\ \hline \mathbf{v} & t \end{array}\right) \quad \text{or} \quad \left(\begin{array}{c|c} W_{ab} & u_{a} \\ \hline v_{a} & t \end{array}\right). \qquad (4.4)$$

The second of these uses the notation of three-dimensional vector analysis to write the dyadic $\mathbf{W}$ and vectors $\mathbf{u}$ and $\mathbf{v}$ in an index-free form. (Dyadic notation is summarized in the Appendix.) In the third expression, (a) is an arbitrary three-dimensional Cartesian coordinate system. If this is other than the system (κ) formed by the first three coordinates of the spacetime system (α), then the equality in (4.4) is symbolic of the precise relations

$$u_a = \sum_{\kappa=1}^{3} A_{a\kappa} T^{\kappa}_{.4},$$

etc., obtained from (4.3) by a Cartesian coordinate transformation. These last two forms will also be written when the spacetime coordinates (α) are a general rectilinear system, but in that case it is necessary to specify separately the inertial reference frame in which the decomposition is to be performed.

In the same manner we shall write

$$u_{\alpha} = (\mathbf{u}, \phi) \qquad (4.5)$$

to denote the splitting of an affine vector u_{α} into a Cartesian vector $\mathbf{u}$ and a scalar ϕ. In this notation the contravariant vector which results when the index is raised is given by

$$u^{\alpha} = (\mathbf{u}, -\phi/c^2) \qquad (4.6)$$

in special relativity and by $\quad u^{\alpha} = (\mathbf{u}, 0) \qquad (4.7)$

in the Newtonian theory, as follows from (3.11) and (3.20) respectively.

4b The Newtonian limit $c \to \infty$

A problem which is closely related to these $(3 + 1)$-dimensional decompositions is that of the nature of the limiting process by which the Newtonian theory can be recovered from the relativistic theory. In a formal sense many Newtonian results can be obtained from relativistic ones by letting $c \to \infty$. In particular, the Galilean boosts (1-6.7) can be obtained from the Lorentz boosts (1-7.6) in this way. But what is the meaning of this limit? The value of c may be changed by a change of the unit of time. As this unit is made larger, so c gets larger. So also do all other velocities. But in the formal limit $c \to \infty$ it is assumed that other velocities, such as v in (1-7.6), remain finite. If such velocities are held numerically fixed while c is increased, it means physically that the unit of time is enlarged and that the physical velocities measured in a fixed reference frame and with respect to fixed units are reduced proportionately. The actual limit cannot be achieved in physical terms, but its theoretical meaning is clear. It is the theory that appears to hold when all velocities v are so small in comparison with the speed of light that only the lowest order terms in any expansion in powers of the dimensionless ratio v/c need to be retained.

Would it not thus be more correct to compare Newtonian and relativistic results by performing this expansion in powers of v/c, and so avoiding the transition to the unachievable limit? No. Such an expansion is useful when one wishes to obtain relativistic corrections to Newtonian results when v/c is small but not negligible, but when one is interested in only the lowest, i.e. Newtonian, terms, the two procedures are precisely equivalent. The more complicated the situation, the greater is the advantage gained by the limit procedure and the more difficult becomes the evaluation of relativistic corrections.

The description of the limit process given above appears to pick a uniquely preferred reference frame, since there is only one frame in which $v/c \to 0$. So how does the limiting theory possess an invariance group, the Galilean group, especially one which differs from the Lorentz group that is the invariance group for all finite values of c? This is most easily seen by a visualizable geometric analogy. Let $Oxyz$ be a fixed Cartesian coordinate system and let $S(r)$ be the surface of the sphere of radius r centred at $(0, 0, -r)$. This sphere passes through O for all r, and is invariant under the three-parameter group $\mathscr{R}$ of rotations about its centre. As $r \to \infty$, $S(r)$ tends to the (x, y) plane Π whose symmetry group $\mathscr{E}$ still has three parameters but which differs from $\mathscr{R}$. This happens because we consider only those points on the sphere which remain at a finite distance from O as $r \to \infty$, and consequently only those rotations of the sphere which preserve this finiteness. Now a general rotation may be represented by its Euler angles $\{\alpha, \beta, \gamma\}$, which specify it as a succession of three rotations :
(i)   one through an angle α about the z-axis, followed by
(ii) one through an angle β about the y-axis, followed by
(iii) one through an angle γ about the z-axis again.
The overall rotation preserves the finiteness of distance from O if and only if $\beta = O(r^{-1})$ as $r \to \infty$. The group $\mathscr{E}$ obtained from $\mathscr{R}$ in the limit as $r \to \infty$ thus corresponds to the portion of $\mathscr{R}$ for which β is infinitesi-

mal. This is not the same as the subgroup of $\mathscr{R}$ obtained by setting $\beta = 0$, since even an infinitesimal β has an effect on Π due to the enormous distance through which it acts, i.e. since $r\beta$ may have a nonzero limit as $r \to \infty$ even though $\beta \to 0$.

This argument is easily put in a quantitative form. If $\mathbf{x} \to \mathbf{x}^*$ under the rotation $\{\alpha, \beta, \gamma\}$, then

$$
\begin{aligned}
x^* &= x(\cos\alpha\cos\beta\cos\gamma - \sin\alpha\sin\gamma) - y(\sin\alpha\cos\beta\cos\gamma \\
&\quad + \cos\alpha\sin\gamma) + (z+r)\sin\beta\cos\gamma, \\
y^* &= x(\cos\alpha\cos\beta\sin\gamma + \sin\alpha\cos\gamma) - y(\sin\alpha\cos\beta\sin\gamma \\
&\quad - \cos\alpha\cos\gamma) + (z+r)\sin\beta\sin\gamma, \\
z^*+r &= -x\cos\alpha\sin\beta + y\sin\alpha\sin\beta + (z+r)\cos\beta.
\end{aligned} \tag{4.8}
$$

Now let $r \to \infty$ with x, y, α, γ, and $\delta \equiv r\beta$ remaining finite. Then $z = O(r^{-1})$, so that we obtain as the limiting case of (4.8)

$$
\begin{aligned}
x^* &= x\cos(\alpha+\gamma) - y\sin(\alpha+\gamma) + \delta\cos\gamma, \\
y^* &= x\sin(\alpha+\gamma) + y\cos(\alpha+\gamma) + \delta\sin\gamma, \\
z^* &= z = 0,
\end{aligned} \tag{4.9}
$$

in which α, γ and δ remain as three independent parameters. This is the general symmetry transformation of a plane.

A flaw in the analogy between the above situation and that in relativistic physics is that the surface $S(r)$ is itself a function of r. The correspondence can be improved if we instead consider the action of the rotation group on the whole three-dimensional space. As before, we consider a rotation whose Euler angles are functions of a parameter r. But this time we move the region of interest rather than the centre of rotation. We may do this by letting all rotations be about the origin and considering points for which x and y remain finite but which have $z = O(r)$ as $r \to \infty$. Put $t = z/r$, so that x, y and t all remain finite. Since the distance of such a point from the centre of rotation is also $O(r)$, each such point lies on a sphere whose radius as before is scaled with r. The difference is that we are no longer restricted to a single sphere, as each point defines its own. The rotation $\{\alpha, \beta, \gamma\}$ is now given by (4.8) with $(z+r)$ and (z^*+r) replaced by $z \equiv rt$ and $z^* \equiv rt^*$ respectively. To preserve the finiteness of x, y and t we again need α, γ and $\delta \equiv \beta r$ to remain finite as $r \to \infty$. The limiting transformation is now found to be

$$
\begin{aligned}
x^* &= x\cos(\alpha+\gamma) - y\sin(\alpha+\gamma) + t\delta\cos\gamma, \\
y^* &= x\sin(\alpha+\gamma) + y\cos(\alpha+\gamma) + t\delta\sin y, \\
t^* &= t.
\end{aligned} \tag{4.10}
$$

If (4.10) is compared with (1-3.2), it is seen to have the form of a two-dimensional origin-preserving Galilean transformation. This is not a coincidence. It was seen in §1-7 that in the nonphysical theory with $C < 0$, the time coordinate is geometrically indistinguishable from the spatial coordinates. If one spatial coordinate is suppressed, we then have precisely the above situation. The limit $r \to \infty$ corresponds to the recovery as $|C| \to \infty$ of the Galilean transformations from the transformations between inertial frames of that theory.

The significant case is the analogy with Lorentz transformations. This is in fact very close. The Lorentz boost (1-7.6) may be put in the form

$$x^* = x \cosh \lambda - ct \sinh \lambda, \quad y^* = y, \\ z^* = z, \quad ct^* = -x \sinh \lambda + ct \cosh \lambda, \Bigg\} \tag{4.11}$$

where $\tanh \lambda = v/c$. This is the hyperbolic analogue of a rotation, and a general Lorentz transformation is constructible as a boost (4.11) both preceded and followed by a rotation of the spatial axes. These rotations correspond to the Euler angles α and γ which remain finite in the limit. The Euler angle β is a precise analogue of λ, with $\beta = O(r^{-1})$ as $r \to \infty$ corresponding to $\lambda = O(c^{-1})$ as $c \to \infty$. The coordinates x, y of (4.10) correspond to the three relativistic spatial coordinates while $z \equiv rt$ corresponds to ct. Both coordinates t remain finite in their respective limits.

These examples are particular cases of a general process known as the *contraction* of a group. The underlying idea is always that certain group parameters are restricted to infinitesimal values but are kept significant by a suitable scaling process which gives them a finite action. More details of the general theory of group contractions may be found in the articles of Inonu & Wigner (1953) and Saletan (1961).

We conclude with some limiting relationships between the Newtonian and relativistic fundamental tensors. In the relativistic case, the proper time $\delta\tau$ between two events with timelike separation was defined in §1-7 and is given in the notation (2.10) by

$$\delta\tau^2 = -^{(r)}g_{\alpha\beta}\,\delta x^\alpha \delta x^\beta. \tag{4.12}$$

In the Newtonian case the absolute time interval δt between any two events is given by

$$\delta t = t_\alpha \,\delta x^\alpha, \tag{4.13}$$

as follows trivially from (3.17). These two expressions can be brought closer together in form if (4.13) is squared to give

$$\delta t^2 = t_\alpha t_\beta \, \delta x^\alpha \delta x^\beta. \tag{4.14}$$

If we now remember that it is $\delta\tau/c$, rather than $\delta\tau$ itself, which is in the relativistic theory the true analogue of δt, then the connexion between (4.12) and (4.14) becomes clear. For (4.12) gives

$$(\delta\tau/c)^2 = - \, (^{(r)}g_{\alpha\beta}/c^2) \, \delta x^\alpha \delta x^\beta, \tag{4.15}$$

while it follows from (3.11) and (3.17) that in a fixed inertial frame

$$\lim_{c \to \infty} (- \, ^{(r)}g_{\alpha\beta}/c^2) = t_\alpha t_\beta. \tag{4.16}$$

The connexion between $^{(r)}g^{\alpha\beta}$ and $^{(n)}g^{\alpha\beta}$ is mathematically simpler, but it does not have such an immediate physical interpretation. It follows from (3.11) and (3.13) that in any Minkowskian coordinate system

$$(^{(r)}g^{\alpha\beta}) = \mathrm{diag} \, (1, 1, 1, \, -c^{-2}), \tag{4.17}$$

so that

$$\lim_{c \to \infty} {}^{(r)}g^{\alpha\beta} = {}^{(n)}g^{\alpha\beta}. \tag{4.18}$$

When combined with (4.16), this shows that (3.23) arises from the limit as $c \to \infty$ of the relativistic identity

$$(- \, ^{(r)}g_{\alpha\gamma}/c^2) \, ^{(r)}g^{\gamma\beta} = -c^{-2}A_\gamma^\beta. \tag{4.19}$$

It is the vanishing of the right-hand side of (4.19) in the limit that prevents t_α from being completely determined by $^{(n)}g^{\alpha\beta}$. Its direction is determined, as is shown by (3.24), but there is an indeterminate scale factor. We now see how the single fundamental tensor of special relativity gives rise to a limiting theory which needs two such tensors.

5   Symmetries and alternating tensors

5a   The bracket notations

We have already seen that an invariant restriction can be placed on a tensor of valence $(0, 2)$ by requiring it to be symmetric or antisymmetric. These properties were defined by (2.29) and (2.30), from which it can be seen that in n dimensions $t_{\alpha\beta}$ has $\frac{1}{2}n(n+1)$ linearly independent components if it is symmetric and $\frac{1}{2}n(n-1)$ if it is antisymmetric. These figures compare with the n^2 linearly independent components of an unrestricted tensor of this valence. Any such unrestricted tensor $t_{\alpha\beta}$ gives rise to a symmetric tensor $s_{\alpha\beta}$ and an antisymmetric tensor $a_{\alpha\beta}$ by

$$s_{\alpha\beta} = \tfrac{1}{2}(t_{\alpha\beta}+t_{\beta\alpha}), \quad a_{\alpha\beta} = \tfrac{1}{2}(t_{\alpha\beta}-t_{\beta\alpha}), \tag{5.1}$$

from which $t_{\alpha\beta}$ itself can be recovered:

$$t_{\alpha\beta} = s_{\alpha\beta} + a_{\alpha\beta}. \tag{5.2}$$

Such properties and their generalizations are very important. A simple example of their use has already appeared in the study of fundamental tensors. Before we proceed much further it is thus wise to study them in more detail and to develop a concise notation for describing them. This will lead us to another important type of fundamental tensor, this time of valence $(0, n)$ in an E_n.

The simple definitions (2.29) and (2.30) can be generalized considerably. They extend naturally to a tensor of valence $(2, 0)$, and also to any two indices of the same type on a tensor of total valence higher than two. A tensor may have several such symmetry properties simultaneously, which need not involve disjoint pairs of indices. A tensor is said to be totally symmetric or totally antisymmetric on a set of three or more indices of the same type if it is symmetric or antisymmetric respectively on any pair of indices of that set.

Two questions arise naturally. First, why are the basic properties (2.29) and (2.30) of more significance than, say, $t_{\alpha\beta} = 2t_{\beta\alpha}$? Secondly, is there a decomposition of a tensor of valence $(0, r)$, $r \geqslant 3$, analogous to that given by (5.1) when $r = 2$? The answer to the first question is simple. Suppose

$$t_{\alpha\beta} = kt_{\beta\alpha}. \tag{5.3}$$

Then by relabelling the indices, we see that also

$$t_{\beta\alpha} = kt_{\alpha\beta}, \tag{5.4}$$

which may be combined with (5.3) to give

$$(k^2 - 1)t_{\alpha\beta} = 0. \tag{5.5}$$

Unless $t_{\alpha\beta}$ is identically zero, the only possibilities are thus $k = \pm 1$.

The second question is more difficult. It is easy to construct a totally symmetric and a totally antisymmetric part for any tensor $t_{\alpha\beta...\gamma}$ of valence $(0, r)$. The symmetric part is simply the average over all $r!$ possible permutations of the indices, while the antisymmetric part has a similar construction but with the odd permutations being subtracted instead of added. These are sufficiently important to be given a special notation. The symmetric part is denoted by enclosing the *indices* between round brackets ( ), while the antisymmetric part is similarly denoted with the use of square brackets [ ]. For $t = 3$ we thus write

$$t_{(\alpha\beta\gamma)} = \frac{1}{3!} \left(t_{\alpha\beta\gamma} + t_{\beta\gamma\alpha} + t_{\gamma\alpha\beta} + t_{\alpha\gamma\beta} + t_{\beta\alpha\gamma} + t_{\gamma\beta\alpha} \right) \tag{5.6}$$

and $$t_{[\alpha\beta\gamma]} = \frac{1}{3!} \left(t_{\alpha\beta\gamma} + t_{\beta\gamma\alpha} + t_{\gamma\alpha\beta} - t_{\alpha\gamma\beta} - t_{\beta\alpha\gamma} - t_{\gamma\beta\alpha} \right). \tag{5.7}$$

However, these two parts alone do not give a complete decomposition in the way that (5.1) does. They have respectively $\frac{1}{6}n(n+1)(n+2)$ and $\frac{1}{6}n(n-1)(n-2)$ linearly independent components, which leaves $\frac{2}{3}n(n^2-1)$ of the total of n^3 unaccounted for. The decomposition can be completed by the addition of two further parts, namely

$$t_{(\alpha\beta)\gamma} - t_{(\alpha\beta\gamma)} \quad \text{and} \quad t_{[\alpha\beta]\gamma} - t_{[\alpha\beta\gamma]}, \tag{5.8}$$

each of which can be shown to have $\frac{1}{3}n(n^2-1)$ linearly independent components, but these extra parts are of a more complex nature than (5.6) and (5.7). The involvement of γ in the symmetry properties of either of (5.8) cannot be described by the simple considerations of (5.3), as it involves all three indices simultaneously. In addition, these two parts are not unique. An alternative choice would be

$$t_{\alpha(\beta\gamma)} - t_{(\alpha\beta\gamma)} \quad \text{and} \quad t_{\alpha[\beta\gamma]} - t_{[\alpha\beta\gamma]}, \tag{5.9}$$

each of which can be shown to be a linear combination of the tensors (5.8). The only simple fact is that the sum of the four parts given by (5.6) and (5.7) with (5.8), or with (5.9), is $t_{\alpha\beta\gamma}$ as required. A proper study of such higher order symmetries needs the powerful techniques of the theory of group representations. It will not be followed further here. The interested reader is referred to the fundamental work of Weyl (1946) on the subject.

The bracket notation for symmetry operations is not restricted to tensors; it can be used on any indexed array. It can also be applied to a subset of indices, as in (5.8), and can be formally factorized so that, for example,

$$t_{\alpha[\beta}(u_{\gamma]} + v_{\gamma]}) = t_{\alpha[\beta}u_{\gamma]} + t_{\alpha[\beta}v_{\gamma]}$$

$$= \tfrac{1}{2}\{t_{\alpha\beta}(u_\gamma + v_\gamma) - t_{\alpha\gamma}(u_\beta + v_\beta)\}. \tag{5.10}$$

Vertical lines are used to exclude indices from the operations when the indices to which they are applied are not consecutive, e.g.

$$u_{(\alpha|\beta\gamma|\delta)} = \tfrac{1}{2}(u_{\alpha\beta\gamma\delta} + u_{\delta\beta\gamma\alpha}). \tag{5.11}$$

If an indexed array has p indices and is totally antisymmetric, those elements which have two numerically equal indices must be zero. If the indices range from 1 to n, as for tensors in an E_n, it follows that the array must vanish identically when $p > n$. When $p = n$, the whole array is determined by the element with index values $12\ldots n$ in that order. Although simple, these observations are often very useful.

Two special arrays of this type with $p = n$ are the *Levi-Civita symbols* $\epsilon_{\alpha\beta\ldots\gamma}$ and $\epsilon^{\alpha\beta\ldots\gamma}$, defined by

$$\epsilon_{\alpha\beta\ldots\gamma} = \epsilon_{[\alpha\beta\ldots\gamma]}, \quad \epsilon^{\alpha\beta\ldots\gamma} = \epsilon^{[\alpha\beta\ldots\gamma]}, \quad \epsilon_{12\ldots n} = \epsilon^{12\ldots n} = 1. \tag{5.12}$$

Let us now find the effect of a coordinate transformation on the one linearly independent component of a totally antisymmetric tensor $u_{\alpha\beta\ldots\gamma}$ of valence $(0, n)$ in an E_n. If upright and sloping numerical values are used for coordinate systems (α) and (a) respectively, then (2.17) and (2.18) give

$$\begin{aligned} u_{12\ldots n} &= A_{12\ldots n}^{\alpha\beta\ldots\gamma} u_{\alpha\beta\ldots\gamma} \\ &= n!\, A_{12\ldots n}^{[12\ldots n]} u_{12\ldots n} \\ &= \det\left(A_a^\alpha\right) u_{12\ldots n}. \end{aligned} \tag{5.13}$$

The second line comes from the antisymmetry of $u_{\alpha\beta\ldots\gamma}$, while the third follows from the second by the definition of a determinant. Any such tensor may thus be taken as a fundamental tensor which determines a preferred family of coordinate systems in which

$$u_{12\ldots n} = 1. \tag{5.14}$$

Two members of this family are related by a transformation which satisfies

$$\det\left(A_a^\alpha\right) = 1. \tag{5.15}$$

Such a transformation is said to be *unimodular*.

If $u^{\alpha\beta\ldots\gamma}$ is a totally antisymmetric tensor of valence $(n, 0)$, the method of derivation of (5.13) shows similarly that

$$u^{12\ldots n} = \det\left(A_\alpha^a\right) u^{12\ldots n}. \tag{5.16}$$

But

$$\det\left(A_\alpha^a\right) = \{\det\left(A_a^\alpha\right)\}^{-1} \tag{5.17}$$

since the matrices (A_α^a) and (A_a^α) are inverse to one another. The product $u^{12\ldots n} u_{12\ldots n}$ is thus an affine scalar. If $u_{\alpha\beta\ldots\gamma}$ is given, a unique tensor $u^{\alpha\beta\ldots\gamma}$ can thus be defined by the requirement that

$$u^{12\ldots n} u_{12\ldots n} = 1. \tag{5.18}$$

From (5.12), this can also be expressed as

$$u^{\alpha\beta\ldots\gamma} u_{\kappa\lambda\ldots\mu} = \epsilon^{\alpha\beta\ldots\gamma} \epsilon_{\kappa\lambda\ldots\mu}. \tag{5.19}$$

Totally antisymmetric tensors of valence $(0, n)$ or $(n, 0)$ in an E_n will be called *alternating tensors*. This name is sometimes applied to a totally antisymmetric tensor of any valence, but in view of the importance of such tensors of these particular valences as fundamental tensors, it is useful to have a specific name for them.

5b Oriented affine spaces

Two coordinate systems (α) and (a) are said to have the same orientation if $\det(A_a^\alpha) > 0$ and the opposite orientation if $\det(A_a^\alpha) < 0$. Similarity of orientation is reflexive† by (2.6) and symmetric by (5.17). Since (2.4) implies

$$\det(A_\alpha^A) = \det(A_a^A)\det_\alpha(A_\alpha^a),$$

it is also transitive. It is thus an equivalence relation and so divides the coordinate systems of an E_n into two classes such that all members of the same class have the same orientation. An E_n is said to be *oriented* if one of these classes is picked out as preferred. Its members are said to have positive orientation, while those of the other class are said to have negative orientation. This extends to a general E_n the distinction that is made in three dimensions between right-handed (positive) and left-handed (negative) coordinate systems. An orientation is also assigned to any ordered set of n linearly independent vectors, namely the orientation of a coordinate system for which they are the basis vectors.

It follows from (5.13) that a fundamental alternating tensor provides a natural orientation for an E_n according to the sign of $u_{12...n}$. A set of n vectors $a^\alpha, b^\alpha, ..., c^\alpha$ then has an orientation given by the sign of

$$u_{\alpha\beta...\gamma}\, a^\alpha b^\beta ... c^\gamma.$$

This quantity is zero if and only if the vectors are linearly dependent, as its value is proportional to the determinant which has the components of these vectors as its columns. The corresponding result for covariant vectors is also true. Conversely, an oriented E_n without additional structure determines an alternating tensor up to a positive scalar factor by the requirement that $u_{12...n} > 0$ in a positively oriented coordinate system, but it cannot distinguish one of these as particularly significant. The additional information contained in a fundamental alternating tensor, besides that of orientation, will be studied in the next section.

Our next result will be to show that an oriented R_n does contain enough structure to determine a particular alternating tensor, so that in contrast to the situation in an E_n, the only addition that a fundamental alternating tensor makes to an R_n is to give it a definite orientation. Let

$$^{(\alpha)}g \equiv \det g_{\alpha\beta}, \tag{5.20}$$

† A binary relation '$\sim$' is said to be reflexive if $a \sim a$ for all a, symmetric if $a \sim b$ implies $b \sim a$, and transitive if $a \sim b$ and $b \sim c$ together imply $a \sim c$. If it has all three properties it is said to be an *equivalence relation*.

where the prefix (α) on the left-hand side denotes the coordinate system in which the determinant is to be evaluated. This notation follows that of (2.1). If the determinant of (3.22) is evaluated for the case given by (3.26), it shows that

$$^{(a)}g = (\det A_a^\alpha)^{2\,(a)}g, \qquad (5.21)$$

from which

$$\sqrt{|^{(a)}g|} = |\det A_a^\alpha|\sqrt{|^{(a)}g|}. \qquad (5.22)$$

As the modulus signs around $\det A_a^\alpha$ are redundant for a transformation between two positively oriented coordinate systems, comparison of (5.22) with (5.13) shows that there exists a unique alternating tensor $\eta_{\alpha\beta\ldots\gamma}$ which satisfies

$$\eta_{\alpha\beta\ldots\gamma} = \sqrt{|^{(a)}g|}\,\epsilon_{\alpha\beta\ldots\gamma} \qquad (5.23)$$

in every such coordinate system. In a negatively oriented system the right-hand side of (5.23) acquires a minus sign. It follows from (4.1) and (5.23) that $\epsilon_{\alpha\beta\ldots\gamma}$ itself behaves as a Cartesian alternating tensor in the preferred coordinate systems of an oriented Euclidean R_n.

We have available two methods for the construction of a contravariant tensor from $\eta_{\alpha\beta\ldots\gamma}$. One of these is by (5.19), which gives

$$\eta^{\alpha\beta\ldots\gamma} = |^{(a)}g|^{-\frac12}\epsilon^{\alpha\beta\ldots\gamma} \qquad (5.24)$$

in any positively oriented coordinate system. The other is by raising the indices. Now the quantity

$$\epsilon^{\alpha\beta\ldots\gamma}g_{\alpha\kappa}g_{\beta\lambda}\cdots g_{\gamma\mu}$$

is totally antisymmetric in its free indices, so that it is a multiple of $\epsilon_{\kappa\lambda\ldots\mu}$. The proportionality factor is its value when $\kappa = 1$, $\lambda = 2$, ..., $\mu = n$, namely $^{(a)}g$. It thus follows from (5.23) and (5.24) that

$$\eta^{\alpha\beta\ldots\gamma}g_{\alpha\kappa}g_{\beta\lambda}\cdots g_{\gamma\mu} = (\mathrm{sgn}\,^{(a)}g)\,\eta_{\alpha\beta\ldots\gamma}, \qquad (5.25)$$

where

$$\mathrm{sgn}\,x = \begin{cases} 1 & \text{if } x > 0 \\ 0 & \text{if } x = 0. \\ -1 & \text{if } x < 0 \end{cases} \qquad (5.26)$$

This is self-consistent since (5.21) shows the sign of $^{(a)}g$ to be invariant under a coordinate trasformation. For the same reason the prefix (α) in (5.25) may be omitted. In fact, it can be seen from the definitions of §3 that in an R_n of signature s,

$$\mathrm{sgn}\,g = (-1)^{(n-s)/2}. \qquad (5.27)$$

However, (5.25) also shows that the two η-tensors are related by the usual raising and lowering rule in R_n only when $g > 0$. If $g < 0$, there is a sign discrepancy. Under these conditions we shall adopt both (5.23) and (5.24), thus adding these tensors to the set, which already contains A^α_β and the metric tensor, to which the raising and lowering processes are not applied. This convention is not universal. Some authors adopt only one of (5.23) and (5.24), and raise and lower indices in the usual way.

Although Newtonian spacetime is not an R_n, when it is oriented it too determines a preferred fundamental alternating tensor. To see this, it is only necessary to note that for the general Galilean transformation (3.15),

$$\det(A^a_\alpha) = \det L = \pm 1. \tag{5.28}$$

Since only the positive sign can occur for a transformation which preserves orientation, there is a unique pair of alternating tensors $^{(n)}\eta_{\alpha\beta\gamma\delta}$ and $^{(n)}\eta^{\alpha\beta\gamma\delta}$ such that

$$^{(n)}\eta_{1234} = {}^{(n)}\eta^{1234} = 1 \tag{5.29}$$

in all positively oriented coordinate systems which satisfy (3.17) and (3.21). The prefix (n) here stands for 'Newtonian' as in (3.21), rather than for a particular coordinate system as in (5.20). Once again the $\cdot$ raising rule for indices has to be suspended, for $^{(n)}\eta_{\alpha\beta\gamma\delta}$ gives zero if its indices are raised with $^{(n)}g^{\alpha\beta}$. However, there is a result which plays a somewhat similar role in a Newtonian spacetime to that played by (5.25) in an R_n. With the prefixes (n) omitted, this is that

$$\eta_{\alpha\beta\gamma\delta} g^{\alpha\kappa} g^{\beta\lambda} g^{\gamma\mu} = \eta^{\kappa\lambda\mu\nu} t_\nu t_\delta. \tag{5.30}$$

It is easily verified in the coordinate systems in which (3.17), (3.21) and (5.29) all hold, and as it is properly constructed, it must thus hold in all coordinate systems.

The kernel symbol η will be reserved for the particular alternating tensors defined above. We now return to a general E_n with a fundamental alternating tensor, and shall revert to the kernel u for the proof of an identity which connects the alternating tensors with the unit tensor. The tensor

$$A^{[\alpha\beta\cdots\gamma]}_{\kappa\lambda\cdots\mu]} \equiv A^{[\alpha}_\kappa A^\beta_\lambda \cdots A^{\gamma]}_\mu \tag{5.31}$$

of valence (n, n) shares with the left-hand side of (5.19) the property of being totally antisymmetric on both its contravariant and its covariant indices. This is easily seen to be the case even though the antisymmetrization is only explicit on the contravariant indices. The two tensors must thus be proportional. To find the constant of proportionality, put $\alpha = \kappa = 1$, $\beta = \lambda = 2, \ldots, \gamma = \mu = n$. This gives 1 in (5.19), while in (5.31) it gives

$$(n!)^{-1} \det (A_\kappa^\alpha) \equiv (n!)^{-1},$$

as in the derivation of (5.13). Hence

$$u^{\alpha\beta\ldots\gamma}u_{\kappa\lambda\ldots\mu} = n! \, A_{\kappa\lambda\ldots\mu}^{[\alpha\beta\ldots\gamma]}. \tag{5.32}$$

More important than (5.32) itself is the identity which results when the final $(n - p)$ indices of each type are contracted. To state and prove this concisely, let $\alpha, \ldots, \beta, \gamma, \delta$ and $\kappa, \ldots, \lambda, \mu, \nu$ now each be a string of p indices, and let $\tau, \ldots, \rho$ be a further $(n - p)$ indices. Then

$$u^{\alpha\ldots\beta\gamma\delta\tau\ldots\rho}u_{\kappa\ldots\lambda\mu\nu\tau\ldots\rho} = p!(n-p)! \, A_{\kappa\ldots\lambda\mu\nu}^{[\alpha\ldots\beta\gamma\delta]}. \tag{5.33}$$

It is proved by downward induction on p, starting from (5.32) which is the case $p = n$. The details will be given in full as an illustration of the power of the bracket notation. With a little practice, only the steps that are written here need be written at all, although the reader to whom this is a new notation may wish to add some intermediate steps. The beauty of the method is that such intermediate steps are often very lengthy, while those written down are concise.

Contraction of the right-hand side of (5.33) on δ and ν gives an expression which can be partially expanded as follows:

$$p!(n-p)! \, A_{\kappa\ldots\lambda\mu\delta}^{[\alpha\ldots\beta\gamma\delta]} = (p-1)! \, (n-p)! \{ A_{\kappa\ldots\lambda\mu\delta}^{[\alpha\ldots\beta\gamma]\delta} - (p-1) \, A_{[\kappa\ldots\lambda\,\mu]\delta}^{[\alpha\ldots\beta|\delta|\gamma]} \}. \tag{5.34}$$

Since δ is now being treated differently from the $(p-1)$ indices $\alpha, \ldots, \beta, \gamma$, the $p!$ terms in the antisymmetrization have been grouped into p sets of $(p-1)!$ terms according to the position of the superscript index δ. Of these p sets, only one is essentially different from the rest. This is the one in which a factor $A_\delta^\delta = n$ occurs. It has been written down first. The remaining $(p-1)$ sets have the superscript δ vertically over one of the $(p-1)$ indices $\kappa, \ldots, \lambda, \mu$. Instead of writing them all individually, they can be combined together as shown. The antisymmetrization over $\kappa, \ldots, \lambda, \mu$ which would have been redundant in the other terms of (5.34) has to be put in explicitly in this term to ensure that all these indices come under δ in their turn.

We next observe that the validity of the identities

$$\left. \begin{aligned} A_{\kappa\ldots\lambda\mu\delta}^{\alpha\ldots\beta\gamma\delta} &= n A_{\kappa\ldots\lambda\mu}^{\alpha\ldots\beta\gamma} \\ A_{\kappa\ldots\lambda\mu\delta}^{\alpha\ldots\beta\delta\gamma} &= A_{\kappa\ldots\lambda\mu}^{\alpha\ldots\beta\gamma} \end{aligned} \right\} \tag{5.35}$$

and

is not affected when any *free* indices are symmetrized or antisymmetrized. It is this restriction to free indices that prevents the contracted form of (5.33) from being dealt with immediately. It arises because the justification for such operations not affecting the validity of an equation is that the result is a sum of equations, each obtained from the original by a relabelling of free indices. Although dummy indices can also be relabelled, such relabelling must be carried out simultaneously on both occurrences of the dummy index. This does not happen when dummy indices are included within bracket operations. If (5.35) is now used in (5.34), the result can be substituted back into (5.33) to give

$$u^{\alpha...\beta\gamma\delta\tau...\rho}u_{\kappa...\lambda\mu\delta\tau...\rho} = (p-1)!\,(n-p+1)!\,A^{[\alpha...\beta\gamma]}_{\kappa...\lambda\mu}. \qquad (5.36)$$

This is (5.33) with p replaced by $(p-1)$, which thus holds for all p by induction.

6 Integration and orientation for surfaces in E_n

6a Definitions

Let ϕ be a scalar field in an E_n in which S is a parametrizable p-surface. If we wish to define an integral of ϕ over S, the simplest method is to choose a particular parametrization $\lambda_1, ..., \lambda_p$ with range $U \subset R^p$ and to evaluate

$$\int ... \int_U \phi \, d\lambda_1 ... d\lambda_p.$$

However, the result will depend on the choice of parametrization. If $\mu_1, ..., \mu_p$ is another choice with range $V \subset R^p$, the well-known rule for changing the variables in a multiple integral states that

$$\int ... \int_U \phi \, d\lambda_1 ... d\lambda_p = \int ... \int_V \phi |J| \, d\mu_1 ... d\mu_p, \qquad (6.1)$$

where

$$J = \det\left(\frac{\partial \lambda_i}{\partial \mu_j}\right) \qquad (6.2)$$

is the Jacobian of the transformation.

If we wish the result to be independent of the parametrization, we must be prepared to use a more complicated definition. The solution is to define an integral whose value is a tensor. If the summation convention is adopted for the latin parameter labels i, j, k which range from 1 to p as well as for the greek tensor indices which range from 1 to n, then

$$\frac{\partial x^\alpha}{\partial \mu_i} = \frac{\partial x^\alpha}{\partial \lambda_j} \frac{\partial \lambda_j}{\partial \mu_i}. \qquad (6.3)$$

From the product of p such equations, it follows that

$$\frac{\partial x^{[\alpha}}{\partial \mu_1} \cdots \frac{\partial x^{\gamma]}}{\partial \mu_p} = \frac{\partial x^{[\alpha}}{\partial \lambda_i} \cdots \frac{\partial x^{\gamma]}}{\partial \lambda_k} \frac{\partial \lambda_i}{\partial \mu_1} \cdots \frac{\partial \lambda_k}{\partial \mu_p}$$

$$= p!\,\frac{\partial x^{[\alpha}}{\partial \lambda_1} \cdots \frac{\partial x^{\gamma]}}{\partial \lambda_p} \frac{\partial \lambda_{[1}}{\partial \mu_1} \cdots \frac{\partial \lambda_{p]}}{\partial \mu_p}$$

$$= J\,\frac{\partial x^{[\alpha}}{\partial \lambda_1} \cdots \frac{\partial x^{\gamma]}}{\partial \lambda_p} \tag{6.4}$$

by a similar chain of reasoning to that of the proof of (5.13). This may be combined with (6.1) to give

$$\int \cdots \int_U \phi\,\frac{\partial x^{[\alpha}}{\partial \lambda_1} \cdots \frac{\partial x^{\gamma]}}{\partial \lambda_p}\,d\lambda_1 \ldots d\lambda_p = \operatorname{sgn} J \int \cdots \int_V \phi\,\frac{\partial x^{[\alpha}}{\partial \mu_1} \cdots \frac{\partial x^{\gamma]}}{\partial \mu_p}\,d\mu_1 \ldots d\mu_p. \tag{6.5}$$

The left-hand side of (6.5) is thus a totally antisymmetric tensor of valence $(p, 0)$ which is invariant under any change of parametrization of S for which $J > 0$. This suggests that we define an orientation for S in an analogous manner to that for E_n, based on the division of its parametrizations into two classes such that any two members of the same class are related by a transformation with positive Jacobian. As in the previous case, S is said to be *oriented* if one of these classes is chosen and labelled as positive. An orientation for S induces an orientation in the tangent p-plane at any point, considered as an E_p. The ordered set $\{\partial x^\alpha/\partial \lambda_i\}$ of p linearly independent vectors in this E_p is simply given the orientation of the parametrization $\{\lambda_i\}$ of S. If $\{\lambda_i\}$ is any positively oriented parametrization, we write

$$dS^{\alpha\cdots\gamma} = p!\,\frac{\partial x^{[\alpha}}{\partial \lambda_1} \cdots \frac{\partial x^{\gamma]}}{\partial \lambda_p}\,d\lambda_1 \ldots d\lambda_p \tag{6.6}$$

and call it the *contravariant surface element* for S. The factor $p!$ is for later convenience. The left-hand side of (6.5), multiplied by $p!$, can now be written unambiguously as

$$\int_S \phi\,dS^{\alpha\cdots\gamma}, \tag{6.7}$$

as it depends only on ϕ and the oriented p-surface S. The special case $p = 1$, which is an integral along a curve C, is known as a line integral and is written as

$$\int_C \phi\,dx^\alpha. \tag{6.8}$$

An orientation for C simply consists of a direction for traversing the curve.

At the beginning of this section, S was taken as a parametrizable p-surface. Let us now consider how the definition of the integral (6.7) can be extended to allow S to be a more general smooth p-surface. Suppose that S is constructed from k overlapping parametrizable p-surfaces S_i, $1 \leqslant i \leqslant k$. We need to evaluate (6.7) for each S_i, and then add the results, making sure that each overlap region is included only once. Now for each separate integral to be well defined, each S_i must be given an orientation. If the contribution from an overlap between S_i and S_j, say, is to be independent of whether it is considered as part of S_i or of S_j, their orientations in this region must be the same. We thus define an orientation for S itself to be an assignment of an orientation to each S_i such that in every overlap region the several orientations all agree. If such an assignment is possible, it has no real dependence on the particular decomposition chosen for S. For if S' is any other parametrizable p-surface wholly contained in S, the orientations which are induced in S' by each of the S_i with which it overlaps are necessarily consistent. It is also clear that a choice of orientation for S_1 determines the orientations of all the other S_i, so that as before, there are only two possible orientations for any (connected) smooth p-surface. However, there exist surfaces for which no consistent assignment is possible. A well-known example in three dimensions is the Möbius band. Such surfaces are said to be *non-orientable*. In conclusion, we see that the integral (6.7) is well defined for any oriented smooth p-surface S, but that there exist non-orientable surfaces for which it cannot be given any consistent meaning. Non-orientable surfaces have little practical significance in physics, and we shall seldom need to mention them again.

If S is an orientable p-surface in an E_n, it is sometimes convenient to make a correspondence between the two possible orientations of E_n and those of S, even though no specific choice of orientation has been made for either. Such a correspondence is known as an *outer orientation* for S. For extra clarity, the orientation defined previously may be called an *inner* orientation. Since an inner orientation for S induces an inner orientation in each tangent p-plane to S, this is so also for outer orientations. No significant generality is thus lost if attention is now confined to the special case of a p-plane Π. It will be shown that an outer orientation for Π can be specified by an inner orientation for any $(n-p)$-plane Π' which intersects Π in a point.

Let $\{a^\alpha, ..., b^\alpha\}$ be an ordered set of $(n-p)$ linearly independent vectors in Π' with positive orientation. Then if $K_1 \equiv \{c^\alpha, ..., d^\alpha\}$ is an ordered set of p linearly independent vectors in Π, the composite set $K_2 \equiv \{a^\alpha, ..., b^\alpha, c^\alpha, ..., d^\alpha\}$ is also linearly independent. If it were not, Π and Π' would intersect in some s-plane, $s \geqslant 1$. The outer orientation for Π that is associated with the inner orientation of Π' is that in which K_2 has the same inner orientation in E_n that K_1 has in Π. Note that in K_2, the vectors of Π' precede those of Π. This is an arbitrary but standard convention. If n is even and p is odd, the opposite outer orientation would result if the vectors of Π were put first in K_2. The most common use of outer orientations is when $p = n-1$ and E_n itself is oriented. The $(n-p)$-plane Π' is then a straight line which does not lie in Π. An outer, and hence also an inner, orientation in the hyperplane Π may then be specified simply by a direction along this line. This is much simpler to give than the $(n-1)$ vectors which would otherwise be necessary to describe this inner orientation.

These rather abstract considerations have a familiar analogue in three dimensions. We have already seen that an inner orientation for a line L is a direction along L. An outer orientation for L corresponds to a direction of rotation about L. The orientation (right- or left-handedness) of the whole space which is determined when L has both an inner and an outer orientation is then the handedness of the screwthread that moves in the positive direction along L when rotated in the positive sense about L. Similarly, an inner orientation for a plane Π is a direction of rotation within Π while an outer orientation is a direction for crossing Π. With this description of its inner orientation, an ordered pair of vectors in Π is positively oriented if the first vector can be made parallel (as distinct from antiparallel) to the second by a rotation in the positive sense through an angle less than π. Figure 1 illustrates both situations. In each case it is the right-handed orientation for the whole space which is determined by the illustrated inner and outer orientations together. An outer orientation may be given to a curved surface by specifying one of the two directions for its normal vector field as positive. In particular, a closed surface has a natural outer orientation determined by the outward direction for its normal vector field.

The use of a screwthread to define an orientation may be extended to an E_n, but it is not very useful there. If λ is a positively oriented parameter on an oriented curve C in E_n, and if the n vectors $d^i x^\alpha/d\lambda^i$, $i = 1, 2, ..., n$ are linearly independent at some point of C, there is a

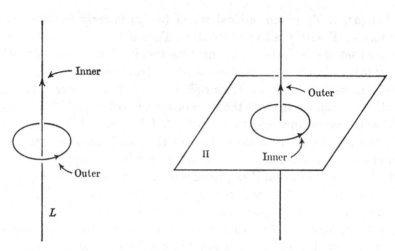

Figure 1. Orientations for a line and a plane in three dimensions.

well-defined orientation for the E_n in which they form a positively oriented set. The other feature mentioned above that has not yet been given an analogue in E_n is the natural outer orientation possessed by a closed surface, i.e. by a surface which bounds a volume. This has an important generalization. Let S be a bounded p-surface in an E_n whose boundary is a $(p-1)$-surface, S' say. We shall show that there is a well-defined correspondence between the inner orientations of S and those of S', although this is an outer orientation for S' only when $p = n$. Let z be a point of S'. Then the tangent $(p-1)$-plane Π' to S' at z lies in the tangent p-plane Π to S at z. If Π is considered as an E_p, the outward direction away from S is an outer orientation for Π' as a hyperplane in this E_p. This induces a correspondence between the inner orientations first of Π and Π', and hence also of S and S' as required. If S is itself oriented, the oriented $(p-1)$-surface consisting of S' with the inner orientation induced from that of S by this correspondence is denoted by ∂S.

6b Stokes' Theorem

There is a theorem that connects integrals over S and ∂S which generalizes the theorems of Stokes and Gauss in three-dimensional vector analysis. This also is known as Stokes' Theorem, and it states that for any scalar field ϕ,

$$\int_S \partial_\alpha \phi \, dS^{\alpha\beta\ldots\delta} = \int_{\partial S} \phi \, dS^{\beta\ldots\delta}. \tag{6.9}$$

The surface elements on the left- and right-hand sides have p and $(p-1)$ indices respectively. We preface its proof with some remarks concerning the result itself. It is valid, and is often required, for surfaces S and ∂S which satisfy a weaker requirement than smoothness, known as piecewise smoothness. A smooth p-surface was defined in §2b as one which can be constructed from a finite number of overlapping parametrizable p-surfaces. If the parametrizable pieces do not overlap, but instead are such that any two which meet have a region of their bounding $(p-1)$-surfaces in common, the resulting p-surface is said to be *piecewise smooth*. The distinction is that between the surface of a sphere, which is smooth, and that of a cube, which is piecewise smooth. An orientation for a piecewise smooth surface is an assignment of an orientation to each piece in such a way that two pieces which have a common region of boundary induce opposite orientations on that region. This agrees with the situation for smooth p-surfaces, as if a join at a boundary is actually smooth, then the outward direction from one piece is the inward direction into the other. A consistent orientation on the whole surface thus indeed corresponds to opposite orientations on the common boundary.

If (6.9) holds for a parametrizable surface S, then it holds also for any bounded orientable piecewise smooth surface. To see this, it is only necessary to add the results for each piece separately. Any region of the boundary of one piece which is not a part of the overall boundary will give a contribution to the right-hand side of (6.9) that will be cancelled by the corresponding contribution from the adjacent piece. This cancellation is ensured by the opposite orientations of the boundary region in the two cases. The resultant right-hand side is thus an integral over the overall boundary, as required.

We shall prove (6.9) for the special case where S can be parametrized as

$$a_i < \lambda_i < b_i, \quad i = 1, 2, ..., p. \tag{6.10}$$

Such a surface is a distorted p-dimensional cube, and its boundary is piecewise smooth. As any bounded piecewise smooth p-surface with piecewise smooth boundary can be built from such distorted cubes, the above discussion shows that the result must hold also for the general case. An illustration of this building process is given in figure 2. This shows how a plane region whose boundary is a smooth closed curve can be built from five distorted squares. It also illustrates the above remarks concerning the orientations of common boundary regions.

To prove (6.9) for the case (6.10), the left-hand side is written out

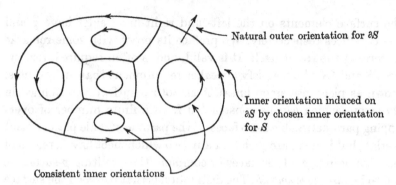

Natural outer orientation for ∂S

Inner orientation induced on ∂S by chosen inner orientation for S

Consistent inner orientations

Figure 2. Building a region with a smooth boundary from distorted squares.

explicitly with the aid of (6.6). Without loss of generality the parameters which satisfy (6.10) may be chosen with positive orientation. It will also be necessary to label one further index in (6.9) explicitly, and so the string $\alpha\beta\ldots\delta$ will be written as $\alpha\beta\gamma\ldots\delta$. Then as in the proof of (5.33), the dummy index α may be separated out from the bracket operation of (6.6) to give

$$\int_S \partial_\alpha \phi \, dS^{\alpha\beta\gamma\ldots\delta} = (p-1)! \int_{a_1}^{b_1} d\lambda_1 \ldots \int_{a_p}^{b_p} d\lambda_p$$

$$\times \sum_{i=1}^{p} (-1)^{i+1} \frac{\partial\phi}{\partial x^\alpha} \frac{\partial x^\alpha}{\partial\lambda_i} \frac{\partial x^{[\beta}}{\partial\lambda_1} \frac{\partial x^\gamma}{\partial\lambda_2} \ldots \frac{\partial x^{\delta]}}{\partial\lambda_p}. \quad (6.11)$$

On the right-hand side the string $\beta\gamma\ldots\delta$ contains $(p-1)$ indices, and in the corresponding sequence of derivatives the one with respect to λ_i is omitted. The factor $(-1)^{i+1}$ takes account of the change of order in the antisymmetrization. Since

$$\frac{\partial\phi}{\partial x^\alpha} \frac{\partial x^\alpha}{\partial\lambda_i} = \frac{\partial\phi}{\partial\lambda_i}, \quad (6.12)$$

the integrand on the right-hand side of (6.11) may be put in the form

$$\sum_{i=1}^{p} (-1)^{i+1} \frac{\partial}{\partial\lambda_i} \left\{ \phi \frac{\partial x^{[\beta}}{\partial\lambda_1} \frac{\partial x^\gamma}{\partial\lambda_2} \ldots \frac{\partial x^{\delta]}}{\partial\lambda_p} \right\} - \phi \sum_{\substack{1 \leqslant i,\, j \leqslant p \\ i \neq j}} \epsilon_{ij} \frac{\partial^2 x^{[\beta}}{\partial\lambda_i \partial\lambda_j} \frac{\partial x^\gamma}{\partial\lambda_1} \ldots \frac{\partial x^{\delta]}}{\partial\lambda_p},$$

$$(6.13)$$

where
$$\epsilon_{ij} = \begin{cases} (-1)^{i+j} & \text{if } 1 \leqslant j < i \leqslant p, \\ 0 & \text{if } j = i, \\ (-1)^{i+j+1} & \text{if } 1 \leqslant i < j \leqslant p. \end{cases} \quad (6.14)$$

The string $\gamma\ldots\delta$ in the final term of (6.13) contains $(p-2)$ indices, and both λ_i and λ_j are to be omitted from the corresponding sequence of derivatives. The factor ϵ_{ij} incorporates both the factor $(-1)^{i+1}$ in (6.11) and the additional factor which results from bringing the λ_j-derivative out of its standard order in the derivative sequence.

It follows from (6.14) that

$$\epsilon_{ij} = -\epsilon_{ji}. \tag{6.15}$$

Since the second-order derivative in (6.13) is symmetric in i and j, and the double summation is made independently over all values of i and j, the terms in this double sum cancel in pairs. The final term in (6.13) is thus identically zero. If the remaining first term is put back into (6.11), the integration over λ_i may be performed explicitly to give

$$(p-1)! \sum_{i=1}^{p} (-1)^{i+1} \int_{a_1}^{b_1} d\lambda_1 \ldots \int_{a_{i-1}}^{b_{i-1}} d\lambda_{i-1} \int_{a_{i+1}}^{b_{i+1}} d\lambda_{i+1}$$

$$\ldots \int_{a_p}^{b_p} d\lambda_p \left[\phi \frac{\partial x^{\beta}}{\partial \lambda_1} \ldots \frac{\partial x^{\delta}}{\partial \lambda_p} \right]_{\lambda_i=a_i}^{\lambda_i=b_i}. \tag{6.16}$$

As the parameters $\lambda_i, \lambda_1, \lambda_2, \ldots, \lambda_{i-1}, \lambda_{i+1}, \ldots, \lambda_p$ in that order form a set with orientation $(-1)^{i+1}$ for S, it follows that $\lambda_1, \lambda_2, \ldots, \lambda_{i-1}, \lambda_{i+1}, \ldots, \lambda_p$ have orientation $(-1)^{i+1}$ on $\lambda_i = b_i$ but $(-1)^i$ on $\lambda_i = a_i$. This is because λ_i is increasing in the outward direction on the first of these faces but in the inward direction on the second. But the corresponding factors of (-1) are precisely those which already occur in (6.16). As all the $2p$ faces of the distorted cube are integrated over in (6.16), this expression is thus precisely the right-hand side of (6.9), which completes the proof.

The integral (6.7) is a tensor of valence $(p, 0)$ and is defined for any oriented p-surface S in an E_n, irrespective of whether the E_n possesses any further structure. The E_n itself need not even be oriented. In this general situation it is the only possible type of surface integral. However, if the E_n has a fundamental alternating tensor $u_{\alpha \ldots \gamma}$ then a second type of surface integral can be constructed, whose value is a totally antisymmetric tensor of valence $(0, n-p)$. This is

$$\int_S \phi \, dS_{\kappa \ldots \lambda}, \tag{6.17}$$

where the covariant surface element for a p-surface S has $(n-p)$ indices and is defined in terms of (6.6) by

$$dS_{\kappa \ldots \lambda} = \frac{1}{p!} u_{\kappa \ldots \lambda \alpha \ldots \beta} \, dS^{\alpha \ldots \beta}. \tag{6.18}$$

When $p = n$ the integral is over a region of E_n and is scalar valued. The covariant surface element (6.18) then has no indices. It is called instead the volume element on the E_n, and will be written as dV. The corresponding contravariant element will be written as $dV^{\alpha\ldots\beta}$, so that

$$dV = \frac{1}{n!} u_{\alpha\ldots\beta} dV^{\alpha\ldots\beta}. \qquad (6.19)$$

In this case the coordinates themselves may be used as parameters, so that we can put $\lambda_1 = x^1, \ldots, \lambda_n = x^n$ in (6.6). When combined with (6.19) this gives

$$dV = u_{12\ldots n} dx^1 dx^2 \ldots dx^n, \qquad (6.20)$$

so that in the preferred coordinate systems determined by (5.14),

$$\int \phi \, dV = \int \ldots \int \phi \, dx^1 \ldots dx^n. \qquad (6.21)$$

This justifies the name 'volume element'. The volume $V(R)$ of any region R of E_n is naturally defined by

$$V(R) = \int_R dV, \qquad (6.22)$$

and from (6.21) is seen to be necessarily positive.

The orientation and the measure of volume $V(R)$ contain between them the full information present in the fundamental alternating tensor, since we can deduce from (6.22), (6.19) and (5.32) that

$$u^{\alpha\ldots\gamma} = \frac{1}{V(R)} \int_R dV^{\alpha\ldots\gamma}. \qquad (6.23)$$

We saw in § 5b that an orientation for E_n determines a fundamental alternating tensor up to a positive scalar factor. It now follows from (6.23) that the volume measure $V(R)$ alone determines such a tensor up to a sign, since the sign of the integral in (6.23) depends on the orientation. It is convenient to turn this round, and to say that an E_n *possesses a volume element* if it has a fundamental alternating tensor which is specified only up to a factor ± 1. The concepts of orientation and volume element for an E_n are then independent of one another, and the E_n has a unique fundamental alternating tensor if and only if it possesses both of them. The results of § 5b show that any R_n possesses a volume element, even when it is not oriented.

To see how the orientation has disappeared from the definition of dV, it is only necessary to consider how it occurs in (6.19). If the sign of $u_{\alpha\ldots\beta}$ is changed, this reverses the orientation of E_n and hence also the sign of $dV^{\alpha\ldots\beta}$. The product thus remains unaltered. In the same way, the covariant surface element (6.18) is unaltered if both the sign of $u_{\alpha...\beta}$ and the orientation of S are changed together. The covariant surface element for a p-surface S thus depends only on the volume element in E_n and the *outer* orientation of S. This dependence on the outer orientation may be made more explicit as follows. If z is a point of S, let Π' be an oriented $(n-p)$-plane through z which meets the tangent plane to S at z only at that point. Let $\{a^\alpha, ..., b^\alpha\}$ be an ordered set of $(n-p)$ vectors at z which lie in Π' and have positive orientation. It now follows from (6.6) and (6.18) that if S has the outer orientation determined by Π', and if the differentials $d\lambda_i$ are considered as positive, then

$$a^\alpha...b^\beta \, dS_{\alpha...\beta} > 0. \tag{6.24}$$

The inverse of (6.18) may be obtained by multiplication with $u^{\kappa...\lambda\gamma...\delta}$ and the use of (5.33). It is

$$dS^{\gamma...\delta} = \frac{1}{(n-p)!} u^{\kappa...\lambda\gamma...\delta} dS_{\kappa...\lambda}. \tag{6.25}$$

With the help of (6.18) and (6.25) and the further use of (5.33), a number of alternative forms can be found for Stokes' theorem (6.9). One commonly used version is that when $p = n$, and S is consequently a region R of E_n, then

$$\int_R \partial_\alpha \phi \, dV = \int_{\partial R} \phi \, dS_\alpha. \tag{6.26}$$

The surface ∂R is a closed hypersurface whose outer orientation is given by the outward direction away from R. If n^α is an outwardly directed vector field on ∂R, (6.24) gives for this case

$$n^\alpha dS_\alpha > 0. \tag{6.27}$$

The result (6.26) is the most natural generalization to E_n of Gauss' Theorem, one form of which states, in the usual notation of three-dimensional vector analysis, that

$$\int_R \nabla \phi \, dV = \int_{\partial R} \phi \, \mathbf{dS}. \tag{6.28}$$

To recover the three-dimensional form of Stokes' theorem from (6.9) we must consider precisely the case $n = 3$, $r = 2$. This gives

$$\int_S u^{\alpha\beta\gamma} \partial_\gamma \phi \, dS_\beta = \int_{\partial S} \phi \, dx^\alpha. \tag{6.29}$$

In a Cartesian coordinate system in a Euclidean R_3, it follows from

(4.1) and (5.24) that $u^{\alpha\beta\gamma} = \epsilon^{\alpha\beta\gamma}$. Equation (6.29) can then be written as

$$\int_S d\mathbf{S} \times \nabla\phi = \int_{\partial S} \phi\, d\mathbf{x}, \qquad (6.30)$$

which is one of the forms of Stokes' theorem, as required.

The definition (6.22) for the volume of a region R of E_n has a natural analogue for a region Σ of a hyperplane Π, when Π has an outer orientation. A *vector area* S_α for Σ can then be defined by

$$S_\alpha = \int_\Sigma dS_\alpha. \qquad (6.31)$$

If $a^\alpha, ..., b^\alpha$ are $(n-1)$ linearly independent vectors which lie in Π, it follows from (6.31) with (6.18) and (6.6) that

$$S_\alpha = ku_{\alpha\beta\cdots\gamma}\, a^\beta...b^\gamma \qquad (6.32)$$

for some scalar k, and hence that

$$S_\alpha a^\alpha = ... = S_\alpha b^\alpha = 0. \qquad (6.33)$$

The vector u_α in the form (2.33) for the equation of Π may thus be taken to be S_α. It was seen in §2b that the vector u_α could be represented by the ordered pair of hyperplanes (2.33) and (2.35). We now see that when the E_n has a volume element, an alternative representation of u_α is by an outer oriented region Σ of the single hyperplane (2.33) which has u_α as its vector area. The connexion between the two representations is as follows. If Π_1 and Π_2 are the hyperplanes (2.33) and (2.35) respectively, then (6.27) shows that the outer orientation of Σ corresponds to the direction from Π_1 to Π_2. The size of Σ is such that the hypercylindrical region of E_n which is generated by any parallel translation of Σ from Π_1 to Π_2 has unit volume.

7 Spacetime geometry in special relativity

In this and the following section we shall consider those geometric properties of spacetime that depend on the specific fundamental tensors of special relativity and Newtonian physics. This section will treat the special relativistic situation, and for simplicity the prefix (r) will be omitted from the corresponding metric tensor of (3.13). Similarly, g and the alternating tensors $\eta_{\alpha\beta\gamma\delta}$ and $\eta^{\alpha\beta\gamma\delta}$ will be the quantities defined in terms of this metric tensor by (5.20), (5.23) and (5.24), and indices will be raised and lowered with it without explicit mention.

The classification of intervals which was given in §1-7 has a natural extension to vectors. A vector a^α is said to be spacelike, null or timelike according as

$$a_\alpha a^\alpha \equiv g_{\alpha\beta} a^\alpha a^\beta \equiv g^{\alpha\beta} a_\alpha a_\beta \qquad (7.1)$$

is positive, zero or negative. The scalar $|a_\alpha a^\alpha|^{\frac{1}{2}}$ is known as the *length* of the vector, so that a null vector has zero length. The proper time or proper distance, as appropriate, between two events is thus the length of their relative position vector. A vector of unit length is called a unit vector. Note the different usage between this and the unit tensor, which is a specific tensor of valence $(1,1)$. Two nonzero vectors a^α and b^α are said to be *orthogonal* if their inner product $a^\alpha b_\alpha$ is zero. Null vectors are thus self-orthogonal.

Choose now a specific inertial frame, and restrict consideration to its natural coordinate systems. In the notation of (4.6) let us put

$$a^\alpha = (\mathbf{a}, p) \quad \text{and} \quad b^\alpha = (\mathbf{b}, q). \qquad (7.2)$$

Since

$$a^\alpha a_\alpha = \mathbf{a}^2 - c^2 p^2, \qquad (7.3)$$

we see that a^α is spacelike, null or timelike according as

$$|\mathbf{a}| >, = \text{or} < c|p|.$$

This shows that if a^α and b^α are both timelike or null, then

$$|\mathbf{a} \cdot \mathbf{b}| \leqslant |\mathbf{a}|\,|\mathbf{b}| \leqslant c^2 |pq|. \qquad (7.4)$$

But as we also have

$$a^\alpha b_\alpha = \mathbf{a} \cdot \mathbf{b} - c^2 pq, \qquad (7.5)$$

a^α and b^α can only be orthogonal if both equalities hold in (7.4) and if $\mathbf{a} \cdot \mathbf{b}$ and $c^2 pq$ have the same sign. This is easily seen to imply that both vectors must be null, and that they must be mutually parallel or antiparallel. Hence in an orthogonal pair of vectors, either one vector is spacelike or they are both null and proportional to one another.

When a^α and b^α are orthogonal, we also have

$$(a^\alpha + b^\alpha)(a_\alpha + b_\alpha) = a^\alpha a_\alpha + b^\alpha b_\alpha. \qquad (7.6)$$

If each is either spacelike or null, this shows that their sum is likewise. In this case also, it can only be null if both vectors are null, which can only happen if they are proportional. It follows from this that one cannot have more than three mutually orthogonal spacelike vectors. Four would form a basis set, which is impossible as no timelike vector can be expressed as a linear combination of orthogonal spacelike vectors, as shown above. A basis set of mutually orthogonal vectors must thus contain three spacelike vectors and one timelike vector. If such a set of vectors is used as the basis vectors of a coordinate system, the corresponding metric tensor components form a diagonal matrix which can be brought to the form (3.11) by a suitable rescaling of the axes.

Two timelike vectors are said to have the same time-orientation if $a^\alpha b_\alpha < 0$ and the opposite time-orientation if $a^\alpha b_\alpha > 0$. This formulation shows that the property is well defined, but it does not make clear its significance. However, it can be seen from (7.2) and (7.5) that the sign of $a^\alpha b_\alpha$ is opposite to that of $a^4 b^4$. The time-orientation of the two vectors is thus the same or opposite according as a^4 and b^4 have the same or opposite signs. It follows easily from this that similarity of time-orientation is an equivalence relation (cf. the corresponding situation for similarity of orientation in § 5b). Hence all timelike vectors can be divided into two classes such that any two members of the same class have the same time-orientation. The sum of two time-like vectors with the same time-orientation is also a timelike vector with this common time-orientation, but that of two timelike vectors with opposite time-orientations is not necessarily even timelike.

The mathematical structure with which we have been dealing so far is an R_4 of signature 2. This does not possess the distinction that is present in physics between past and future, but we can now see how this may be added. It is necessary to choose one of the two time-orientation classes and to label its vectors as future-pointing and those of the other class as past-pointing. An R_4 of signature 2 for which this has been done is said to be *time-oriented*. It is easily seen from the definitions of § 1-7 that a Lorentz transformation between two Minkowskian coordinate systems is orthochronous if and only if their timelike basis vectors have the same time-orientation. If the systems instead have the same orientation then the transformation is said to be *proper*.

An important property of the Lorentz transformations which has so far been mentioned only incidentally is that they form a group, in the sense of abstract algebra, known as the (full) Lorentz group. This is easily seen to be true for the transformations between the members of any preferred family of coordinate systems which is specifiable by one or more fundamental tensors. The full Lorentz group is not topologically connected, since those transformations which reverse either the orientation of a Minkowskian coordinate system or the time-orientation of its timelike basis vector cannot be obtained from the identity by a continuous deformation. The transformations which can be so obtained are in fact precisely those which are both proper and orthochronous. They form a subgroup of the full Lorentz group known naturally as the proper orthochronous Lorentz group. These transformations may also be characterized as those which preserve both the direction of time and the orientation of the *spatial* coordinates. The stated result then follows since any future-pointing unit vector may be continuously deformed into any other, and any two right-handed spatial coordinate systems may also be obtained from one another by continuous deformation.

By convention, the R_4 of spacetime is both oriented and time-oriented, since attention is normally restricted to coordinate systems in which the time coordinate increases into the future and in which the spatial coordinates are right-handed. This is very different from the question of whether the laws of physics themselves determine a particular orientation or time-orientation in spacetime. On a cosmological scale the expansion of the universe provides an arrow for the direction of time, while in thermodynamics such an arrow is given by the Second Law. This suggests, but does not prove, that there may be an asymmetry between the two directions of time in some of the fundamental laws of physics. However, no fundamental law outside of the domain of quantum physics has yet shown such an asymmetry if the second law of thermodynamics itself is excluded from the classification 'fundamental' due to its statistical nature. On a quantum level the situation is somewhat different. If the so-called *PCT* theorem is valid then no quantum law can ever provide an absolute distinction between the two time-orientations of spacetime. But evidence is growing that none of the three operations P (parity inversion), C (charge conjugation) and T (time-reversal) separately, nor any product of two of them, leaves invariant all the quantum laws. If this is true, then there are fundamental quantum laws which can be used to determine an orientation for spacetime, and also laws which provide an association between the two time-orientations and the distinction between particles and antiparticles.

It is inconvenient to develop the laws of physics without ever making definitions which depend on an arbitrarily assigned orientation or time-orientation, especially when dealing with integrals. However, the above discussion shows that the behaviour of equations under a change of convention is important. It is thus best to keep any such dependence as simple as possible. When there is a choice, definitions will be linked to a time-orientation in preference to an orientation, as of the two, this distinction appears the most fundamental. It is certainly the one most evident in everyday experience.

Let us now turn from the classification of vectors to that of hyperplanes. A hyperplane Π is said to be spacelike if every displacement in Π is spacelike. Now if the equation of Π is

$$n_\alpha(x^\alpha - y^\alpha) = 0, \qquad (7.7)$$

where y is a fixed point, then the vector n^α is orthogonal to every displacement in Π. The results on orthogonality obtained above show that Π is thus spacelike if and only if n^α is timelike. It should be noted that the existence of n^α as a contravariant vector is dependent on the presence of a spacetime metric. As a contravariant vector it determines a unique direction which is called the normal direction to Π, and in consequence the vector itself is said to be *normal* to Π. The same name is not attached to the covariant form n_α which exists in any E_n as the geometric association with a direction is absent.

This inverse association between the natures of Π and its normal is continued by calling Π null or timelike if n^α is null or spacelike respectively. If Π is null, it contains its normal direction. Every other direction in Π is then spacelike. If Π is timelike, of any three mutually orthogonal directions in Π, one is timelike and two are spacelike. The fact that the normal to a null hyperplane is also tangent to it shows that care is necessary in extrapolating from Euclidean geometry to the geometry of spacetime.

As was mentioned in § 1-9, a curve in spacetime is also known as a world line. It is said to be spacelike, null or timelike if its tangent vector at every point has that character. Similarly, if all the tangent hyperplanes to a hypersurface have the same character, then that character is also attached to the hypersurface itself. Any future-pointing timelike direction provides any spacelike hypersurface with an outer orientation. All integrals over spacelike hypersurfaces evaluated with the covariant surface element dS_α will be assumed to be taken with this outer orientation unless the contrary is stated explicitly. Note that the vector dS^α itself is then past-pointing, to accord with (6.27).

The spacetime metric enables the vector area S_α of a region Σ of any non-null hyperplane to be expressed as a product of a scalar magnitude and a unit vector. When Σ is spacelike, the significance of the magnitude may be found from an evaluation of S_α in a Minkowskian coordinate system which has x^4 constant on Σ. The three coordinates x^α, $\alpha = 1, 2, 3$ may then be used as parameters on Σ, but they are negatively oriented with respect to the inner orientation determined by the future-pointing outer orientation. This puts a minus sign on

the right-hand side of (6.6). If this equation is then combined with
(6.31), (6.18) and the definition (5.23) of the appropriate alternating
tensor, it gives

$$S_\alpha = cA\delta_\alpha^4, \quad A = \iiint_\Sigma dx^1 dx^2 dx^3. \tag{7.8}$$

When taken together with (4.17) this shows that the length of the
vector S_α is A, which is the volume of the three-dimensional region Σ
as measured by an observer to whom Σ lies in a hyperplane of simultaneity.

The world line traced out by the motion of a particle is necessarily
timelike as there must exist at any instant a Minkowskian coordinate
system in which the particle is instantaneously at rest. The future-
pointing unit tangent vector v^α to this world line is known as the *velocity* of the particle. It is given by $v^\alpha = dx^\alpha/d\tau$ if τ is the proper time
along the path of the particle measured so as to increase into the
future. The velocity is sometimes called the four-velocity to disting-
uish it from the three-dimensional velocity vector $\mathbf{v} \equiv d\mathbf{x}/dt$ which is
defined in each inertial frame. Since

$$dx^\alpha/dt = (\mathbf{v}, 1), \tag{7.9}$$

the constant of proportionality between v^α and dx^α/dt may be found
from the scaling condition $v^\alpha v_\alpha = -1$. This gives

$$v^\alpha = c^{-1}\gamma\, dx^\alpha/dt, \quad \gamma = (1 - \mathbf{v}^2/c^2)^{-\frac{1}{2}}. \tag{7.10}$$

It follows from (7.8) that the direction of S^α is that of the four-velocity
of the observer who sees the whole of Σ simultaneously.

If δx^α is the relative position vector of two events in spacetime, the
quadratic invariant Φ which is given in tensorial form by (3.14)
determines either the proper time or proper distance between the
events as appropriate. But these quantitites are only the observed
time or observed distance to a special class of observers, namely those
that see the events as spatially coincident or as simultaneous respectively. Let us now see how a tensorial form can be given to both the
time δt and the distance δr between the events as measured by an
arbitrary observer with velocity u^α. These scalar quantities are determined by the expressions

$$\delta t = \delta x^4, \quad (\delta r)^2 = (\delta x^1)^2 + (\delta x^2)^2 + (\delta x^3)^2 \tag{7.11}$$

which hold in those Minkowskian coordinate systems in which
$u^\alpha = c^{-1}\delta_4^\alpha$. Since this implies $u_\alpha = -c\delta_\alpha^4$, the equations

$$\delta t = -c^{-1}u_\alpha\,\delta x^\alpha, \quad (\delta r)^2 = (g_{\alpha\beta} + u_\alpha u_\beta)\,\delta x^\alpha\,\delta x^\beta \tag{7.12}$$

are valid in these systems. Since they are correctly constructed according to the rules of §1, they thus also hold in all other coordinate systems, as required. We see that when u^α is orthogonal to δx^α then $\delta t = 0$ and $(\delta r)^2 = \Phi$, while if u^α is parallel to δx^α then $\delta r = 0$ and $(\delta t)^2 = -c^{-2}\Phi$, so that the special cases determined by Φ do follow from (7.12). Equations (7.12) may also be combined with (7.10) to give a tensorial form for the speed of a particle with four-velocity v^α relative to a frame with four-velocity u^α. We find that

$$\gamma = -u_\alpha v^\alpha, \quad \gamma^2 \mathbf{v}^2 = c^2(g_{\alpha\beta} + u_\alpha u_\beta)\, v^\alpha v^\beta. \qquad (7.13)$$

Since $v^\alpha v_\alpha = -1$, the second of equations (7.13) can be simplified with the aid of the first back down to the definition of γ given in (7.10). However, there is an alternative form for this equation, and for the expression for δr in (7.12), that will be useful later. This involves the tensor

$$g^{\cdot\cdot\gamma\delta}_{\alpha\beta} \equiv A^{\gamma\delta}_{[\alpha\beta]}, \qquad (7.14)$$

which is antisymmetric on each index pair and satisfies

$$g^{\cdot\cdot\gamma\delta}_{\alpha\beta}\, g^{\cdot\cdot\epsilon\zeta}_{\gamma\delta} = g^{\cdot\cdot\epsilon\zeta}_{\alpha\beta}. \qquad (7.15)$$

The new notation is needed for something apparently adequately described by the right-hand side of (7.14) as it is most commonly used with its indices lowered. In this form it is given by

$$g_{\alpha\beta\gamma\delta} = \tfrac{1}{2}(g_{\alpha\gamma} g_{\beta\delta} - g_{\alpha\delta} g_{\beta\gamma}) \qquad (7.16)$$

and is known as the *bivector metric tensor*. Then since $u^\alpha u_\alpha = -1$, (7.12) implies

$$(\delta r)^2 = -2g_{\alpha\beta\gamma\delta}\, \delta x^\alpha u^\beta \delta x^\gamma u^\delta. \qquad (7.17)$$

There is a corresponding form for the second equation of (7.13). The tensor (7.14) can also be expressed with the use of (5.33) as

$$g^{\cdot\cdot\gamma\delta}_{\alpha\beta} = \tfrac{1}{4}\eta_{\alpha\beta\kappa\lambda}\, \eta^{\gamma\delta\kappa\lambda}. \qquad (7.18)$$

If combined with (5.25), this shows that

$$g_{\alpha\beta\gamma\delta} = -\tfrac{1}{4}\eta_{\alpha\beta\kappa\lambda}\, \eta_{\gamma\delta\mu\nu}\, g^{\kappa\mu} g^{\lambda\nu}. \qquad (7.19)$$

The significance of this result will be seen in the next section.

8   Spacetime geometry in Newtonian physics

8a Intrinsic theory

In this section we shall first consider some geometric properties of Newtonian spacetime in its own right, and shall then consider their relationship to the relativistic situation of the preceding section. The prefix (r) will again be attached to all relativistic variables, so that $g^{\alpha\beta}$, t_α and $\eta_{\alpha\beta\gamma\delta}$ can unambiguously denote the Newtonian variables of (3.21), (3.17) and (5.29).

Contravariant vectors, which represent directions, have a natural division into spacelike vectors a^α which satisfy $t_\alpha a^\alpha = 0$, and timelike vectors b^α which satisfy $t_\alpha b^\alpha \neq 0$. A vector is spacelike if and only if a representative line segment appears simultaneous to any observer, which is a meaningful statement since simultaneity is absolute in Newtonian physics. The timelike vectors may be further subdivided into future-pointing and past-pointing classes according as $t_\alpha b^\alpha$ is positive or negative, so that a definite time-orientation is already present in the fundamental tensors in contrast to the relativistic situation. This is because the Galilean group is the analogue of the orthochronous Lorentz group rather than of the full Lorentz group. Null vectors have no Newtonian analogue.

A hyperplane Π is naturally said to be spacelike only if it is a surface of constant absolute time t. If it is represented by equation (7.7), this is so if and only if n_α is parallel to t_α. Such a covariant vector will be said to be timelike. Due to (3.23) and the raising convention of §3, this can also be expressed as $n^\alpha = 0$. If $n^\alpha \equiv g^{\alpha\beta}n_\beta \neq 0$, n^α is necessarily spacelike and so we shall say that n_α is spacelike. In this case Π is said to be timelike. The only spacelike hypersurfaces are the hyperplanes of simultaneity, but there can be curved timelike hypersurfaces.

The analogue of (7.8) is that the vector area S_α of a region of a spacelike hyperplane satisfies $S_\alpha = At_\alpha$, where A is the Euclidean volume of the region. This suggests that for any timelike covariant vector $n_\alpha = kt_\alpha$, the scalar $|k|$ should be considered as its magnitude. The vector can also be called future-oriented† or past-oriented according as $k > 0$ or $k < 0$. We shall adopt these definitions, but it means that the magnitude of a spacelike covariant vector n_α must be defined separately. This is naturally taken to be $|g^{\alpha\beta}n_\alpha n_\beta|^{\frac{1}{2}}$, which is the length of the three-vector $\mathbf{n}$ in the decomposition

$$n_\alpha = (\mathbf{n}, k) \qquad (8.1)$$

in the notation (4.5). In the decomposition of a timelike vector, $\mathbf{n} = \mathbf{0}$. The three-vector $\mathbf{n}$ is uniquely determined by n_α irrespective of the frame in which the decomposition is made, since (8.1) implies

$$n^\alpha = (\mathbf{n}, 0). \qquad (8.2)$$

† Covariant vectors don't point.

Two spacelike covariant vectors m_α, n_α will be said to be orthogonal if these corresponding three-vectors are orthogonal. This is so if and only if

$$m^\alpha n_\alpha \equiv g^{\alpha\beta} m_\alpha n_\beta = 0. \qquad (8.3)$$

Due to the irreversibility of the index raising operation, these definitions for the magnitude of a covariant vector do not extend immediately to contravariant vectors. In particular, any contravariant vector which is obtained from a covariant one by raising the index is spacelike. Now the general spacelike vector n^α has the form (8.2), which may be obtained in this way from all covariant vectors of the form (8.1), for any value of k. But the magnitude of (8.1) is independent of k. It is thus consistent to define the length of a spacelike contravariant vector as the magnitude of *any* spacelike covariant vector from which it is obtainable by raising the index. It is easily verified that the length of the spacelike position vector δx^α which connects two simultaneous events is in fact the distance between those events, which we should expect of any reasonable definition. The length of a timelike contravariant vector b^α is naturally taken as $|b^\alpha t_\alpha|$, so that the length of any timelike position vector is the absolute time interval between the events concerned. Orthogonality of two spacelike contravariant vectors m^α and n^α is again defined by (8.3) for any appropriate n_α, which agrees with the usual Euclidean concept. We now have enough geometric language to characterize the basis vectors of any Galilean coordinate system. Both the contravariant and covariant sets consist of three mutually orthogonal unit spacelike vectors together with one unit timelike vector which is future-pointing or future-oriented as appropriate.

Of the four combinations of spacelike and timelike with covariant and contravariant, the case of the spacelike contravariant vectors stands out from the other three for the clumsiness of the definitions given above. It is the only one in which a direct four-dimensional definition was not given for its magnitude. Such a definition can be given, as we shall now see, but we shall approach the problem indirectly via a related problem. The tangent vector to the world line of an observer, or any moving particle, is necessarily timelike. As in the relativistic case, when it is taken to be of unit length and future-pointing, it is called the (four-)velocity of the observer or particle. This corresponds to parametrization by absolute time in accordance with (7.9). We shall obtain a tensor expression for the distance δr between two events with relative position vector δx^α, as measured by an observer with velocity u^α. Only in the special case of simultaneous events will this be independent of u^α. The result will be the Newtonian analogue of the second equation of (7.12). The analogue of the first equation is (4.13), as the absolute time interval is the observed time interval to all observers. Only the distance is relative in the Newtonian theory.

The central role in this expression is played by the tensor

$$h_{\alpha\beta} \equiv \tfrac{1}{2}\eta_{\alpha\kappa\lambda\mu}\,\eta_{\beta\nu\rho\sigma}\,g^{\kappa\nu}g^{\lambda\rho}u^\mu u^\sigma. \tag{8.4}$$

This rather complicated quantity has a simple value in any positively oriented Galilean coordinate system which has u^α as its timelike contravariant basis vector. In such a system $u^\alpha = \delta^\alpha_4$ and in addition (5.29), (3.17) and (3.21) all hold. Hence with the notation (5.12),

$$h_{\alpha\beta} = \tfrac{1}{2}\epsilon_{\alpha\kappa\lambda 4}\,\epsilon_{\beta\nu\rho 4}\,g^{\kappa\nu}g^{\lambda\rho}. \tag{8.5}$$

Since $\epsilon_{\alpha\beta\gamma\delta}$ vanishes if two indices are equal, $g^{\kappa\nu}$ can be replaced in (8.5) by $\delta^{\kappa\nu}$ as the two only differ when $\kappa = \nu = 4$. But it is trivially true that

$$\epsilon_{\beta\nu\rho 4}\delta^{\kappa\nu}\delta^{\lambda\rho} = \delta_{\beta\sigma}\,\epsilon^{\sigma\kappa\lambda 4}, \tag{8.6}$$

so that (8.5) is equivalent to

$$h_{\alpha\beta} = \tfrac{1}{2}\delta_{\beta\sigma}\,\epsilon_{\alpha\kappa\lambda 4}\,\epsilon^{\sigma\kappa\lambda 4}. \tag{8.7}$$

The product of the two ϵ-symbols can now be expanded with the use of (5.19) and (5.33) to give finally

$$h_{\alpha\beta} = \delta_{\alpha\beta} - \delta_{\alpha 4}\,\delta_{\beta 4}. \tag{8.8}$$

Hence $h_{\alpha\beta}\,\delta x^\alpha \delta x^\beta = (\delta x^1)^2 + (\delta x^2)^2 + (\delta x^3)^2 = (\delta r)^2. \tag{8.9}$

The required tensorial expression is thus

$$(\delta r)^2 = h_{\alpha\beta}\,\delta x^\alpha \delta x^\beta, \tag{8.10}$$

which holds in every coordinate system. It should be noted that $h_{\alpha\beta}$ does not depend on the conventional choice of orientation since it contains a product of two alternating tensors, each of which changes sign under a change of orientation.

If the index β in (8.4) is raised by multiplication with $g^{\beta\gamma}$, the result can be simplified with the use of (5.30) and (5.33) to give

$$h_\alpha{}^\gamma = A^\gamma_\alpha - t_\alpha u^\gamma, \tag{8.11}$$

and hence also $h^{\alpha\gamma} = g^{\alpha\gamma}. \tag{8.12}$

The connexion with our earlier discussion of length is now clear. If

a^α and b^α are two spacelike vectors, there exist vectors a_α and b_α from which they can be obtained by raising the indices. It follows from (8.12) that

$$k \equiv h_{\alpha\beta} a^\alpha b^\beta = g^{\alpha\beta} a_\alpha b_\beta. \tag{8.13}$$

The left-hand equality shows that k is independent of the particular choice of a_α and b_α, while the right-hand one shows that k is independent of u^α. It thus depends only on a^α, b^α and $g^{\alpha\beta}$. This enables the length of a^α to be defined unambiguously as $\sqrt{(h_{\alpha\beta} a^\alpha a^\beta)}$, and orthogonality of a^α and b^α to be characterized by $k = 0$. Since the development from (8.5) to (8.9) is purely for the purposes of interpretation, we thus have a completely tensorial development of length and orthogonality for spacelike contravariant vectors, as required.

It is possible to go even further if we define

$$g_{\alpha\beta\gamma\delta} \equiv -\tfrac{1}{4} \eta_{\alpha\beta\kappa\lambda} \eta_{\gamma\delta\mu\nu} g^{\kappa\mu} g^{\lambda\nu}. \tag{8.14}$$

The factor $-\tfrac{1}{4}$ is to make (8.14) formally identical to the relativistic result (7.19). Then since $t_\alpha u^\alpha = 1$, (8.13) can be rewritten with the aid of (8.4) and (8.14) as

$$(2g_{\gamma\alpha\beta\delta} a^\gamma b^\delta - kt_\alpha t_\beta) u^\alpha u^\beta = 0. \tag{8.15}$$

This can hold for all timelike unit vectors u^α only if the part of the bracketed tensor which is symmetric in α and β vanishes. We thus have

$$2g_{\gamma(\alpha\beta)\delta} a^\gamma b^\delta = kt_\alpha t_\beta, \tag{8.16}$$

which defines the scalar product k of a^α and b^α without any mention of extraneous variables. The fact that a^α and b^α must be spacelike for k to be well defined still enters as it is only in this case that the left-hand side is proportional to $t_\alpha t_\beta$ at all. The tensor (8.14) also enables us to write (8.10) as

$$(\delta r)^2 = -2g_{\alpha\beta\gamma\delta} \delta x^\alpha u^\beta \delta x^\gamma u^\delta, \tag{8.17}$$

which is formally the same as (7.17).

8b The Newtonian situation as a limiting case

This analogy is actually more than merely formal. It underlies the limiting process by which the Newtonian results can be deduced from the relativistic ones. Let us now consider this in more detail. It was seen in §4b that variation of c corresponds to a change in the unit of time, and that consequently for a given physical system all three-dimensional velocities dx/dt will increase indefinitely as $c \to \infty$. The actual limiting transition involves changing the physical system with

c so that these speeds remain finite in the limit. As this cannot hold simultaneously in all inertial frames, one particular frame K has to be picked out as that frame in which these speeds do indeed remain finite. Let $v^\alpha(c)$ be the four-velocity of a material particle which is being so scaled. From (7.10), v^α is related to the three-velocity $\mathbf{v}$ in K by

$$v^\alpha = (c^{-1}\gamma\mathbf{v}, c^{-1}\gamma), \quad v_\alpha = (c^{-1}\gamma\mathbf{v}, -c\gamma). \tag{8.18}$$

If $\mathbf{v}$ remains finite as $c \to \infty$, this gives

$$cv^\alpha \to (\mathbf{v}, 1), \quad c^{-1}v_\alpha \to (0, -1). \tag{8.19}$$

To avoid writing a large number of prefixes (r) and (n), let us now denote all Newtonian variables by an overbar, e.g. $\bar{g}^{\alpha\beta}$. Then

$$\bar{v}^\alpha = (\mathbf{v}, 1) \quad \text{and} \quad \bar{t}_\alpha = (0, 1), \tag{8.20}$$

so that (8.19) can be written in the tensorial form

$$cv^\alpha \to \bar{v}^\alpha, \quad c^{-1}v_\alpha \to -\bar{t}_\alpha. \tag{8.21}$$

Note that the limit of $v^\alpha v_\alpha = -1$ is thus $\bar{v}^\alpha \bar{t}_\alpha = 1$, so that the Newtonian unit vector condition is the limit of the corresponding relativistic condition.

Since $\sqrt{(-g)} = c$ in any Minkowskian coordinate system, it follows from (5.23) and (5.29) that

$$c^{-1}\eta_{\alpha\beta\gamma\delta} \to \bar{\eta}_{\alpha\beta\gamma\delta}, \tag{8.22}$$

while in the present notation (4.16) and (4.18) read

$$c^{-2}g_{\alpha\beta} \to -\bar{t}_\alpha \bar{t}_\beta, \quad g^{\alpha\beta} \to \bar{g}^{\alpha\beta}. \tag{8.23}$$

When these are taken together with (7.19) and (8.14), they show that

$$c^{-2}g_{\alpha\beta\gamma\delta} \to \bar{g}_{\alpha\beta\gamma\delta}. \tag{8.24}$$

This cannot be obtained directly from (7.16) as there is a cancellation of the c^4 behaviour that is present in each separate term of its right-hand side.

We can now give further mathematical foundation to the argument used in §4b to show that the chosen frame K does not have a preferred status in the limiting theory. Consider for example the relativistic expressions for δt given by (7.12), and for δr given by (7.17). Take for u^α any velocity which is scaled with c in the manner of (8.21), so that it represents a general inertial frame of the limiting theory. Then these expressions tend to well-defined limits as $c \to \infty$, namely the Newtonian results (4.13) and (8.17) respectively, whose Galilean invariance is shown by the isotropy of the coefficient tensors t_α and $g_{\alpha\beta\gamma\delta}$ under Galilean coordinate transformations.

Two more significant results that can be given a tensorial form and so removed from their restriction to K are

$$\gamma \to 1, \quad c^2(\gamma - 1) \to \tfrac{1}{2}\mathbf{v}^2, \tag{8.25}$$

which follow from the definition (7.10) for γ. If u^α and v^α are velocities which each satisfy (8.21), it follows immediately that

$$u^\alpha v_\alpha \to -1. \tag{8.26}$$

By the first of equations (7.13), this generalizes the first of (8.25). Now the right-hand side of the identity

$$g_{\alpha\beta} + v_\alpha v_\beta = 2g_{\alpha\gamma\delta\beta} v^\gamma v^\delta, \tag{8.27}$$

which follows from (7.16), has by (8.21) and (8.24) a well-defined· limit as $c \to \infty$. We thus obtain

$$g_{\alpha\beta} + v_\alpha v_\beta \to 2\bar{g}_{\alpha\gamma\delta\beta} \bar{v}^\gamma \bar{v}^\delta. \tag{8.28}$$

This will later be of importance in its own right. For present purposes its significance is that it implies, on multiplication by $c^2 u^\alpha u^\beta$, that

$$c^2\{1 - (u_\alpha v^\alpha)^2\} \to 2\bar{g}_{\alpha\beta\gamma\delta} \bar{u}^\alpha \bar{v}^\beta \bar{u}^\gamma \bar{v}^\delta. \tag{8.29}$$

This is not quite the analogue of (8.25), as from (7.13) we see that when u^α is the velocity of K then

$$\gamma - 1 = -(1 + u_\alpha v^\alpha). \tag{8.30}$$

However we have identically that

$$1 - (u_\alpha v^\alpha)^2 = (1 - u_\alpha v^\alpha)(1 + u_\alpha v^\alpha), \tag{8.31}$$

while from (8.26) we get $(1 - u_\alpha v^\alpha) \to 2$.

Equation (8.29) thus implies

$$c^2(1 + u_\alpha v^\alpha) \to \bar{g}_{\alpha\beta\gamma\delta} \bar{u}^\alpha \bar{v}^\beta \bar{u}^\gamma \bar{v}^\delta, \tag{8.32}$$

which is the required result. Its relationship to the second of equations (8.25) may be completed by consideration of (8.17) for the special case $\delta x^\alpha = \bar{v}^\alpha$. Then δr will be the distance travelled in unit time by a particle with four-velocity $\bar{v}^\alpha$ as measured in a frame with four-velocity $\bar{u}^\alpha$, i.e. it will be the speed of the particle relative to the frame. When this is combined with (8.30) we obtain (8.25) as a special case of (8.32), as required.

3
Foundations of dynamics

1 The free motion of particle systems

1a Conservation laws

The mathematical developments of the preceding chapter have given us the tools needed for a study of the properties of matter in the absence of gravitation. The present chapter lays the foundations of this study. Since the properties of Newtonian and relativistic spacetime have been treated in a unified fashion, the distinction between the two resting on their fundamental tensors, it will be possible to develop the corresponding theories of matter in a similarly unified manner. This is a considerable help when one wishes to compare the two theories and to relate them to one another.

In classical physics, dynamics is generally divided into two parts. There is the dynamics of point particles, and there is continuum dynamics. The latter itself has two main branches, the studies of fluids and of elastic media. If one takes a macroscopic viewpoint, matter is seen as being essentially continuous. Continuum dynamics is then the fundamental branch, and particle dynamics is an idealization which can be deduced from it to treat very small bodies. A collision between two particles can be analyzed in detail if necessary by considering their deformation on impact and the distribution of the forces that they exert on one another at their surface of mutual contact.

But consider what happens as a conceptual microscope is taken to a typical continuum such as a gas. The first change occurs when its molecular structure is seen. At this level the macroscopic continuum properties are seen to be statistical averages over molecular collisions which are governed by the classical laws of particle dynamics. These averaging processes are studied in the kinetic theory of gases. As the power is increased, the molecules themselves are seen to have a complex internal structure. The constituent atoms are soon seen, with their nuclei and surrounding electrons. The laws here are those of quantum mechanics. The molecular collisions of the kinetic theory are thus also analyzable in detail, but by a very different process to that involved in the analysis of a collision between macroscopic particles. The speeds of both macroscopic particles and of molecules in gases are such that under normal conditions, the Newtonian laws are a good approximation to the relativistic ones. This ceases at the next level which is revealed by even higher magnification. Here we reach the subnuclear particles, which are studied primarily through their behaviour in high speed collisions. Provided that the relativistic laws are used, even these collisions are found to obey the classical laws of particle dynamics. At this level the question has been raised of whether or not it is even meaningful to ask for a detailed description of the internal processes which occur during a collision. Certainly physics as it is at present cannot supply one. But this does not mean that such particles are free from internal structure. Experiments in which high energy electrons are scattered by nucleons indicate that nucleons have quite a complicated internal structure.

Let us see now what significance all this has for the development of dynamics. The above remarks are, of course, all made with hindsight, but this does not prevent them from shedding light on the logical development of the subject. They reveal first of all that particle dynamics is more fundamental than continuum dynamics and so it should be studied first. They also show that one should not enquire too closely as to just what is meant by a particle, or by a collision. There are a number of levels at which these terms are meaningful. The laws of particle dynamics are a set of laws which are common to all levels and which restrict the possible consequences of a given collision without determining that outcome completely.

At every level the particles have internal structure if examined sufficiently closely. This structure is controlled by additional laws which depend on the level, and which place additional restrictions on the effects of a collision. The laws of quantum mechanics prevent complete determinism at anything other than the macroscopic level. At every level collisions can occur which have drastic consequences, in that the number of particles involved need not be the same both before and after a collision. Macroscopically, bodies can coalesce or fragment. At a molecular level, chemical reactions can occur. At a subnuclear level, particles can be created and annihilated. But internal structure can also be altered in other ways, in which the identity of the particles is not destroyed. An atom may change from its ground state to an excited state, or a macroscopic particle may be deformed. It is assumed in particle dynamics that enough is known about the particles to enable their state to be described, so that such changes can be recognized. A quantity which depends on the internal state of the particle but not on its motion is known as a *function of state*. The internal structure of a particle may change even when the particle is isolated. The most obvious example is a particle in a state of continuous vibration, but it can also happen in numerous other ways. A function of state which remains constant during any change in an isolated particle will be called an *isolate invariant*.

There is one more point to be made before we turn to the mathematical formulation of the laws. This is a consistency requirement. Due to the uncertainty as to just what constitutes a particle, it should be possible to regard a group of particles close together and with similar velocities as a single particle. If this group is dispersed by impact with another particle, it could then be treated as a two-particle problem in which fragmentation occurs. For the theory to be self-consistent, such alternative views of the same situation must lead to the same result. As the theory will be based on the idealization of point particles, i.e. particles of zero spatial extent which trace out a world line in spacetime, this need for self-consistency provides a useful check on the validity of the idealization. Such a group of particles being treated as a single particle will be called a *composite*.

The terminology needed to describe the kinematics of particle motion was developed in §§2-7 and 2-8a. From these we recall that the world line of any particle is timelike, and that its future-pointing unit tangent vector $v^\alpha \equiv dx^\alpha/d\tau$ is known as the velocity of the particle. Here, τ is the absolute time in the Newtonian theory and the proper time in the relativistic theory. The first assertion of particle dynamics is that there exists an isolate invariant m for every particle, known as its *mass*, which in some sense measures the quantity of matter in the particle. From this, the *momentum vector* p^α is defined by

$$p^\alpha = mv^\alpha. \tag{1.1}$$

For an isolated particle p^α is constant, since m is an isolate invariant and v^α is constant by the Principle of Inertia. If a particle is moving under external influences, those influences may thus be given a quantitative measure by the *force vector* F^α which is defined by

$$F^\alpha = dp^\alpha/d\tau. \tag{1.2}$$

The preceding statements lack precision until some method is given which enables m to be determined. This is provided by the *law of conservation of momentum*, which states that in any collision, the sum of the momenta of the particles involved is the same immediately after the collision as it was immediately before the collision. It will be seen in the subsequent development how far-reaching is this apparently very simple law. For our immediate purpose, it is clear that it provides in principle the required method for the quantitative comparison of masses. Once a unit of mass is chosen, the concepts of mass, momentum and force then all become well defined.

Consider now a system of particles which move freely except for mutual collisions. Let Σ be any hypersurface with an outer orientation. As was seen in § 2-6a, this corresponds to an inner orientation for any line which crosses Σ. Then the momentum flux $P^\alpha(\Sigma)$ across Σ is defined by

$$P^\alpha(\Sigma) = \Sigma\,(\pm p^\alpha), \qquad (1.3)$$

where the sum is taken over all points of intersection of particle paths with Σ, and the sign of each term is that of the orientation with which p^α crosses Σ. If Σ is taken as a closed hypersurface which surrounds one collision, and if the outward direction is taken as positively oriented, then the signs in (1.3) will be positive for the outgoing particles and negative for the incoming particles. The corresponding momenta will have the same values at Σ as they have immediately before or after the collision, since the particles move freely between collisions. The law of conservation of momentum thus states that $P^\alpha(\Sigma)$ is zero. But the interior of any closed hypersurface may be subdivided into regions each of which contains just one collision. If the corresponding results are added for the bounding hypersurfaces of each region, it follows that the momentum flux across any closed hypersurface is zero.

An important special case of this result arises when Σ is taken as illustrated in figure 3. Here, Σ_1 and Σ_2 are two spacelike hypersurfaces which give cross sections of the region of spacetime occupied by the colliding particles, and Σ_3 is a cylindrical timelike hypersurface which joins their boundaries. Their outer orientations are indicated by the arrows marked ' $+$ '. Since these are not consistent, the condition of zero total flux over the closed hypersurface formed by Σ_1, Σ_2 and Σ_3 together is

$$P^\alpha(\Sigma_1) - P^\alpha(\Sigma_2) + P^\alpha(\Sigma_3) = 0. \qquad (1.4)$$

But $P^\alpha(\Sigma_3) = 0$ since no particle world lines meet Σ_3. Hence

$$P^\alpha(\Sigma_1) = P^\alpha(\Sigma_2). \qquad (1.5)$$

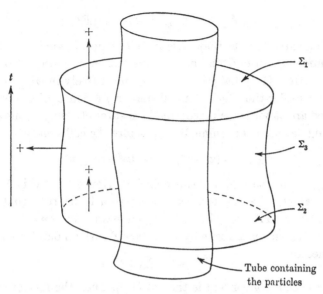

Figure 3. Hypersurfaces and orientations used in the definition of the total
momentum of a system of colliding particles.

Any hypersurface such as Σ_1 or Σ_2 will be called a *spacelike section* of
the system. We thus see that if Σ is any spacelike section of the system
with a future-pointing outer orientation, then $P^\alpha(\Sigma)$ is independent of
the choice of Σ. Its constant value P^α is called the *total momentum* of the
system. For such a Σ, all the signs in the sum (1.3) are positive so that
P^α is, like each separate summand, a future-pointing timelike vector.
By analogy with (1.1), the length of P^α is known as the *total mass* of the
system. It will be denoted by M, and we shall write $P^\alpha = MV^\alpha$ where
V^α is a unit vector. The inertial frame whose velocity is V^α is known as
the *zero-momentum frame*, as in that frame the three-vector formed
from the spatial components of P^α vanishes. For reasons that will be
seen in § 1*b*, it is also known as the *centre-of-mass frame*, but we shall
keep to the former name.

For a system of point particles, conservation of momentum implies
also conservation of angular momentum, which in spacetime form is
defined with respect to a fixed point z in *spacetime*. Specifically, the
angular momentum about z of a particle at x with momentum p^α is an
antisymmetric tensor $j^{\alpha\beta}$ of valence $(2,0)$ defined by

$$j^{\alpha\beta} = (x^\alpha - z^\alpha)\,p^\beta - (x^\beta - z^\beta)\,p^\alpha. \tag{1.6}$$

Since $dx^\alpha/d\tau = v^\alpha$, when (1.2) is satisfied we find that

$$dj^{\alpha\beta}/d\tau = G^{\alpha\beta}, \tag{1.7}$$

where $\qquad G^{\alpha\beta} = (x^\alpha - z^\alpha) F^\beta - (x^\beta - z^\beta) F^\alpha.$ $\qquad$ (1.8)

The terms involving v^α that arise in (1.7) cancel in view of (1.1). The antisymmetric tensor $G^{\alpha\beta}$ is known as the *moment* of the force F^α about z. It follows from (1.7) that $j^{\alpha\beta}$ is constant for a freely moving particle.

Now consider the effect of a collision on $j^{\alpha\beta}$. Since all the particles involved are at the same place x at the instant of collision, the sum of their angular momenta immediately before the collision is

$$\sum j^{\alpha\beta} = (x^\alpha - z^\alpha) \sum p^\beta - (x^\beta - z^\beta) \sum p^\alpha. \qquad (1.9)$$

But $\sum p^\alpha$ is unchanged by the collision. Hence so also is $\sum j^{\alpha\beta}$, as claimed above. We thus have an exact parallel with the situation for momentum, and it is developed in the same way. The flux $J^{\alpha\beta}(z, \Sigma)$ of angular momentum across a hypersurface Σ with an outer orientation is defined by
$$J^{\alpha\beta}(z, \Sigma) = \sum (\pm j^{\alpha\beta}), \qquad (1.10)$$

where the sign convention is that of (1.3). Then the flux of angular momentum across any closed hypersurface is zero. If Σ is chosen as a spacelike section of the system, it follows that $J^{\alpha\beta}(z, \Sigma)$ is independent of the choice of Σ. This common value $J^{\alpha\beta}(z)$ is known as the *total angular momentum* of the system about z. If y is an alternative reference point, equations (1.3), (1.6) and (1.10) show that for any hypersurface Σ,

$$J^{\alpha\beta}(z, \Sigma) - J^{\alpha\beta}(y, \Sigma) = (y^\alpha - z^\alpha) P^\beta(\Sigma) - (y^\beta - z^\beta) P^\alpha(\Sigma). \quad (1.11)$$

1b Centroid lines

It seems reasonable to consider the unit vector $V^\alpha = P^\alpha/M$ as a mean velocity for the system. However, for this to be satisfactory there should be some world line parallel to V^α which lies within the system and which in some sense describes its dynamical centre. With the ultimate aim of finding such a line, we shall now consider in some detail the implications of (1.11).

If (1.11) is applied to the total angular momentum $J^{\alpha\beta}(z)$ by choosing Σ appropriately, it implies that the vector S_α defined by

$$S_\alpha \equiv \tfrac{1}{2}\eta_{\alpha\beta\gamma\delta} V^\beta J^{\gamma\delta} \qquad (1.12)$$

has the same value whether $J^{\gamma\delta}$ is evaluated about y or about z. It is thus independent of the choice of reference point, and so must represent a part of the total angular momentum which is internal to the system. It satisfies
$$S_\alpha V^\alpha = 0 \qquad (1.13)$$

and can be seen from (1.6) to vanish for a system which consists only of a single particle. It is called the *spin vector*, and is one half of a decomposition of $J^{\alpha\beta}(z)$ into two pieces. The other part carries all the dependence on z, and describes the position and motion of the system relative to z. To find this second part, we first deduce from (1.12) and (2-5.33) that

$$\eta^{\alpha\beta\gamma\delta} S_\alpha = J^{\beta\gamma} V^\delta + 2V^{[\beta}J^{\gamma]\delta}. \tag{1.14}$$

Then if q_α is any timelike vector such that

$$q_x V^\alpha = 1, \tag{1.15}$$

it follows from (1.14) on multiplication by q_δ that

$$J^{\alpha\beta} = \eta^{\alpha\beta\gamma\delta} S_\gamma q_\delta + 2M^{[\alpha} V^{\beta]}, \tag{1.16}$$

where
$$M^\alpha(z, q_\beta) = J^{\alpha\beta}(z) q_\beta. \tag{1.17}$$

The vector q_α determines a field of parallel spacelike hyperplanes $q_\alpha x^\alpha = $ constant, one through every point of spacetime. There is precisely one point X on each of these such that

$$M^\alpha(X, q_\beta) = 0. \tag{1.18}$$

To see this it is only necessary to multiply (1.11) by q_β and to use (1.15). If z is a fixed point, the required point X on the hyperplane

$$(x^\alpha - z^\alpha) q_\alpha = k \tag{1.19}$$

is then found to be
$$X^\alpha = z^\alpha + kV^\alpha + M^\alpha(z, q_\beta)/M. \tag{1.20}$$

This is consistent with (1.19) since (1.17) and the antisymmetry of $J^{\alpha\beta}$ imply

$$M^\alpha q_\alpha = 0. \tag{1.21}$$

Two significant results follow from (1.20). First, it shows that the locus of these points as k varies is a straight line l parallel to V^α. Secondly, it can be rewritten with the aid of (1.19) as

$$M^\alpha(z, q_\beta) = q_\beta\{(X^\alpha - z^\alpha) P^\beta - (X^\beta - z^\beta) P^\alpha\}. \tag{1.22}$$

This is precisely the form that M^α would have for a single particle of mass M with world line l. Such a particle would be moving freely since l is straight. Its velocity would be V^α, and it follows from (1.6) and (1.20) that its angular momentum about z would be $2M^{[\alpha} V^{\beta]}$. Let us now write

$$L^{\alpha\beta} = 2M^{[\alpha} V^{\beta]}, \quad S^{\alpha\beta} = \eta^{\alpha\beta\gamma\delta} S_\gamma q_\delta, \tag{1.23}$$

so that $$J^{\alpha\beta}(z) = L^{\alpha\beta}(z, q_\gamma) + S^{\alpha\beta}(q_\gamma)$$ (1.24)

by (1.16). Then once q_α is chosen, we have the following situation. The tensor $L^{\alpha\beta}$ is the angular momentum about z that a single particle would have if it were moving freely with momentum P^α along the world line l determined by (1.18). The tensor $S^{\alpha\beta}$ is independent of z, but equals the total angular momentum of the system about any point of l. Hence, provided that l actually lies within the system, it can be said to describe its dynamical centre in a natural sense. We shall call it the *centroid line* with respect to q_α. $L^{\alpha\beta}$ and $S^{\alpha\beta}$ are known as the *orbital* and *spin* angular momentum tensors respectively. The vector M^α is called the *dipole moment* of the system about z. Its significance in determining l can be seen most directly from (1.19) and (1.20) with $k = 0$.

Recall now that in the definition of $J^{\alpha\beta}(z)$ from (1.10), the hypersurface Σ needed to specify the summation can be chosen to be any spacelike section of the system. If it is taken as the hyperplane (1.19), (1.10) shows with (1.6) and (1.17) that

$$M^\alpha(z, q_\beta) = \Sigma\{(x^\alpha - z^\alpha)\, p^\beta q_\beta\} - kP^\alpha.$$ (1.25)

But in view of (1.15) we also have

$$\Sigma p^\beta q_\beta = M.$$ (1.26)

This enables (1.25) and (1.20) to be combined to give

$$X^\alpha = \Sigma p^\beta q_\beta x^\alpha / \Sigma p^\gamma q_\gamma.$$ (1.27)

Now each p^α is timelike and future-pointing. Since q^β is also timelike, all the inner products $p^\alpha q_\alpha$ have the same sign, which is positive in view of (1.15). Hence (1.27) expresses X^α as a weighted mean with positive coefficients of the individual position vectors x^α. The point X in which l meets Σ must thus lie within any closed convex 2-surface in Σ which encloses all the particles of the system. This settles the one point left outstanding above, as it gives a precise sense in which the centroid line l lies within the system.

We must now consider the dependence of l on q_α. Here for the first time the Newtonian and relativistic theories diverge. In the Newtonian theory all timelike vectors are proportional to t_α, and since the proportionality is fixed by (1.15), the only possibility is

$$q_\alpha = t_\alpha.$$ (1.28)

Equations (1.26) and (1.27) may then be simplified with the use of (1.1) to give $$M = \Sigma m, \quad MX^\alpha = \Sigma m x^\alpha,$$ (1.29) which shows that the centroid line is precisely the world line traced out by the centre of mass of the system, in the usual Newtonian sense.

In the relativistic case, each choice of q_α is particularly convenient in one particular inertial frame, namely the frame in which the hyperplanes (1.19) are surfaces of constant time. This occurs when the frame velocity is antiparallel to q^α. Of the family of parallel centroid lines which thus exist corresponding to different q^α, one may be picked out as particularly significant. This is the one that corresponds to the zero-momentum frame. It is given by

$$q_\alpha = -V_\alpha \qquad (1.30)$$

and is known as the *centre-of-mass line*. It is the only one that is distinguished by a property of the system rather than of the observer.

Another geometrical construction determined by the system alone is the *centroid tube* which is formed by all the centroid lines taken together. This is of interest since it places a lower bound on the system's size, as it must lie entirely within the system. Let us now investigate it in more detail. If z is any point on the centre-of-mass line, we see from (1.16) and (1.30) that

$$J^{\alpha\beta}(z) = -\eta^{\alpha\beta\gamma\delta}S_\gamma V_\delta. \qquad (1.31)$$

If this is used in (1.17) and the result substituted into (1.20), it gives

$$X^\alpha - z^\alpha = M^{-1}\eta^{\alpha\beta\gamma\delta}S_\beta q_\gamma V_\delta + kV^\alpha. \qquad (1.32)$$

As q_α and k vary, X generates the whole centroid tube. This shows that for systems in which S_α is zero, all the centroid lines coincide and the tube is degenerate.

Even when S_α is nonzero, (1.32) and (1.13) imply

$$S_\alpha(X^\alpha - z^\alpha) = 0, \qquad (1.33)$$

so that the tube lies in the timelike hyperplane through the centre-of-mass line which is orthogonal to S^α. In this case, if it were not for the restriction that q_α has to be timelike, X would generate the whole of this hyperplane. To see this, let d^α be any vector such that

$$S_\alpha d^\alpha = 0. \qquad (1.34)$$

Then if we put
$$X^\alpha - z^\alpha = d^\alpha \qquad (1.35)$$

in (1.32), we can use (2-5.33) to solve that equation together with the constraint (1.15) to give
$$k = -d^\alpha V_\alpha \qquad (1.36)$$

and
$$q_\alpha = (M/S^2)\eta_{\alpha\beta\gamma\delta} d^\beta V^\gamma S^\delta - V_\alpha + \theta S_\alpha, \qquad (1.37)$$

where θ is arbitrary and $S^2 \equiv S_\alpha S^\alpha$.   (1.38)

Since (1.13) shows that S_α is either zero or spacelike, S^2 is strictly positive unless S_α itself is zero.

It follows from (1.37), (2-5.25) and (2-5.33) that

$$q_\alpha q^\alpha = (M/S)^2 (g_{\alpha\beta} + V_\alpha V_\beta) d^\alpha d^\beta - 1 + \theta^2 S^2. \qquad (1.39)$$

There thus exist timelike values of q_α if and only if

$$(g_{\alpha\beta} + V_\alpha V_\beta) d^\alpha d^\beta < (S/M)^2, \qquad (1.40)$$

the geometrical significance of which follows from (1-7.12). We see that a line in the hyperplane (1.33) which is parallel to the centre-of-mass line l_0 lies within the centroid tube provided that its perpendicular distance from l_0 is less than S/M. This tube is thus quite descriptive of the motion. It is a cylindrical region of a timelike hyperplane, the normal to which gives the direction of the spin vector, while the axis and radius of the cylinder give the centre-of-mass line and the spin/mass ratio respectively.

2 Mass and energy in Newtonian physics

2a Three-dimensional formulation

An important concept in Newtonian physics which occurs in, but transcends, dynamics is that of energy. Since energy in dynamics is dependent on the choice of inertial frame, we must begin a study of it by changing from the spacetime viewpoint of the preceding section to the separate space and time of a particular inertial frame. In this section only the Newtonian theory will be considered. The corresponding relativistic development will be given in the next section. A problem of terminology arises when both three- and four-dimensional physical variables are considered simultaneously, as the four-dimensional variables are often named after their three-dimensional analogues. Where confusion may arise, these names will be prefixed by 'three-' or 'four-' as appropriate, but these prefixes will be omitted when the meaning is clear from the context.

When a fixed spacetime origin O is chosen, the coordinates x^α form the position four-vector with respect to O which will be decomposed as $x^\alpha = (\mathbf{x}, t)$. The notation is that of (2-4.6). The corresponding relationship between the four-velocity $v^\alpha = dx^\alpha/dt$ and the three-velocity $\mathbf{v} = d\mathbf{x}/dt$ is then

$$v^\alpha = (\mathbf{v}, 1). \qquad (2.1)$$

The four-momentum p^α can now be decomposed as

$$p^\alpha = (\mathbf{p}, m), \quad \mathbf{p} = m\mathbf{v}, \tag{2.2}$$

where m is the mass of the particle by (1.1), and the three-vector $\mathbf{p}$ is known as its three-momentum. Conservation of four-momentum is thus expressed in each inertial frame by the separate laws of conservation of three-momentum and of mass.

In the general systems of particles considered in the preceding section, no distinction was made between collisions in which coalescence or fragmentation occurs and those in which each particle retains its separate identity. Newtonian physics makes such a distinction. In keeping with Newton's conception of mass as a measure of quantity of matter, it is assumed that the mass of a particle remains constant as long as that particle retains its identity. This is consistent with, but is a refinement of, the conservation of mass in identity-preserving collisions. It also restricts the possible effects of an external force. The assumption already made that m is an isolate invariant leaves open the possibility that m might be changed by the action of such a force. This is now prohibited. In the notation of (1.2), it shows that all forces F^α must satisfy

$$F^\alpha t_\alpha = 0. \tag{2.3}$$

In a particular inertial frame we write

$$F^\alpha = (\mathbf{F}, 0), \tag{2.4}$$

so that

$$d\mathbf{p}/dt = \mathbf{F}, \tag{2.5}$$

and we call $\mathbf{F}$ the three-force.

Another class of collisions of particular significance consists of those that do not change the internal states of the particles involved. This is much more restrictive than mere preservation of identity. In these it is found that kinetic energy is conserved, where the kinetic energy of a particle of mass m and speed $v \equiv |\mathbf{v}|$ is defined to be $\frac{1}{2}mv^2$. There is an important point of interpretation to be made here. 'No change of internal state' must be taken to mean that the internal states of the particles immediately after the collision are the same as their states immediately before it. Nothing is implied about the states *during* the collision. This is consistent with the general comment made earlier that one must not look too closely at the nature of collisions.

When collisions are considered in which internal states are changed, it is necessary to transcend dynamics in order to get a law of conservation of energy. It is then found that every particle possesses a function of state U, known as its *internal energy*, which is an isolate invariant and is such that the sum of the kinetic and internal energies of all the particles involved in any collision is left unchanged. For each particle the sum

$$E \equiv \tfrac{1}{2}mv^2 + U \qquad (2.6)$$

is known as its *total energy*. Collisions which leave unchanged the internal energy of each individual particle are said to be *elastic*. All others are said to be *inelastic*. Elastic collisions do not necessarily leave unchanged the internal states of the particles, although such state-preserving collisions are necessarily elastic.

The nature of the variables which determine U depends on the conceptual level at which the theory is being considered. At a molecular level, U is a function of the quantum numbers which specify the state. At a macroscopic level, the all-inclusive law of conservation of energy becomes the First Law of Thermodynamics. The internal energy then depends on temperature as well as on the mechanical variables needed to specify a state of deformation of the body.

Let us now consider the change in the energy of a particle caused by a continuously acting force, rather than by a collision. If its internal state is not affected, the change will be solely in its kinetic energy and is given from (2.2) and (2.5) by

$$\frac{d}{dt}\,(\tfrac{1}{2}mv^2) = \mathbf{v}\cdot\mathbf{F}. \qquad (2.7)$$

The right-hand side is known as the *rate of working* of the force $\mathbf{F}$. It is indeed a rate if the work performed by $\mathbf{F}$ when its point of application is displaced by an amount $\delta\mathbf{r}$ is defined to be $\mathbf{F}\cdot\delta\mathbf{r}$. The work done by the force then equals the increase in energy of the particle on which it acts.

Some external influences will affect U. In this case, since U is an isolate invariant, the rate of change

$$G \equiv dU/dt \qquad (2.8)$$

is as much a measure of those influences as is the force vector F^α defined by (1.2). The scalar G will be referred to simply as the *rate of supply of energy*. Detailed consideration of G must involve a study of the internal structure of the particle, which depends greatly on the conceptual level at which one works. For a macroscopic particle this is closely related to the dynamics of continuous media, which will be treated at a later stage. For the moment it is enough to remark that in this macroscopic case, G may be nonzero either because mechanical forces are deforming the particle, or because heat is being supplied to it.

An influence which changes the velocity of a particle without affecting its internal state has the same effect relative to an observer as would a suitable change in the velocity of that observer. Such an influence is thus, in some sense, inconsequential. It would have $\mathbf{F} \neq 0$ but $G = 0$. In contrast, an influence with $\mathbf{F} = 0$ but $G \neq 0$ would not affect the state of motion of the particle, but would change its internal state in a manner which could not be compensated for by any change in the observer's own situation. For this reason, influences with nonzero G are considered to have an inherently different nature to those with zero G. The latter will be said to be given by a *pure force*, while if $G \neq 0$ we shall speak of an *impure force*. An influence which has $G \neq 0$ but $\mathbf{F} = 0$ will be called a *pure energy source*.

It must be noted that although we are developing quite a lot of language with which to discuss forces, it will remain largely empty until further physical properties are ascribed to the concepts that are being named. It would be possible within the Newtonian theory to do so at the present stage. Newton's Third Law, which states that the forces of action and reaction are equal and opposite, provides all that is required to give real content to these concepts. However, nothing comparable can be done in relativity theory outside of the dynamics of continuous media. This is because the lack of absolute simultaneity prevents such equality from being meaningful unless both forces act at the same point in space as well as in time. As the Newtonian theory is being developed for comparison with the relativistic theory rather than for its own sake, only those developments will be made which have a relativistic parallel.

2b Four-dimensional formulation

Let us now consider how the law of conservation of energy can be expressed in spacetime form. If u^α is the four-velocity of the reference frame, it follows from §2-8 that

$$\mathbf{v}^2 = -2g_{\alpha\beta\gamma\delta}\, u^\alpha v^\beta u^\gamma v^\delta. \tag{2.9}$$

Since also $u^\alpha t_\alpha = 1$, (2.6) can thus be written as

$$E = E_{\alpha\beta}\, u^\alpha u^\beta, \tag{2.10}$$

where

$$E_{\alpha\beta} = mg_{\alpha\gamma\delta\beta}\, v^\gamma v^\delta + U t_\alpha t_\beta. \tag{2.11}$$

The symmetric tensor $E_{\alpha\beta}$, which is independent of any choice of reference frame, will be called the *energy tensor* of the particle. Its complete set of components in any inertial frame may be written down easily if it is first noted that (2-8.14) and (2-5.30) imply

$$g_{\dot\alpha\beta\dot\gamma}{}^\delta = A^\delta_{[\alpha} t_{\beta]} t_\gamma, \qquad (2.12)$$

so that $E^{\ \beta}_{\dot\alpha} = \tfrac{1}{2}m(A^\beta_\alpha - t_\alpha v^\beta).$ (2.13)

We then see that $E_{\alpha\beta} = \begin{pmatrix} \tfrac{1}{2}m\mathbf{I} & -\tfrac{1}{2}\mathbf{p} \\ -\tfrac{1}{2}\mathbf{p} & E \end{pmatrix}$ (2.14)

in the notation of (2-4.4), where $\mathbf{I}$ is the unit dyadic whose Cartesian components are $\delta_{\kappa\lambda}$.

It follows from (2.10) that conservation of energy in every inertial frame implies conservation of each component of $E_{\alpha\beta}$. We now see that the other laws that this requires are the already known ones of mass and three-momentum conservation, so that all five conservation laws may be combined into the single requirement that the energy tensor $E_{\alpha\beta}$ be conserved. In a system of particles which move freely except for mutual collisions, this can be formulated as in the preceding section as a statement that the net flux of $E_{\alpha\beta}$ through any closed hypersurface is zero.

The total energy tensor of such a system can also be expressed in the single-particle form (2.11). To see this, let $P^\alpha = MV^\alpha$ be the total four-momentum as in §1a. Then since $(v^\alpha - V^\alpha)$ is spacelike, where v^α is the four-velocity of an individual particle of the system, it follows from (2-8.13) and (2-8.16) that

$$mg_{\alpha\gamma\delta\beta}(v^\gamma - V^\gamma)(v^\delta - V^\delta) = mt_\alpha t_\beta g_{\epsilon\gamma\delta\zeta} V^\epsilon v^\gamma v^\delta V^\zeta. \qquad (2.15)$$

If the brackets on the left-hand side are expanded and the result summed over any spacelike section of the system, we then find that

$$\Sigma E_{\alpha\beta} = Mg_{\alpha\gamma\delta\beta} V^\gamma V^\delta + U^* t_\alpha t_\beta, \qquad (2.16)$$

where $U^* = \Sigma(E_{\alpha\beta} V^\alpha V^\beta).$ (2.17)

The right-hand side of (2.16) is the energy tensor that a particle of mass M and internal energy U^* would have if it were travelling along the centre-of-mass line of the system. Equation (2.17) shows that U^* is the total energy of the system, both kinetic and internal, measured in the zero-momentum frame. If the system is considered as a composite particle, this checks the self-consistency of the theory. Note that U^* can be nonzero even when the internal energies of each individual particle are zero, so that the internal energy contribution to (2.11) is necessary for this check to work.

Finally, let us find the effect of external influences on the energy tensor of a single particle. It follows from (1.2), (2.8) and (2.11) that

$$\frac{d}{dt}E_{\alpha\beta} = 2g_{\alpha\gamma\delta\beta}v^{(\gamma}F^{\delta)} + Gt_\alpha t_\beta. \tag{2.18}$$

But if $h_{\alpha\beta}$ is defined as in (2-8.4) with u^α taken to be the particle velocity v^α, then in virtue of (2-8.11) and (2.3) we have

$$F^\alpha = g^{\alpha\beta}h_{\beta\gamma}F^\gamma. \tag{2.19}$$

When this is substituted into (2.18), the result may be simplified with the use of (2.12) to give

$$\frac{d}{dt}E_{\alpha\beta} = -t_{(\alpha}F_{\beta)}, \tag{2.20}$$

where

$$F_\alpha \equiv h_{\alpha\beta}F^\beta - Gt_\alpha. \tag{2.21}$$

The notation (2.21) is consistent as it implies

$$g^{\alpha\beta}F_\alpha = F^\beta \tag{2.22}$$

in accordance with the convention on raising indices. We shall call F_α the *covariant force vector*. It is more comprehensive than F^α, since it fully describes an impure force. The rate of supply of energy is given by

$$G = -F_\alpha v^\alpha, \tag{2.23}$$

so that a pure force is characterized by

$$F_\alpha v^\alpha = 0. \tag{2.24}$$

In a particular inertial frame it may be seen that

$$F_\alpha = (\mathbf{F}, -\mathbf{v}\cdot\mathbf{F} - G). \tag{2.25}$$

The decompositions (2.14) and (2.25) confirm that the $\alpha = \beta = 4$ component of (2.20) corresponds to the sum of (2.7) and (2.8), as is to be expected.

3   Mass and energy in the relativistic theory

3a   *The relativistic unification of mass and energy*

The terminology used by a relativistic observer in a particular inertial frame is based on that of the Newtonian theory. The parallel is kept as close as possible by the insertion of appropriate factors of c, to preserve the independence of the units of length and time. This differs from the usual procedure for spacetime variables, where the introduction of c would be very artificial. So although we saw in §1 that identical equations often hold between spacetime variables in the Newtonian and relativistic theories, since proper time is measured in units of length, analogous variables in the two theories may differ by factors of c. In the three-dimensional formulation we shall find greater differences between the equations, but similarly named variables in the two theories will strictly correspond to one another.

The relativistic analogue of (2.1) is, from (2-7.10),

$$v^\alpha = (c^{-1}\gamma \mathbf{v}, c^{-1}\gamma) \tag{3.1}$$

where
$$\mathbf{v} = d\mathbf{x}/dt, \quad \gamma = (1 - \mathbf{v}^2/c^2)^{-\frac{1}{2}}. \tag{3.2}$$

To have a three-momentum $\mathbf{p}$ which is conserved and which agrees with (2.2) in the limit $c \to \infty$, we must thus put

$$p^\alpha = (c^{-1}\mathbf{p}, m\gamma/c), \quad \mathbf{p} = m\gamma \mathbf{v}. \tag{3.3}$$

By analogy with (2.2), the coefficient of proportionality $m\gamma$ between $\mathbf{p}$ and the three-velocity $\mathbf{v}$ is called the *apparent mass* of the particle. When any confusion might arise between m and $m\gamma$, m is called its *rest mass*. Since $m\gamma = cp^4$, apparent mass is conserved in collisions.

With these definitions we have kept the formal parallel with the Newtonian theory. Conservation of four-momentum is expressed in each inertial frame by the separate laws of conservation of three-momentum and of apparent mass. However, there is one important difference between the Newtonian and relativistic situations. The presence of the factor γ shows that in contrast to the Newtonian mass, apparent mass is not a function of state. Let us consider the consequences of this for a collision which does not alter the internal states of the particles involved. Since rest mass is a function of state, and indeed is an isolate invariant, the rest masses of all the particles will be unchanged by the collision. Hence both $\sum m$ and $\sum m\gamma$ will be unchanged. This has an interesting implication for the Newtonian theory, as we see from (2-8.25) that in the limit $c \to \infty$,

$$c^2(\textstyle\sum m\gamma - \sum m) \to \sum \tfrac{1}{2}m\mathbf{v}^2. \tag{3.4}$$

It follows that kinetic energy must be conserved in such collisions, which is a correct Newtonian result but one which had to be added as a quite separate hypothesis in the systematic development of the Newtonian theory itself.

In a general collision within the Newtonian theory, kinetic energy alone is not conserved. Hence, by the reverse of the above argument, within the relativistic theory a general collision can neither preserve the rest masses of the individual particles, nor even conserve their sum. This behaviour of rest mass resembles that of internal energy rather than that of mass in the Newtonian theory. In view of (3.4), it is convenient to define a relativistic kinetic energy T for a particle of rest mass m by

$$T = mc^2(\gamma - 1). \qquad (3.5)$$

Conservation of apparent mass can then be expressed by the constancy of

$$\sum T + c^2 \sum m. \qquad (3.6)$$

If mc^2 is called the rest energy of the particle, (3.6) shows that the sum of the kinetic and rest energies of the particles involved in a collision is constant. This shows explicitly the parallel between rest mass and Newtonian internal energy, and for this reason the law of conservation of apparent mass is more commonly called the relativistic law of conservation of energy.

3b The relativistic distinction between mass and energy

It is now clear that the Newtonian concepts of mass and energy are both different aspects of the same relativistic quantity. This suggests that any energy-carrying field such as the electromagnetic field has as much right to be considered as matter as does the ordinary substantial matter that is usually designated as such. Before we can see why the mass and energy aspects separate in the Newtonian limit, we must consider what are the properties of substantial matter that so clearly distinguish it in everyday experience from such energy-carrying fields. At a phenomenological level there are primarily two such properties. One is the localizability of substantial matter, and the other is that the dominant contribution to the mass is inert. The former is responsible for point particles being approximately realizable, and for the conception of a particle as identifiable. The concept of identity has already been invoked several times. In the theory of continuous media it leads to a particular material element being sufficiently recognizable to be followed during its motion, so that the velocity of the material is everywhere well defined. Let us now turn our attention to the latter property.

More specifically, it is that for a given particle, the variation in mass over the range of states which is being considered is small in comparison with its total mass in any of these states. Let one such state be chosen as a reference state, and let m_0 be its mass. If the relativistic internal energy U of a state of mass m relative to the reference state is defined by

$$U = (m - m_0)c^2, \tag{3.7}$$

then in any identity-preserving collision the law of conservation of energy can be expressed as the conservation of kinetic plus internal energy. In this way the contribution $c^2 \sum m_0$ to (3.6) is removed from consideration. This is useful as this contribution, which is constant, is normally far larger than either of the variable contributions $\sum T$ and $\sum U$. We shall call m_0 the *inert mass* of the particle.

Such a separation of inert mass and internal energy can be made at any conceptual level at which it is useful. Let us consider a few examples. In nuclear physics, the natural reference state for a nucleus is that in which all the nucleons are entirely separated. The internal energy is then known as the binding energy of the nucleus. It is a useful concept even though it may be of the order of one per cent of the rest energy. In atomic physics, on the other hand, the nucleus is treated as inert and attention is centred on the electronic structure of the atoms. The natural reference state now is that in which the nucleus is in its ground state and the electrons are widely separated from it and from one another. Relative to this, the binding energies of atomic states are at most of the order of 10^{-6} of the rest energy. This illustrates the dependence of the internal energy concept on the situations being considered, as atomic and nuclear physics both treat the same physical system but they concern different ranges of states.

Another significant point is also brought out by these examples. The reference states have been so chosen that inert mass is conserved also in those collisions which come within the scope of the theory in question but which do not preserve identity. These are the nuclear reactions and chemical reactions respectively. This can be achieved also at a macroscopic level if the reference state of a particle is taken as that in which the atoms are all in their ground states but are widely separated from one another. This is more suitable than the totally ionized reference state of atomic physics as the inner electron shells are inert for most macroscopic processes. The only conceptual level of interest which remains is that of the physics of elementary particles. Here, internal energy is not a useful concept as the interchange that occurs between kinetic and rest energies in particle collisions and decays is too large.

We see that at every level except this last one, there exists an inert mass which is conserved in all collisions, and which is constant as long as a particle retains its identity. This fact will be referred to as the *law of conservation of inert mass*. It enables the dominant inert mass contribution to be omitted from the conserved energy in all collisions, not only in the identity-preserving ones. It, rather than the law of conservation of apparent mass, is the true analogue of the Newtonian law of conservation of mass. It is less fundamental than conservation of four-momentum in that it is, at any conceptual level, a phenomenological manifestation of details of the internal structure of the particles that are outside the scope of the theory at that level. Inert mass thus transcends dynamics in the same way as does energy in the Newtonian theory.

The dependence of the kinetic energy T on the four-velocity u^α of the reference frame is easily expressed. If the definition (3.5) is combined with (2-7.13), it gives

$$T = -mc^2(v^\alpha u_\alpha + 1). \tag{3.8}$$

If this is combined with (1.1) and the definition (3.7) of U, the total energy

$$E \equiv T + U \tag{3.9}$$

can then be expressed as

$$E = c^2(m\gamma - m_0) = -c^2(p^\alpha u_\alpha + m_0). \tag{3.10}$$

For a system of particles, the flux $M_0(\Sigma)$ of inert mass through a hypersurface Σ is defined as

$$M_0(\Sigma) = \Sigma(\pm m_0), \tag{3.11}$$

with the same interpretation and sign convention as in (1.3). The flux of inert mass through a closed hypersurface is then zero, whether or not the system moves freely between collisions. If Σ is a spacelike section of the system with future-pointing outer orientation, $M_0(\Sigma)$ is independent of the choice of Σ and is known as the total inert mass of the system. Suppose now that the particles do move freely between collisions, and that their total four-momentum is $P^\alpha = MV^\alpha$. If the system is treated as a single composite particle, its internal energy U^* is given by

$$U^* = c^2(M - M_0). \tag{3.12}$$

Since $M = -P^\alpha V_\alpha$, this can also be written as

$$U^* = -c^2\Sigma(p^\alpha V_\alpha + m_0), \tag{3.13}$$

where the sum may be taken over the instantaneous values in any inertial frame. Comparison of (3.13) with (3.10) shows that U^* is the total kinetic and internal energy of the system, measured in the zero-momentum frame. This is the relativistic analogue of the result (2.17).

3c The non-relativistic limit

Let us now formalize the transition to the Newtonian limit. As in § 2-8b, overbars will be used to denote Newtonian variables. To get the known Newtonian results, it is clearly necessary that inert mass m_0 and internal energy U must separately remain finite as $c \to \infty$. Since the limiting process is a lengthening of the unit of time, there is no difficulty concerning m_0. It can be considered as independent of c and identified directly with the Newtonian mass $\overline{m}$. But U causes problems. For U to remain finite in the limit, the corresponding contribution U/c^2 to the rest mass must tend to zero. For heat energy, which is a macroscopic form of energy due to particle motion at a microscopic level, this occurs naturally. The speeds at a microscopic level must be scaled just as are all other speeds, so that they remain finite in the limit. Hence so also will their contribution to U. But the discrete energy levels of atomic and nuclear structure are very different. There is no adjustable parameter at our disposal which can vary these. Consequently, at an atomic level, Newtonian theory can only be regarded as a valid approximation if the variation in rest mass over the range of states under consideration is no greater than the possible error in experimental measurements. The orders of magnitude of binding energies given above show that this excludes nuclear theory, but it is acceptable in atomic and molecular physics except for very precise calculations.

As quantum phenomena really lie outside the scope of this book, we shall from now on confine attention to the macroscopic situation. The formal Newtonian limit is designed to reproduce the results which hold approximately when v/c and $U/(m_0 c^2)$ are small. It thus requires

$$m_0 = \overline{m}, \quad U = c^2(m - m_0) \to \overline{U}, \tag{3.14}$$

from which also

$$m \to \overline{m}. \tag{3.15}$$

These are the primary assumptions concerning mass and energy. They concern scalars, and hence have no dependence on any choice of inertial frame. It follows immediately from (3.3), (3.5), (3.9) and (2.6) that in a fixed frame we also have

$$\mathbf{p} \to \overline{\mathbf{p}}, \quad T \to \tfrac{1}{2}\overline{m}\overline{\mathbf{v}}^2, \quad E \to \overline{E}, \tag{3.16}$$

as is to be expected.

The basic assumption which governs the limiting behaviour of four-vectors is (2-8.21). If the four-velocities of both particle and reference frame satisfy this, then (2-8.21) and (2-8.32) give

$$cp^\alpha \to \overline{p}^\alpha, \quad c^{-1}p_\alpha \to -\overline{m}l_\alpha, \quad p^\alpha u_\alpha \to -\overline{m}, \left.\rule{0cm}{0.5cm}\right\}$$
$$c^2(p^\alpha u_\alpha + m) \to \overline{m}\overline{g}_{\alpha\beta\gamma\delta}\,\overline{u}^\alpha\overline{v}^\beta\overline{u}^\gamma\overline{v}^\delta. \quad\quad (3.17)$$

The last of these may be combined with (3.14) and (2.11) to show that

$$-c^2(p^\alpha u_\alpha + m_0) \to \overline{E}_{\alpha\beta}\,\overline{u}^\alpha\overline{u}^\beta. \quad\quad (3.18)$$

This shows how the Newtonian energy tensor arises from the limiting form of the expression (3.10) for the relativistic energy E. It completes the recovery of the dynamical variables of §2 from their relativistic counterparts.

A noticeable feature of the relativistic theory as compared with the Newtonian theory is the inversion of the roles of mass and energy. It is energy rather than inert mass which combines with three-momentum to form a four-vector. Similarly, it is inert mass rather than energy which transcends dynamics. One result of this is that the force vector F^α defined by (1.2) in the relativistic theory is more analogous to the Newtonian covariant force vector F_α of (2.21) than to the contravariant F^α of the Newtonian equation (1.2). This tends to hide the distinction between a pure and an impure force in the relativistic theory. Nevertheless, such a distinction still exists and it has the same significance and as much importance as in the Newtonian theory.

The guideline is that a pure force is one which leaves U, and hence also m, unchanged, while a pure energy source leaves v^α unchanged. Since $v^\alpha v_\alpha = -1$, it follows by differentiation that

$$v_\alpha\, dv^\alpha/d\tau = 0. \quad\quad (3.19)$$

When used in conjunction with (1.1) and (1.2), this shows that

$$dm/d\tau = -v_\alpha F^\alpha \quad\quad (3.20)$$

and

$$m\, dv^\alpha/d\tau = F^\alpha + v^\alpha v_\beta F^\beta. \quad\quad (3.21)$$

A pure force is thus characterized by

$$v_\alpha F^\alpha = 0 \quad\quad (3.22)$$

while for a pure energy source F^α has the form

$$F^\alpha = Kv^\alpha. \quad\quad (3.23)$$

The closeness of the analogies with (2.24), and with the case $F^\alpha = 0$

of (2.21), is evident. We see from (3.20) and (3.23) that

$$dm/d\tau = K. \tag{3.24}$$

It follows from (3.20) and (3.21) that an impure force F^α may be resolved into a pure force P^α and a pure energy source Kv^α with

$$P^\alpha = F^\alpha + v^\alpha v_\beta F^\beta, \quad K = -v_\beta F^\beta. \tag{3.25}$$

The three-dimensional force vector $\mathbf{F}$ and energy supply variable G are defined in terms of P^α and K respectively in such a way that the Newtonian equations

$$d\mathbf{p}/dt = \mathbf{F}, \quad dE/dt = G \tag{3.26}$$

remain valid for a pure force and a pure energy source respectively. With the use of (3.1) and (3.10) it may be seen that this requires

$$P^\alpha = (c^{-2}\gamma\mathbf{F}, P^4), \quad G = c^3 K. \tag{3.27}$$

Note that even with a pure energy source, not all the energy supplied goes into U. The second of equations (3.26) implies that the division is

$$dT/dt = (1 - \gamma^{-1})G, \quad dU/dt = \gamma^{-1}G. \tag{3.28}$$

The value of P^4 can be seen from (3.22) and (3.27) to be

$$P^4 = c^{-4}\gamma\mathbf{v}\cdot\mathbf{F}. \tag{3.29}$$

The equations which correspond to (3.26) for a general four-force can now be written down from (1.2). They are

$$d\mathbf{p}/dt = \mathbf{F} + c^{-2}G\mathbf{v} \tag{3.30}$$

and

$$dE/dt = \mathbf{v}\cdot\mathbf{F} + G. \tag{3.31}$$

As in the Newtonian theory, $\mathbf{v}\cdot\mathbf{F}$ is known as the rate of working of the force $\mathbf{F}$. The term $c^{-2}G\mathbf{v}$ in (3.30) has no Newtonian analogue. It expresses the change in momentum caused by the change in rest mass which G produces, and is another manifestation of the inertia of energy.

The Newtonian limit of the force variables is easily found. The main result follows on combining (3.18) with (1.2), (2.20) and the limit

$$cd/d\tau \to d/d\bar{t} \tag{3.32}$$

to give

$$c^3 F_\alpha u^\alpha \to \overline{F}_\alpha \overline{u}^\alpha. \tag{3.33}$$

Since u^α is arbitrary and $cu^\alpha \to \overline{u}^\alpha$ from (2-8.21), we must thus have

$$c^2 F_\alpha \to \overline{F}_\alpha. \tag{3.34}$$

It follows from (3.34) and (2-8.23) that

$$c^2 F^\alpha \to \overline{F}^\alpha. \qquad (3.35)$$

If (2-8.21) applied to v^α is combined with (3.34) and (3.35), it gives

$$cF^\alpha v_\alpha \to -\overline{F}^\alpha \overline{\iota}_\alpha, \quad c^3 K = -c^3 F_\alpha v^\alpha \to -\overline{F}_\alpha \overline{v}^\alpha = \overline{G}, \qquad (3.36)$$

the consistency of which yields (2.3). The corresponding three-dimensional forms which now follow with the aid of (3.27) and (2.4) are the expected results

$$\mathbf{F} \to \overline{\mathbf{F}}, \quad G \to \overline{G}. \qquad (3.37)$$

4   The free motion of a continuum

4a   *Consequences of momentum conservation*

At a microscopic level, a gas which is free from external forces may be regarded as composed of a large number of particles which move freely between collisions. If one does not look that closely at it, one sees only certain average properties. Some of these, such as the velocity of the gas, are direct averages of corresponding microscopic properties. Others, such as temperature and viscosity, have no immediate microscopic analogues. It is possible in both Newtonian and relativistic physics to deduce the macroscopic properties of a gas from a study of the averaging processes involved. But this does not remove the need for a more phenomenological study of gases, as the identification of appropriate averages with pressure, temperature, etc. depends on a knowledge of the macroscopic properties of these variables, which can only come from such a study. In this book only the phenomenological theory will be treated. Reviews of the kinetic theory approach in both the special and general theories of relativity have been given by Ehlers (1971) and Stewart (1971).

The laws of conservation of momentum and angular momentum, together with that of energy in the Newtonian theory and inert mass in the relativistic theory, form the starting point of phenomenological continuum dynamics. For the kinetic theory model of a gas, their validity follows immediately from the corresponding laws in particle dynamics. It is assumed that they hold universally. This immediately broadens the scope of the resulting theory to include liquids and solids, whose relativistic statistical physics has not yet been developed. In particle dynamics the appropriate variables were each defined first for an individual particle. By summation, a corresponding flux across a hypersurface was then constructed for the system as a whole. Continuum mechanics clearly has to take the flux as the basic concept as by its very nature it must not look at the individual particles at a microscopic level. The laws then state that for a continuum which is free from external forces, the net fluxes across any closed hypersurface are zero. This version of the laws was seen to hold for a system of particles, but there it was a derived form. With only the concept of flux at our disposal, it must now become the fundamental version.

Let z be a fixed point in spacetime, and u_α be a fixed covariant vector. If Π is the hyperplane

$$u_\alpha(x^\alpha - z^\alpha) = 0, \tag{4.1}$$

it follows from the results of § 2-6b that any covariant vector parallel to u_α is the vector area of some outer-oriented region of Π. There thus exists a continuously variable region $\Sigma(\lambda)$ of Π depending on a real parameter λ, which always contains z and whose vector area is λu_α. As $\lambda \to 0$, $\Sigma(\lambda)$ shrinks on to the point z. The momentum flux $P^\alpha(\Sigma)$ across $\Sigma(\lambda)$ will be a continuous function of λ, and in a medium homogeneous in space and time it would be proportional to λ. The essential hypothesis for a continuous medium is that a small enough region may be regarded as homogeneous. The limit

$$p^\alpha(z, u_\beta) = \lim_{\lambda \to 0} P^\alpha(\Sigma(\lambda))/\lambda \tag{4.2}$$

thus always exists, and as indicated by the notation it depends only on z and u_β. It satisfies

$$p^\alpha(z, ku_\beta) = \lim_{\lambda \to 0} P^\alpha(\Sigma(k\lambda))/\lambda = kp^\alpha(z, u_\beta). \tag{4.3}$$

Choose now a fixed rectilinear coordinate system (α) and let Π^γ, $\gamma = 1, 2, 3, 4$, be hyperplanes such that the γth coordinate is constant on Π^γ. Then provided that the components of u_α are all nonzero, the five intersecting hyperplanes Π^γ and Π enclose a bounded region R of spacetime. This is a four-dimensional generalization of a triangle and tetrahedron. Label the five faces as Σ^γ and Σ respectively, and give them the outer orientation determined by the outward direction from R. If the Π^γ are suitably chosen and considered as functions of λ, it is possible to ensure that Σ contains z and has vector area λu_α. Now an application of (2-6.26) with $\phi = 1$ shows that the total vector area of the boundary ∂R is zero. Hence if the vector area of the face which lies in Π^γ is $S_\alpha^{(\gamma)}$, we must have

$$\lambda u_\alpha + \sum_\gamma S_\alpha^{(\gamma)} = 0. \tag{4.4}$$

But the construction of Π^γ ensures that $S_\alpha^{(\gamma)}$ is parallel to the γth covariant basis vector $e_\alpha^{(\gamma)}$ of the coordinate system. Equation (4.4) thus implies

$$S_\alpha^{(\gamma)} = -\lambda u_\gamma e_\alpha^{(\gamma)}, \qquad (4.5)$$

where the summation convention is suspended for bracketed indices.

It follows from (4.2), (4.3) and (4.5) that

$$\lim_{\gamma \to 0} P^\alpha(\Sigma^\gamma(\lambda))/\lambda = -u_\gamma p^\alpha(z, e_\beta^{(\gamma)}). \qquad (4.6)$$

But by conservation of momentum,

$$P^\alpha(\Sigma) = -\sum_\gamma P^\alpha(\Sigma^\gamma). \qquad (4.7)$$

If this is divided by λ and the limit $\lambda \to 0$ taken, (4.2) and (4.6) enable the result to be expressed as

$$p^\alpha(z, u_\beta) = T^{\alpha\gamma}(z) u_\gamma \qquad (4.8)$$

where

$$T^{\alpha\gamma}(z) = p^\alpha(z, e_\beta^{(\gamma)}). \qquad (4.9)$$

The two-index quantity $T^{\alpha\gamma}(z)$ thus completely determines the momentum flux across any sufficiently small hyperplane through z. Since $p^\alpha(z, u_\beta)$ is a contravariant vector for all choices of u_β, the quotient rule of §2-2a enables us to deduce from (4.9) that $T^{\alpha\gamma}$ is a tensor of valence (2,0). It is to a continuous medium what p^α is to a particle. In relativity theory it is variously known as the *energy tensor*, the *energy–momentum tensor* and the *stress–energy tensor*. It will be shown below that it describes all these quantities. In the Newtonian theory 'energy' has to be replaced by 'mass' in these names.

In the notation of (4.2) we now have

$$P^\alpha(\Sigma(\lambda)) = \lambda u_\beta T^{\alpha\beta} + o(\lambda). \qquad (4.10)$$

If this is combined with the definition of an integral as a limit of a sum, it shows that the momentum flux across an arbitrary outer-oriented hypersurface Σ is given by

$$P^\alpha(\Sigma) = \int_\Sigma T^{\alpha\beta} dS_\beta. \qquad (4.11)$$

When Σ is the boundary ∂R of a region R of spacetime, conservation of momentum thus requires

$$\int_{\partial R} T^{\alpha\beta} dS_\beta = 0. \qquad (4.12)$$

The left-hand side of (4.12) is not quite in a form to which the version

(2-6.26) of Stokes' Theorem can be applied. However, this can easily be corrected. Let a_α and b_α be arbitrary constant vectors. If (2-6.26) is applied to the scalar field $\phi = T^{\alpha\beta} a_\alpha b_\beta$, it gives

$$a_\alpha b_\beta \int_R \partial_\gamma T^{\alpha\beta} dV = a_\alpha b_\beta \int_{\partial R} T^{\alpha\beta} dS_\gamma. \qquad (4.13)$$

But the vectors a_α and b_β are arbitrary. Equation (4.13) can thus only hold for all choices if

$$\int_R \partial_\gamma T^{\alpha\beta} dV = \int_{\partial R} T^{\alpha\beta} dS_\gamma. \qquad (4.14)$$

A similar argument may be used to justify the replacement of ϕ in (2-6.26) by an arbitrary tensor field. If (4.14) is contracted between β and γ, the right-hand side vanishes by (4.12). But the region R is arbitrary. The integrand of the integral over R must thus vanish identically, which gives

$$\partial_\beta T^{\alpha\beta} = 0. \qquad (4.15)$$

This is the differential equation which expresses the law of conservation of momentum.

4b Consequences of angular momentum conservation

Consider now the flux $J^{\alpha\beta}(z, \Sigma)$ across an arbitrary hypersurface Σ of angular momentum about a fixed point z. It will be assumed that the dependence of $J^{\alpha\beta}(z, \Sigma)$ on z given by (1.11) is a universal characteristic of angular momentum. If this or something similar were not the case, the laws of conservation of angular momentum about different points would be independent both of one another and of the law of conservation of momentum. This is physically unacceptable as it gives too many restrictions on the motion. Since (1.11) must hold for a continuum describable by kinetic theory, it is thus almost forced upon us to adopt it also for an arbitrary continuous medium.

In virtue of (4.11), the integral

$$L^{\alpha\beta}(z, \Sigma) = \int_\Sigma \{(x^\alpha - z^\alpha) T^{\beta\gamma}(x) - (x^\beta - z^\beta) T^{\alpha\gamma}(x)\} dS_\gamma \qquad (4.16)$$

also has this dependence. The difference

$$S^{\alpha\beta}(\Sigma) = J^{\alpha\beta}(z, \Sigma) - L^{\alpha\beta}(z, \Sigma) \qquad (4.17)$$

is thus independent of z. The arguments which led to the existence of a tensor field $T^{\alpha\beta}$ satisfying (4.15) can now be repeated for $S^{\alpha\beta}(\Sigma)$ with one refinement. In the notation of (4.2), it follows from (4.16) that

$$\lim_{\lambda \to 0} L^{\alpha\beta}(z, \Sigma(\lambda))/\lambda = 0. \tag{4.18}$$

This is because z lies on $\Sigma(\lambda)$ for all λ. Hence if $s^{\alpha\beta}(z, u_\gamma)$ is defined by

$$s^{\alpha\beta}(z, u_\gamma) = \lim_{\lambda \to 0} S^{\alpha\beta}(\Sigma(\lambda))/\lambda \tag{4.19}$$

by analogy with (4.2), then it is also true that

$$s^{\alpha\beta}(z, u_\gamma) = \lim_{\lambda \to 0} J^{\alpha\beta}(z, \Sigma(\lambda))/\lambda. \tag{4.20}$$

Conservation of angular momentum leads from (4.20) to the existence of a tensor field $U^{\alpha\beta\gamma}(z)$ which satisfies

$$U^{\alpha\beta\gamma} = U^{[\alpha\beta]\gamma} \tag{4.21}$$

and is such that $s^{\alpha\beta}(z, u_\gamma) = U^{\alpha\beta\gamma}(z)\, u_\gamma. \tag{4.22}$

It is known as the *spin tensor* of the matter. Equation (4.22) may then be used with (4.19) to show that for an arbitrary Σ,

$$S^{\alpha\beta}(\Sigma) = \int_\Sigma U^{\alpha\beta\gamma}\, dS_\gamma. \tag{4.23}$$

The vanishing of $J^{\alpha\beta}(z, \partial R)$ for an arbitrary region R of spacetime now gives

$$\partial_\gamma U^{\alpha\beta\gamma} = 2T^{[\alpha\beta]} \tag{4.24}$$

as the differential expression of the law of conservation of angular momentum.

In a kinetic theory in which molecules are treated as structureless point particles, the limit (4.20) is necessarily zero and so the spin tensor vanishes. There are occasions, however, when it is necessary to go deeper than this and to take into account the finite size of the molecules. The molecules themselves may then have an angular momentum of rotation which can give a contribution to $J^{\alpha\beta}$ that remains nonzero when the limit (4.20) is taken. A nonzero $U^{\alpha\beta\gamma}$ also arises naturally for matter described by a Lagrangian field theory. For these reasons it would be unrealistic to ignore the possibility of a nonzero spin tensor. However, it will be shown below that for macroscopic purposes a theory with nonzero spin tensor may normally be replaced by one in which the spin tensor is zero.

The vanishing of $U^{\alpha\beta\gamma}$ for a kinetic theory based on structureless particles shows that the similarity between the decompositions (1.24) and (4.17) is only superficial. The two tensors $S^{\alpha\beta}$ are analogous only to the extent that they describe in both cases an angular momentum which is free from any choice of reference point. They differ significantly in their origin. That of (1.24) is specifically a large-scale property of a whole system which depends for its existence on the nonzero size of the system. That of (4.17) is seen from (4.20), (4.22) and (4.23) to have its origin in the small-scale properties of a material element.

4c The symmetric energy–momentum tensor

The tensor
$$T^{\alpha\beta\gamma} \equiv \tfrac{1}{2}(U^{\beta\gamma\alpha} - U^{\alpha\beta\gamma} + U^{\alpha\gamma\beta}) \qquad (4.25)$$

which satisfies
$$T^{\alpha\beta\gamma} = T^{\alpha[\beta\gamma]} \qquad (4.26)$$

provides an alternative description of the spin since (4.25) can be inverted to give
$$U^{\alpha\beta\gamma} = -2T^{[\alpha\beta]\gamma}. \qquad (4.27)$$

With (4.24) this implies
$$T^{[\alpha\beta]} + \partial_\gamma T^{[\alpha\beta]\gamma} = 0, \qquad (4.28)$$

so that the tensor $*T^{\alpha\beta}$ defined by
$$*T^{\alpha\beta} \equiv T^{\alpha\beta} + \partial_\gamma T^{\alpha\beta\gamma} \qquad (4.29)$$

is symmetric:
$$*T^{\alpha\beta} = *T^{(\alpha\beta)}. \qquad (4.30)$$

It follows from (4.15) and (4.26) that it also satisfies
$$\partial_\beta *T^{\alpha\beta} = 0. \qquad (4.31)$$

The two properties (4.30) and (4.31) are those which (4.15) and (4.24) give for the energy–momentum tensor of matter with zero spin.

Consider now the quantities
$$*P^\alpha(\Sigma) = \int_\Sigma *T^{\alpha\beta} \, dS_\beta \qquad (4.32)$$

and
$$*J^{\alpha\beta}(z, \Sigma) = \int_\Sigma \{(x^\alpha - z^\alpha) *T^{\beta\gamma} - (x^\beta - z^\beta) *T^{\alpha\gamma}\} \, dS_\gamma, \qquad (4.33)$$

which would be the momentum and angular momentum fluxes of such matter. A version of Stokes' Theorem which may be obtained from (2-6.9) with the aid of (2-6.18) is

$$2\int_\Sigma dS_{[\alpha} \, \partial_{\beta]} \phi = \int_{\partial\Sigma} \phi \, dS_{\alpha\beta}. \qquad (4.34)$$

If this is used on the contributions to (4.32) and (4.33) which arise from the final term of (4.29), it is found that

$$*P^\alpha(\Sigma) = P^\alpha(\Sigma) + \tfrac{1}{2} \int_{\partial\Sigma} T^{\alpha\beta\gamma} \, dS_{\beta\gamma} \qquad (4.35)$$

and $$*J^{\alpha\beta}(z, \Sigma) = J^{\alpha\beta}(z, \Sigma) + \int_{\partial\Sigma} (x^{[\alpha} - z^{[\alpha}) \, T^{\beta]\gamma\delta} \, dS_{\gamma\delta}. \qquad (4.36)$$

If Σ is taken as the boundary ∂R of a spacetime region R, it is a closed surface and hence itself has no boundary, i e. $\partial\Sigma = 0$. In this case the final terms of (4.35) and (4.36) are absent. This shows that the starred momentum and angular momentum fluxes are conserved in virtue of conservation of the unstarred fluxes, a result which has already been proved in differential form. Another situation where these final terms vanish is when the 2-surface $\partial\Sigma$ lies outside the matter distribution, so that $T^{\alpha\beta\gamma}$ vanishes on it. This happens when Σ is a spacelike section of the distribution, so that the starred and unstarred fluxes also give the same value for the total momentum and angular momentum. They thus differ only in the precise distribution of the fluxes within the body.

For small bodies, particle dynamics gives an operational meaning to the total momentum of a body. However, as yet in our theoretical development no prescription has been given even in principle for the measurement of the momentum distribution within a body. It is not sufficient to remove a material element from the body, to measure its mass m, and to calculate its momentum from this as mv^α. Removal and isolation of the element changes its state of stress, which will upset the validity of the measurements. If the element is removed but its state of stress is maintained by a surrounding membrane, then it is not isolated and it simply becomes part of a different body. We are then back where we started. It will indeed be seen shortly that internal stress contributes significantly to the momentum flux. Only when the theory has been developed further and more contact made with observable quantities can the precise meaning of the momentum distribution be determined.

The significance of this is that it is not meaningful to ask which of the starred or unstarred fluxes are the correct ones. They will each lead to a consistent theory which will specify how they may be measured, at least in principle. However, the two theories will not be at quite the same level. A theory capable of distinguishing between $L^{\alpha\beta}$ and $S^{\alpha\beta}$ will need to delve more deeply into the origin of intrinsic spin than will one content with uniting them into a single flux $J^{\alpha\beta}$.

The occurrence mentioned above of a well-defined $S^{\alpha\beta}$ in kinetic theory when the nonzero size of the molecules is taken into account is a good example of this. Since macroscopic phenomenological physics seldom needs to separate $L^{\alpha\beta}$ and $S^{\alpha\beta}$, it can work solely with the tensor $*T^{\alpha\beta}$ of (4.29) instead of with the more detailed pair of tensors $T^{\alpha\beta}$ and $U^{\alpha\beta\gamma}$ (or $T^{\alpha\beta\gamma}$).

When all three are under consideration, $*T^{\alpha\beta}$ is known in relativistic theory as the *Belinfante–Rosenfeld* or *symmetrized* energy–momentum tensor, as it was introduced in Lagrangian field theory independently by Belinfante (1940) and Rosenfeld (1940). Otherwise, the unqualified term 'energy–momentum tensor' (or in the Newtonian theory, mass–momentum tensor) is used for $*T^{\alpha\beta}$. The momentum and angular momentum fluxes are then correspondingly taken as $*P^{\alpha}(\Sigma)$ and $*J^{\alpha\beta}(z, \Sigma)$ of (4.32) and (4.33) respectively. This will be the situation from now on in this book, and the distinguishing asterisks will thus be omitted.

In conclusion, for macroscopic purposes a material continuum which is free from external forces is described by a tensor field $T^{\alpha\beta}$ known in the relativistic theory as the energy–momentum tensor and in the Newtonian theory as the mass–momentum tensor. This satisfies

$$\partial_{\beta} T^{\alpha\beta} = 0 \qquad (4.37)$$

in virtue of the conservation of momentum, and

$$T^{\alpha\beta} = T^{\beta\alpha} \qquad (4.38)$$

in virtue of the conservation of angular momentum. It follows from (4.38) that the divergence in (4.37) may be taken with either index of $T^{\alpha\beta}$.

5 Spacetime decomposition of $T^{\alpha\beta}$

It has already been mentioned that $T^{\alpha\beta}$ describes the internal stresses within the matter, in addition to the distribution of four-momentum from which it was defined. Let us now consider how this comes about. For this purpose the material four-velocity v^{α} must be introduced. As discussed in § 3b, this is considered to be well defined as a macroscopic manifestation of the atomic structure of substantial matter. It will be assumed to be timelike, to ensure the existence of a rest frame for any material element. In the following development it will be supposed that the chemical composition of the matter is constant and homogeneous. This eliminates two complications that can occur in inhomogeneous fluids. One is mixing by diffusion, and the other is a change of composition by chemical reaction. Such matter is said to form a *single-component system*.

To every point x_0 in the region W of spacetime occupied by the matter, there corresponds a unique world line $x(\tau)$ such that

$$x(0) = x_0, \quad dx^{\alpha}/d\tau = v^{\alpha}(x), \tag{5.1}$$

known as the *material orbit* through x_0. If Σ_0 is an arbitrary hypersurface, then for each real τ a new hypersurface $\Sigma(\tau)$ may be defined by requiring $x(\tau) \in \Sigma(\tau)$ whenever $x_0 \in \Sigma_0$. When considered as a function of proper/absolute time τ, $\Sigma(\tau)$ is said to be moving with the matter, or to be *comoving*. In a macroscopic sense the matter on $\Sigma(\tau)$ is the same as that on Σ_0 but at a different stage in its evolution with time. The flux across $\Sigma(\tau)$ of inert mass (which can be identified with the Newtonian mass in accordance with §3c) is thus independent of τ. Note that at a microscopic level there may have been diffusion in both directions across the moving boundary $\partial\Sigma$, but this is not recognizable macroscopically due to the assumed chemical homogeneity of the material.

Consider now the region R of spacetime illustrated in figure 4. This is bounded by $\Sigma_1 \equiv \Sigma(\tau_1)$ and $\Sigma_2 \equiv \Sigma(\tau_2)$, where $\tau_2 > \tau_1$, and by the hypersurface Δ formed from the material orbits which join their boundaries Let the outer orientation of Σ_1 and Σ_2 be given by the direction of increasing τ, and let that of Δ be the inward direction into R. Then (2-6.26) gives

$$\int_R \partial_\alpha \phi \, dV = \int_{\Sigma_2} \phi \, dS_\alpha - \int_{\Sigma_1} \phi \, dS_\alpha - \int_\Delta \phi \, dS_\alpha. \tag{5.2}$$

Now any ordered set $\{\lambda_1, \lambda_2, \lambda_3\}$ of parameters for Σ_1 may be extended by the addition of τ to a parametrization $\{\tau, \lambda_1, \lambda_2, \lambda_3\}$ for R. It is only necessary to take the material orbits as curves of constant λ_i. If the former set is taken as defining the positive orientation of Σ_1, then the latter set is positively oriented for R. It thus follows from the definitions (2-6.6), (2-6.18) and (2-6.19) that

$$\int_R \partial_\alpha \phi \, dV = \int_{\tau_1}^{\tau_2} d\tau \int_{\Sigma(\tau)} \partial_\alpha \phi \, v^\beta \, dS_\beta. \tag{5.3}$$

Furthermore, if λ_1 is constant on a portion of the boundary $\partial\Sigma_1$ and increases outwards, then $\{\lambda_2, \lambda_3\}$ is positively oriented on $\partial\Sigma_1$. But this

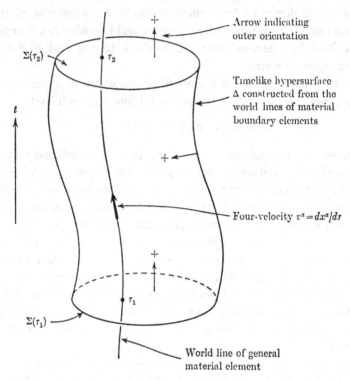

Figure 4. The spacetime hypercylinder formed by the evolution of
a material continuum.

also implies that $\{\tau, \lambda_2, \lambda_3\}$ is positively oriented on Δ. Hence

$$\int_\Delta \phi \, dS_\alpha = \int_{\tau_1}^{\tau_2} d\tau \int_{\partial\Sigma(\tau)} \phi v^\beta \, dS_{\alpha\beta}. \tag{5.4}$$

If (5.3) and (5.4) are substituted back into (5.2) and the result differen-
tiated with respect to τ_2, it gives the useful formula

$$\frac{d}{d\tau} \int_{\Sigma(\tau)} \phi \, dS_\alpha = \int_{\Sigma(\tau)} \partial_\alpha \phi v^\beta \, dS_\beta + \int_{\partial\Sigma(\tau)} \phi v^\beta \, dS_{\alpha\beta}. \tag{5.5}$$

It follows immediately from (4.11), (4.37) and (5.5) that

$$\frac{d}{d\tau} P^\alpha(\Sigma) = \int_{\partial\Sigma} T^{\alpha\beta} v^\gamma \, dS_{\beta\gamma}. \tag{5.6}$$

To interpret this, consider Σ as part of a larger hypersurface Σ^*. If
the exterior of Σ in Σ^* is labelled as Σ', the right-hand side of (5.6)
gives the force exerted by the matter in Σ' on the matter in Σ. The

boundary $\partial\Sigma'$ will be composed of two parts, the inner of which will be $-\partial\Sigma$, i.e. the surface $\partial\Sigma$ but with the opposite orientation. Hence this force is the negative of that exerted by the matter in Σ on the matter in Σ'. Newton's Third Law is thus recovered, for two materials in contact, in a form valid in both the Newtonian and relativistic theories.

It is natural to consider the total force of (5.6) as a sum, with an elemental force

$$dF^\alpha = T^{\alpha\beta}v^\gamma\,dS_{\beta\gamma} \tag{5.7}$$

being exerted on each 2-surface element. The equation for angular momentum which corresponds to (5.6) is easily found to be

$$\frac{d}{d\tau}J^{\alpha\beta}(z,\Sigma) = 2\int_{\partial\Sigma}(x^{[\alpha}-z^{[\alpha})\,dF^{\beta]}. \tag{5.8}$$

The right-hand side is the total moment of these elemental forces defined in accordance with (1.8). Had we retained a nonzero spin tensor $U^{\alpha\beta\gamma}$, this would not have been the case. There would have been an additional moment acting on each surface element, determined by $U^{\alpha\beta\gamma}$ and independent of z.

The development so far has been valid for the Newtonian and relativistic theories equally, but at this point the theories separate. Consider first the relativistic theory. In this, in accordance with (3.25) each elemental force may be resolved into a pure force

$$dP^\alpha = B^\alpha_\beta\,dF^\beta \tag{5.9}$$

where

$$B^\alpha_\beta \equiv A^\alpha_\beta + v^\alpha v_\beta \tag{5.10}$$

and a pure energy source with a rate of supply

$$dK = -v_\alpha\,dF^\alpha. \tag{5.11}$$

Suppose now that dP^α and dK are known for every 2-surface element through a point z. Let us investigate how much information they convey about $T^{\alpha\beta}$. Since $dS_{\alpha\beta}$ is antisymmetric but otherwise unrestricted except in magnitude, it follows from (5.7) that the consequential knowledge of dF^α determines

$$Z^{\alpha\beta\gamma} = T^{\alpha[\beta}v^{\gamma]}. \tag{5.12}$$

As

$$2Z^{\alpha\beta\gamma}v_\gamma = -B^\beta_\gamma T^{\alpha\gamma}, \tag{5.13}$$

we thus see that dP^α and dK determine respectively the tensors

$$\sigma^{\alpha\beta} \equiv -B^\alpha_\gamma B^\beta_\delta T^{\gamma\delta} \quad\text{and}\quad q^\alpha \equiv -v_\beta B^\alpha_\gamma T^{\beta\gamma}. \tag{5.14}$$

The minus signs are to accord with standard usage.

This does not yet prove the converse, that $\sigma^{\alpha\beta}$ and q^{α} respectively determine dP^{α} and dK, but this is in fact true. To see it, expand the right-hand side of the identity

$$T^{\alpha\beta} \equiv (B^{\alpha}_{\gamma} - v^{\alpha}v_{\gamma})(B^{\beta}_{\delta} - v^{\beta}v_{\delta})T^{\gamma\delta}. \qquad (5.15)$$

Since $T^{\alpha\beta}$ is symmetric, this gives

$$T^{\alpha\beta} = \mu v^{\alpha}v^{\beta} + q^{\alpha}v^{\beta} + v^{\alpha}q^{\beta} - \sigma^{\alpha\beta} \qquad (5.16)$$

where $$\mu \equiv v_{\alpha}v_{\beta}T^{\alpha\beta}, \qquad (5.17)$$

which implies $\quad dP^{\alpha} = -\sigma^{\alpha\beta}v^{\gamma}dS_{\beta\gamma}, \quad dK = q^{\alpha}v^{\beta}dS_{\alpha\beta} \qquad (5.18)$
as required.

The tensor $\sigma^{\alpha\beta}$ which describes the pure forces of mutual reaction between adjacent material elements is known as the *stress tensor*. The vector q^{α} describes the flow of energy between such elements by heat conduction, and if present also by electromagnetic radiation. This will be seen in more detail in Chapters 4 and 5. It is known as the *energy flux vector*. It follows from (5.14), (5.10) and (4.38) that

$$v_{\alpha}q^{\alpha} = 0, \quad v_{\beta}\sigma^{\alpha\beta} = 0, \quad \sigma^{\alpha\beta} = \sigma^{\beta\alpha}. \qquad (5.19)$$

When combined with (5.16), these relations imply (5.14) and (5.17). The decomposition of $T^{\alpha\beta}$ is thus determined by (5.16) and (5.19) alone. Under it, the ten linearly independent components of $T^{\alpha\beta}$ are seen to split into the six of $\sigma^{\alpha\beta}$, three of q^{α} and one of μ.

The parallel development of the Newtonian theory is somewhat simpler. Here, in accordance with (2.3) each elemental force must satisfy

$$t_{\alpha}dF^{\alpha} = 0. \qquad (5.20)$$

If $Z^{\alpha\beta\gamma}$ is still defined by (5.12), (5.13) is replaced by

$$2Z^{\alpha\beta\gamma}t_{\gamma} = B^{\beta}_{\gamma}T^{\alpha\gamma} \qquad (5.21)$$

where now $$B^{\alpha}_{\beta} \equiv A^{\alpha}_{\beta} - v^{\alpha}t_{\beta}. \qquad (5.22)$$

Equations (5.20) and (5.21) together imply

$$t_{\beta}B^{\alpha}_{\gamma}T^{\beta\gamma} = 0, \qquad (5.23)$$

so that instead of (5.15) we find that

$$T^{\alpha\beta} = \mu v^{\alpha}v^{\beta} - \sigma^{\alpha\beta} \qquad (5.24)$$

where $$\mu = T^{\alpha\beta}t_{\alpha}t_{\beta}, \quad \sigma^{\alpha\beta} = -B^{\alpha}_{\gamma}B^{\beta}_{\delta}T^{\gamma\delta}. \qquad (5.25)$$

The elemental force is given by

$$dF^{\alpha} = -\sigma^{\alpha\beta}v^{\gamma}dS_{\beta\gamma} \qquad (5.26)$$

in terms of the symmetric stress tensor $\sigma^{\alpha\beta}$ which satisfies

$$t_\beta \sigma^{\alpha\beta} = 0. \tag{5.27}$$

There is no analogue of the energy flux vector. This is in agreement with the inversion of the roles of mass and energy between the two theories which was seen to occur in particle dynamics.

It remains only to interpret μ. Before this is done, it is useful to consider the flux of inert mass and its conservation. Here we revert to a simultaneous treatment of the two theories. The techniques of §4a lead in this case to a flux vector ρ^α such that the total flux $M_0(\Sigma)$ of inert mass across a hypersurface Σ is given by

$$M_0(\Sigma) = \int_\Sigma \rho^\alpha \, dS_\alpha. \tag{5.28}$$

The conservation law $M_0(\partial R) = 0$ then gives the differential form

$$\partial_\alpha \rho^\alpha = 0. \tag{5.29}$$

If (5.29) is used in (5.5), it follows that

$$\frac{d}{d\tau} M_0(\Sigma) = \int_{\partial\Sigma} \rho^\alpha v^\beta \, dS_{\alpha\beta}. \tag{5.30}$$

Now according to the discussion of the second paragraph of this section, there is no net flow of matter into Σ across any element of the boundary $\partial\Sigma$ when Σ moves with the matter. Each elemental contribution $\rho^\alpha v^\beta \, dS_{\alpha\beta}$ to the right-hand side of (5.30) must thus vanish, for all $dS_{\alpha\beta}$. In view of the antisymmetry of $dS_{\alpha\beta}$ this implies

$$\rho^{[\alpha} v^{\beta]} = 0, \tag{5.31}$$

which shows that ρ^α is parallel to v^α. There thus exists a scalar field ρ, known as the *density of inert mass*, such that

$$\rho^\alpha = \rho v^\alpha. \tag{5.32}$$

Hence from (5.29), $$\partial_\alpha(\rho v^\alpha) = 0. \tag{5.33}$$

Return now to the Newtonian theory and let dS_α be the vector area of a spacelike hypersurface element. Such an element has a special significance since it corresponds to constant absolute time. The momentum flux across it is thus the quantity that any observer would consider to be the total momentum of the material element at that time. Since dS_α must be parallel to t_α, it follows from (5.24) and (5.27) that this flux is $(\mu v^\beta \, dS_\beta) v^\alpha$. This is the momentum of a mass $\mu v^\beta \, dS_\beta$

which has velocity v^α. But the mass of the element is also given from (5.32) as $\rho v^\beta \, dS_\beta$. Hence $\mu = \rho$, so that μ in (5.24) is also the mass density. It follows that

$$\rho^\alpha = T^{\alpha\beta} t_\beta, \tag{5.34}$$

so that (5.29) is a consequence of (4.37), as is necessary for the consistency of the interpretation.

In the relativistic theory a spacelike hypersurface element dS_α belongs to the hypersurface of simultaneity of an observer whose velocity is parallel to dS^α. The corresponding momentum flux $T^{\alpha\beta} \, dS_\beta$ thus gives the total momentum of the material element as seen by that particular observer. It can be seen from (5.16) that the contribution to this flux from q^α never vanishes, while that from $\sigma^{\alpha\beta}$ vanishes only when the matter is at rest relative to the observer, in which case dS_α and v_α are parallel. This is to be expected as a consequence of the inertia of energy. A nonzero flux q^α always supplies energy, while the internal stresses described by $\sigma^{\alpha\beta}$ do work only when their point of application moves relative to the observer. These contributions are additional to the convective momentum flux due to the motion of the matter. This is described by the term $\mu v^\alpha v^\beta$ in $T^{\alpha\beta}$, which gives a flux across dS_β of $(\mu v^\beta \, dS_\beta)v^\alpha$ as in the Newtonian case. In accordance with the results of § 3a, the coefficient of v^α should be interpreted as the rest mass of the element. Comparison of its form with (5.28) and (5.32) shows that μ is thus the rest mass density of the matter. In conformity with (3.7), the difference between the densities μ and ρ of rest and inert mass respectively is used to define the *specific internal energy* u of the matter. This is the internal energy per unit inert mass, and is given by

$$\mu = \rho(1 + u). \tag{5.35}$$

The factor of c^2 which would be required for direct correspondence with (3.7) has been omitted for convenience in the developments to be made in Chapter 4.

An equation for the variation of u along a material orbit may be deduced from (4.37), which implies

$$\partial_\beta(v_\alpha T^{\alpha\beta}) = T^{\alpha\beta} \partial_\beta v_\alpha. \tag{5.36}$$

If this is expanded with the use of (5.16) and (5.33), it gives

$$\rho v^\alpha \partial_\alpha u = \sigma^{\alpha\beta} \partial_\alpha v_\beta - \partial_\alpha q^\alpha - v^\alpha q^\beta \partial_\alpha v_\beta \tag{5.37}$$

as required.

6   Three-dimensional interpretation of $T^{\alpha\beta}$

The decomposition of $T^{\alpha\beta}$ given above is analogous to the separation (3.25) of a general force into a pure force and a pure energy source. For a complete interpretation of $T^{\alpha\beta}$ it is necessary to be able to relate the spacetime variables μ, q^{α} and $\sigma^{\alpha\beta}$ to the corresponding three-dimensional variables used by a particular observer. This requires results analogous to those of equations (3.27) to (3.31). The relativistic theory will be considered first.

To an inertial observer, the only significant fluxes are those across hypersurfaces of constant time t. These hypersurfaces are the instantaneous three-dimensional volumes of his reference frame. If Σ is the hypersurface which represents the spatial volume V at time t, then the four-momentum flux $P^{\alpha}(\Sigma)$ gives in accordance with (3.3) the total three-momentum and apparent mass of the matter within V at time t. To express this quantitatively, let $\mathbf{p}$ and ϵ/c^2 be the densities of three-momentum and apparent mass in the usual three-dimensional sense, so that ϵ is the density of total energy including rest energy. Then from (3.3),

$$P^{\alpha}(\Sigma) = \left(c^{-1} \int_V \mathbf{p}\, d^3\mathbf{x}, c^{-3} \int_V \epsilon\, d^3\mathbf{x} \right), \qquad (6.1)$$

where $d^3\mathbf{x} \equiv dx^1 dx^2 dx^3$. But if the spatial coordinates $\mathbf{x}$ are used to parametrize Σ in the evaluation of dS_{α} from (2-6.18) and (2-6.6), we find that

$$dS_{\alpha} = \delta_{\alpha}^4 \eta_{1234}\, d^3\mathbf{x}. \qquad (6.2)$$

It also follows from (2-5.23) and (2-3.11) that in a Minkowskian coordinate system $\eta_{1234} = c$. With the use of (4.11), (6.1) thus gives

$$T^{4\alpha} = T^{\alpha 4} = (c^{-2}\mathbf{p}, c^{-4}\epsilon). \qquad (6.3)$$

The next step in the interpretation is to analyze the dP^{α} and dK of (5.18) in accordance with (3.27). To do this, it is necessary to relate $dS_{\alpha\beta}$ to the vector surface element $d\mathbf{S}$ used in three dimensions. If the direction of $d\mathbf{S}$ is taken as outwards from V, and $dS_{\alpha\beta}$ has the sign appropriate to the boundary of the outer-oriented hypersurface Σ, the results of §2-6 show that

$$dS_{\alpha\beta} = \left(\begin{array}{c|c} 0 & -c\,d\mathbf{S} \\ \hline c\,d\mathbf{S} & 0 \end{array} \right). \qquad (6.4)$$

The notation, as also that used below, is as in (2-4.4).

The corresponding decompositions of q^{α} and $\sigma^{\alpha\beta}$ have to satisfy

(5.19). This shows that they are completely determined by their totally spatial components and hence may be expressed as

$$q^\alpha = (Q, c^{-2} Q \cdot v) \tag{6.5}$$

and

$$\sigma^{\alpha\beta} = \left(\begin{array}{c|c} \Sigma & c^{-2} \Sigma \cdot v \\ \hline c^{-2} \Sigma \cdot v & c^{-4} v \cdot \Sigma \cdot v \end{array} \right). \tag{6.6}$$

It now follows from (5.18) that

$$dK = c^{-3} dG, \quad dP^\alpha = (c^{-2} \gamma \, dF, c^{-4} \gamma v \cdot dF) \tag{6.7}$$

where

$$dG = -c^3 \gamma (Q - c^{-2} Q \cdot vv) \cdot dS \tag{6.8}$$

and

$$dF = c^2 (\Sigma - c^{-2} \Sigma \cdot vv) \cdot dS. \tag{6.9}$$

By (3.27) and (3.29), dF and dG are respectively the contributions to the three-dimensional force acting on V, and the rate of supply of energy to V, arising from material contact at dS. The three-dimensional energy flux vector q and stress dyadic σ are defined so that

$$dG = -q \cdot dS, \quad dF = \sigma \cdot dS. \tag{6.10}$$

They are thus given by

$$q = c^3 \gamma (Q - c^{-2} Q \cdot vv) \tag{6.11}$$

and

$$\sigma = c^2 (\Sigma - c^{-2} \Sigma \cdot vv), \tag{6.12}$$

which may be inverted easily since they imply

$$Q \cdot v = c^{-3} \gamma q \cdot v, \quad \Sigma \cdot v = c^{-2} \gamma^2 \sigma \cdot v. \tag{6.13}$$

It should be noted that σ is not symmetric. However, in virtue of the symmetry of Σ its transpose σ' may be evaluated from (6.12) and (6.13) in terms of σ. We obtain

$$\sigma' = \sigma + c^{-2} \gamma^2 (\sigma \cdot vv - v\sigma \cdot v), \tag{6.14}$$

from which

$$v \cdot \sigma = \gamma^2 (\sigma \cdot v - c^{-2} vv \cdot \sigma \cdot v). \tag{6.15}$$

The above results may be combined with (5.16) and (6.3) to show that

$$T^{\alpha\beta} = \left(\begin{array}{c|c} c^{-2}(vp - \sigma' + qv/c^2) & c^{-4}(\epsilon v + q - v \cdot \sigma) \\ \hline c^{-2} p & c^{-4} \epsilon \end{array} \right). \tag{6.16}$$

This is only one of several equivalent decompositions. However, although it hides the symmetry of $T^{\alpha\beta}$ it is the one most suited to our needs. It enables the conservation equation $\partial_\beta T^{\beta\alpha} = 0$, in this form in which the divergence is taken with the first index of $T^{\alpha\beta}$, to be written in terms of the three-dimensional variables as

$$\frac{\partial \epsilon}{\partial t} + \nabla \cdot (\epsilon \mathbf{v}) = \nabla \cdot (\mathbf{v} \cdot \boldsymbol{\sigma}) - \nabla \cdot \mathbf{q} \qquad (6.17)$$

and

$$\frac{\partial \mathbf{p}}{\partial t} + \nabla \cdot (\mathbf{v}\mathbf{p}) = \nabla \cdot \boldsymbol{\sigma}' - c^{-2}\nabla \cdot (\mathbf{q}\mathbf{v}). \qquad (6.18)$$

The interpretation of (6.17) and (6.18) is assisted by a well-known result in three-dimensional vector analysis which can also be deduced from (5.5). This is that for a volume $V(t)$ of the observer's instantaneous three-space which moves with the matter,

$$\frac{d}{dt}\int_V \phi\, d^3\mathbf{x} = \int_V \left\{ \frac{\partial \phi}{\partial t} + \nabla \cdot (\phi \mathbf{v}) \right\} d^3\mathbf{x}. \qquad (6.19)$$

To deduce it from (5.5) it is only necessary to observe that in the derivation of that equation, no use is made of τ being the proper time. Provided that v^{α} is consistently interpreted as $dx^{\alpha}/d\lambda$, (5.5) thus holds if τ is replaced by an arbitrary parametrization λ of the material orbits. The situation which corresponds to (6.19) is obtained when λ is the coordinate time t and the hypersurfaces $\Sigma(\lambda)$ which represent $V(t)$ lie in the hyperplanes of constant t. The $\alpha = 4$ component of (5.5) then gives (6.19) when use is made of (6.2) and (6.4). With the aid of (6.19) and Gauss' Theorem (2-6.26) in three dimensions, (6.17) may be put in the integral form

$$\frac{d}{dt}\int_V \epsilon\, d^3\mathbf{x} = \int_{\partial V} (\mathbf{v} \cdot d\mathbf{F} + dG). \qquad (6.20)$$

The notation is that of (6.10). Apart from the trivial difference that the separately conserved inert energy has been included in ϵ but not in E, (6.20) is seen to be the equation for a continuum which corresponds to (3.31) for a particle. Similarly, (6.18) in integral form corresponds to (3.30).

The conservation of inert mass, which is given by (5.33), has from (6.19) the three-dimensional form

$$\frac{d}{dt}\int \rho\gamma\, d^3\mathbf{x} = 0. \qquad (6.21)$$

The apparent density of inert mass is thus $\rho\gamma$. The factor of γ arises not from any change in the inert mass of a material element due to its motion, but instead from the Lorentz contraction of its volume. Since lengths in the direction of motion are contracted by a factor of γ^{-1}

while lengths perpendicular to the direction of motion are unaltered, a body which has volume V_0 in its rest frame has the smaller volume $V = V_0/\gamma$ when viewed in a frame in which it is in motion. The density of inert mass, or of any Lorentz scalar, is thus increased by a factor of γ.

The energy and momentum densities ϵ and $\mathbf{p}$ of (6.3) may be evaluated from (5.16) in terms of μ, $\mathbf{q}$ and $\boldsymbol{\sigma}$. It is found that

$$\epsilon = \gamma^2(\mu c^2 + 2c^{-2}\mathbf{q}\cdot\mathbf{v} - c^{-2}\mathbf{v}\cdot\boldsymbol{\sigma}\cdot\mathbf{v}) \tag{6.22}$$

and
$$\mathbf{p} = \gamma^2(\mu\mathbf{v} + 2c^{-4}\mathbf{q}\cdot\mathbf{v}\mathbf{v} - c^{-2}\boldsymbol{\sigma}\cdot\mathbf{v}) + c^{-2}\mathbf{q}, \tag{6.23}$$

which by (6.14) and (6.15) are consistent with the symmetry of the decomposition (6.16) of $T^{\alpha\beta}$. Their quantitative interpretation is more complicated than that of inert mass. In the rest frame of a material element they reduce to $\quad \epsilon = \mu c^2, \quad \mathbf{p} = \mathbf{q}/c^2.$ \tag{6.24}

These confirm that μ is the density of rest mass. In addition they show that an energy flux $\mathbf{q}$ carries an equivalent momentum $\mathbf{q}/c^2$, as is to be expected. Now in a frame in which the matter is in motion, the apparent density of apparent mass will be $\mu\gamma^2$. One factor of γ arises from the Lorentz contraction of the volume, while the other gives the velocity dependence of apparent mass found first in §3a. This explains in a general frame the terms involving μ. The energy flux $\mathbf{q}$ will carry the same momentum as before. But there are in addition contributions to (6.22) and (6.23) involving $\mathbf{q}$ and $\boldsymbol{\sigma}$ which vanish in the rest frame. What is their origin?

This is seen most easily by a simple construction. Suppose that one has a rigid body at rest to which a variable homogeneous stress $\boldsymbol{\sigma}(t)$ is being applied. Since the body does not deform under this stress, μ will not change. The equations of motion (6.17) and (6.18) are trivially satisfied in this rest frame since every term vanishes separately. But if the same body is viewed in a frame in which its velocity is $\mathbf{v}$, the stress will no longer be homogeneous. By the Lorentz transformation (1-7.6), $\boldsymbol{\sigma}$ will be in this frame a function not of t but of $t^* \equiv (t - \mathbf{v}\cdot\mathbf{x}/c^2)$. Since this is the only time-dependent feature of the system, ϵ and $\mathbf{p}$ will similarly be functions of t^*. The equations of motion now reduce to

$$\left. \begin{aligned} \gamma^{-2}\frac{\partial\epsilon}{\partial t} &= -\frac{\partial}{\partial t}(c^{-2}\mathbf{v}\cdot\boldsymbol{\sigma}\cdot\mathbf{v}), \\ \gamma^{-2}\frac{\partial\mathbf{p}}{\partial t} &= -\frac{\partial}{\partial t}(c^{-2}\boldsymbol{\sigma}\cdot\mathbf{v}), \end{aligned} \right\} \tag{6.25}$$

which integrate to give a dependence on $\boldsymbol{\sigma}$ in accordance with (6.22) and (6.23)

A complication arises when the analogous process is applied to the terms in $\mathbf{q}$. The natural analogue is to consider a body with perfect thermal conductivity which has a variable heat flow $\mathbf{q}(t)$ in its rest frame. However in view of (6.24), which will be assumed valid in the rest frame, such a heat flow exerts an external force on the body since it does not satisfy (6.18). This force performs no work in the rest frame, but it does in the frame in which the body is in motion. This must be taken into account in the calculation. If this is done, the dependence on $\mathbf{q}$ can also be recovered.

This completes the interpretation of $T^{\alpha\beta}$ in the relativistic theory. In the Newtonian theory the situation is much simpler. When Σ is a hypersurface of constant absolute time, the surface element dS_α still satisfies (6.2) but now with $\eta_{1234} = 1$ so that

$$dS_\alpha = t_\alpha \, d^3\mathbf{x}. \qquad (6.26)$$

In keeping with (2.2), the densities ρ and $\mathbf{p}$ of mass and three-momentum are defined by

$$P^\alpha(\Sigma) = \left(\int p \, d^3\mathbf{x}, \int \rho \, d^3\mathbf{x} \right). \qquad (6.27)$$

With the use of (4.11), (5.24) and (5.27), this leads to

$$\mu = \rho, \quad \mathbf{p} = \rho\mathbf{v}, \qquad (6.28)$$

in confirmation of the results of §5. In virtue of (5.27) the stress tensor has the simple decomposition

$$\sigma^{\alpha\beta} = \left(\begin{array}{c|c} \boldsymbol{\sigma} & 0 \\ \hline 0 & 0 \end{array} \right) \qquad (6.29)$$

with $\boldsymbol{\sigma}$ symmetric. Since (6.4) again holds if the factor c is omitted, the expression (5.26) for the elemental force becomes

$$dF^\alpha = (\mathbf{dF}, 0), \quad \mathbf{dF} = \boldsymbol{\sigma} \cdot \mathbf{dS}. \qquad (6.30)$$

This is consistent with (2.4) and shows that $\boldsymbol{\sigma}$ is the usual three-dimensional stress tensor. The conservation equation (4.37) now gives

$$\frac{\partial \rho}{\partial t} + \boldsymbol{\nabla} \cdot (\rho\mathbf{v}) = 0 \qquad (6.31)$$

and

$$\frac{\partial}{\partial t}(\rho\mathbf{v}) + \boldsymbol{\nabla} \cdot (\rho\mathbf{v}\mathbf{v}) = \boldsymbol{\nabla} \cdot \boldsymbol{\sigma}. \qquad (6.32)$$

The first is the usual form of the mass conservation equation. With its use the second may be simplified to the standard form

$$\rho \left(\frac{\partial \mathbf{v}}{\partial t} + \mathbf{v} \cdot \boldsymbol{\nabla}\mathbf{v} \right) = \boldsymbol{\nabla} \cdot \boldsymbol{\sigma} \qquad (6.33)$$

for the Newtonian equation of motion of a continuum.

7 Energy in Newtonian continuum mechanics

7a *The Newtonian stress–energy tensor*

In the relativistic theory the energy–momentum tensor $T^{\alpha\beta}$ and the flux vector ρ^α of inert mass give a complete macroscopic description of a continuum for dynamical purposes. However, in the Newtonian theory the description given so far lacks energy variables, and the mass–momentum tensor $T^{\alpha\beta}$ determines ρ^α according to (5.34). It is clear from the discussions of §§2*b* and 4*a* that the spacetime description of energy in a Newtonian continuum leads to a tensor field $T_{\dot\alpha\dot\beta}{}^\gamma$ which determines the tensor flux $E_{\alpha\beta}(\Sigma)$ of energy across a hypersurface Σ by

$$E_{\alpha\beta}(\Sigma) = \int_\Sigma T_{\dot\alpha\dot\beta}{}^\gamma \, dS_\gamma. \tag{7.1}$$

The structure of $T_{\dot\alpha\dot\beta}{}^\gamma$ will now be investigated in more detail.

Its two most immediate properties are its symmetry

$$T_{\dot\alpha\dot\beta}{}^\gamma = T_{\dot\beta\dot\alpha}{}^\gamma \tag{7.2}$$

which follows from that of $E_{\alpha\beta}$, and its conservation equation

$$\partial_\gamma T_{\dot\alpha\dot\beta}{}^\gamma = 0 \tag{7.3}$$

which holds in the absence of external forces. But the main restriction on its structure follows from the relationship between the energy flux $E_{\alpha\beta}(\Sigma)$ and the fluxes $P^\alpha(\Sigma)$ of four-momentum and $M_0(\Sigma)$ of mass. For a system of particles it follows from (2.13) that

$$2E_{\dot\alpha}{}^\beta(\Sigma) = A_\alpha^\beta M_0(\Sigma) - t_\alpha P^\beta(\Sigma). \tag{7.4}$$

To ensure that (7.3) adds only one conserved quantity to the four already known, it will be assumed that (7.4) holds also for a general continuum. Since Σ is arbitrary, (4.11), (5.28) and (5.32) show that (7.4) is equivalent to

$$2T_{\dot\alpha}{}^{\beta\gamma} = \rho A_\alpha^\beta v^\gamma - t_\alpha T^{\beta\gamma}, \tag{7.5}$$

which implies

$$T^{\alpha\beta\gamma} = \tfrac{1}{2}\rho g^{\alpha\beta} v^\gamma. \tag{7.6}$$

Now in the notation of (5.22), it follows from that equation that

$$T_{\dot\alpha\dot\beta}{}^\gamma = B_\alpha^\delta B_\beta^\varepsilon T_{\dot\delta\dot\varepsilon}{}^\gamma + t_{(\alpha} \sigma_{\dot\beta)}^\gamma \tag{7.7}$$

identically, where

$$\sigma_{\dot\alpha}^\beta = (2B_\alpha^\gamma v^\delta + t_\alpha v^\gamma v^\delta) T_{\dot\gamma\dot\delta}{}^\beta. \tag{7.8}$$

But it also follows from (7.6) and (2-8.11), with the u^α of the latter taken as the material velocity v^α, that

$$B_\alpha^\delta B_\beta^\varepsilon T_{\dot\delta\dot\varepsilon}{}^\gamma = h_{\alpha\delta} h_{\beta\varepsilon} T^{\delta\varepsilon\gamma} = \tfrac{1}{2}\rho h_{\alpha\beta} v^\gamma. \tag{7.9}$$

If this is written out explicitly with the use of (2-8.4) and (2-8.14), it can be used to put (7.7) in the form

$$T_{\dot\alpha\dot\beta}{}^\gamma = \rho g_{\alpha\delta\epsilon\beta} v^\delta v^\epsilon v^\gamma + t_{(\alpha}\, \sigma_{\dot\beta)}{}^\gamma. \tag{7.10}$$

The mass–momentum tensor may be evaluated from (7.10) if (2.12) is used to raise the index β to obtain

$$2T_{\dot\alpha}{}^{\beta\gamma} = \rho A_\alpha^\beta v^\gamma - t_\alpha(\rho v^\beta v^\gamma - \sigma^{\beta\gamma}) \tag{7.11}$$

where

$$\sigma^{\beta\gamma} = g^{\beta\delta}\sigma_{\dot\delta}{}^\gamma. \tag{7.12}$$

If (7.11) is compared with (7.5), it is seen that

$$T^{\alpha\beta} = \rho v^\alpha v^\beta - \sigma^{\alpha\beta}. \tag{7.13}$$

Since $T^{\alpha\beta}$ is symmetric, so also is $\sigma^{\alpha\beta}$. It follows from (7.12) and (2-3.23) that

$$t_\alpha \sigma^{\alpha\beta} = 0. \tag{7.14}$$

We have thus recovered (5.24) and (5.27) from (7.7), which demonstrates the consistency of the notation $\sigma_{\dot\alpha}{}^\beta$ with that of the stress tensor $\sigma^{\alpha\beta}$ of §5.

The remaining components of $\sigma_{\dot\alpha}{}^\beta$ may be interpreted with the aid of (b.5), which is valid in the Newtonian theory if τ is taken as usual as the absolute time t. If it is applied to (7.1), in virtue of (7.3) and (7.10) it gives

$$\frac{d}{dt}E_{\alpha\beta}(\Sigma) = -t_{(\alpha}\int_{\partial\Sigma} dF_{\beta)} \tag{7.15}$$

where

$$dF_\alpha = -\sigma_{\dot\alpha}{}^\beta v^\gamma dS_{\beta\gamma}. \tag{7.16}$$

Comparison of (7.15) with (2.20) shows that dF_α is the covariant force acting on an element of the 2-surface $\partial\Sigma$. In accordance with (2.22) and (2.23), this may be resolved into a pure force described by the contravariant dF^α, and a pure energy source described by

$$dG = -v^\alpha dF_\alpha.$$

We see from (7.16) that the pure force is determined by the stress tensor in accordance with (5.26). If the methods of §5 are applied to dG, they show that it is determined by the vector

$$q^\alpha = B_\gamma^\alpha v^\beta \sigma_{\dot\beta}{}^\gamma \tag{7.17}$$

where B_γ^α is defined by (5.22). This vector satisfies

$$dG = q^\alpha v^\beta dS_{\alpha\beta}, \quad q^\alpha t_\alpha = 0. \tag{7.18}$$

As in the relativistic case it is known as the energy flux vector.

There is only one component of $\sigma_\alpha^{\cdot\beta}$ which is not determined by $\sigma^{\alpha\beta}$ and q^α together. To see this, once again let $h_{\alpha\beta}$ be defined by (2-8.4) with v^α replacing u^α. Then it follows from (7.12), (7.14) and (2-8.11) that raising the index α on the tensor

$$\sigma_\alpha^{\cdot\beta} - h_{\alpha\gamma}\,\sigma^{\gamma\beta} \tag{7.19}$$

gives zero. This tensor must thus have the form $t_\alpha\,Q^\beta$ for some vector Q^β. Since $v^\alpha h_{\alpha\gamma} \equiv 0$, multiplication of (7.19) by v^α gives

$$Q^\beta = v^\alpha \sigma_\alpha^{\cdot\beta}. \tag{7.20}$$

Substitution from (5.22) into (7.17) now shows that

$$Q^\alpha = \rho u v^\alpha + q^\alpha \tag{7.21}$$

where the scalar u is defined by

$$\rho u = v^\alpha t_\beta\,\sigma_\alpha^{\cdot\beta}. \tag{7.22}$$

On collecting together these results, we find that

$$\sigma_\alpha^{\cdot\beta} = t_\alpha(\rho u v^\beta + q^\beta) + 2g_{\alpha\gamma\delta\epsilon}\,v^\gamma v^\delta \sigma^{\epsilon\beta}, \tag{7.23}$$

where only u remains to be interpreted.

The final step follows when Σ in (7.1) is taken as spacelike, so that dS_α is parallel to t_α. In this case we find that

$$E_{\alpha\beta}(\Sigma) = \int_\Sigma (g_{\alpha\delta\epsilon\beta}\,v^\delta v^\epsilon + u t_\alpha t_\beta)\,\rho v^\gamma\,dS_\gamma. \tag{7.24}$$

Since $\rho v^\gamma\,dS_\gamma$ is the mass of a material element, the bracket in (7.24) must be the energy tensor per unit mass. If it is compared with (2.11), it is seen that u is thus the specific internal energy.

The tensor $\sigma_\alpha^{\cdot\beta}$ will be called the *Newtonian stress–energy tensor.* Since

$$t_\alpha v^\alpha = 1, \tag{7.25}$$

it follows from (7.10) that

$$v^\alpha v^\beta \partial_\gamma\,T_{\dot\alpha\dot\beta}{}^\gamma = v^\alpha \partial_\beta\,\sigma_\alpha^{\cdot\beta} \tag{7.26}$$

identically. The conservation equation (7.3) thus implies

$$v^\alpha \partial_\beta\,\sigma_\alpha^{\cdot\beta} = 0, \tag{7.27}$$

which is the required energy equation for a Newtonian continuum. It gives information present in (7.3) which is not contained in the equation (4.37) of conservation of four-momentum. Indeed, it gives all such extra information, since (4.37) and (7.27) together imply

(7.3). This is easily verified with the aid of (7.5) and (7.26). In this sense (7.27) is to the Newtonian theory what (5.29) is to the relativistic theory.

If (7.23) is substituted into (7.27), use of (5.33) enables the result to be written as

$$\rho v^\alpha \partial_\alpha u = \sigma_{\cdot\alpha}^{\,\beta} \partial_\beta v^\alpha - \partial_\alpha q^\alpha. \qquad (7.28)$$

This corresponds to the relativistic result (5.37), and it shows clearly that (7.27) is the continuum analogue of (2.8). Because of (7.25), the only part of the expression (7.23) for $\sigma_{\cdot\alpha}^{\,\beta}$ which contributes to the first term on the right-hand side of (7.28) is the part involving $\sigma^{\alpha\beta}$. The three terms of (7.28) thus give the separation of (7.27) into the contributions from u, q^α and $\sigma^{\alpha\beta}$.

The three-dimensional energy flux vector $\mathbf{q}$ in a particular inertial frame is defined, as in the relativistic theory, such that the first of equations (6.10) holds. It may be seen from (7.18) that this requires

$$q^\alpha = (\mathbf{q}, 0). \qquad (7.29)$$

In virtue of (6.29) it now follows from (7.23) that

$$\sigma_{\cdot\alpha}^{\,\beta} = \left(\begin{array}{c|c} \boldsymbol{\sigma} & 0 \\ \hline \rho u \mathbf{v} + \mathbf{q} - \boldsymbol{\sigma}\cdot\mathbf{v} & \rho u \end{array} \right) \qquad (7.30)$$

so that (7.28) has the three-dimensional form

$$\rho \left(\frac{\partial u}{\partial t} + \mathbf{v}\cdot\nabla u \right) = \boldsymbol{\sigma}{:}\nabla\mathbf{v} - \nabla\cdot\mathbf{q} \qquad (7.31)$$

in agreement with the usual treatment of continuum mechanics as given, for example, in Batchelor (1967).

7b Relation to the relativistic description

Let us now consider how these results follow as limits of the corresponding relativistic formulae. If the limits (3.35) and (3.36) are applied to the elemental forces (5.18), they give

$$c^2 dP^\alpha \to \overline{dF^\alpha}, \quad c^3 dK \to \overline{dG}. \qquad (7.32)$$

Here, as in the discussions of limits in preceding sections, overbars denote Newtonian variables. Since η_{1234} has the value c in the natural coordinates of a relativistic inertial frame but 1 in those of a Newtonian frame, the corresponding 2-surface and 3-surface elements $dS_{\alpha\beta}$ and dS_α also differ by a factor of c. Specifically,

$$c^{-1} dS_\alpha = \overline{dS}_\alpha, \quad c^{-1} dS_{\alpha\beta} = \overline{dS}_{\alpha\beta}. \qquad (7.33)$$

If (7.33) and (2-8.21) are taken into consideration, it follows from (5.18), (5.26) and (7.18) that (7.32) implies

$$c^2\sigma^{\alpha\beta} \to \bar{\sigma}^{\alpha\beta}, \quad c^3 q^\alpha \to \bar{q}^\alpha. \tag{7.34}$$

The restrictions of (5.27) and (7.18) arise in the limit from (5.19). Similarly, (3.14) and (3.15) lead to their continuum equivalents

$$\rho \to \bar{\rho}, \quad \mu \to \bar{\rho}, \quad c^2 u = c^2(\mu - \rho)/\rho \to \bar{u}, \tag{7.35}$$

while when (3.17) and (3.18) are applied to the fluxes $P^\alpha(\Sigma)$ and $M_0(\Sigma)$ they give

$$c^2 T^{\alpha\beta} \to \bar{T}^{\alpha\beta}, \quad c T^{\alpha\beta} u_\beta \to -\bar{\rho} v^\alpha \tag{7.36}$$

and

$$-c^3(T^{\alpha\beta} u_\beta + \rho v^\alpha) \to \bar{T}_{\beta\gamma;}{}^\alpha \bar{u}^\beta \bar{u}^\gamma. \tag{7.37}$$

As before, u^α denotes the velocity of the reference frame, which is assumed to satisfy (2-8.21). It is clear from these that (5.24) is the limit of (5.16), that the Newtonian version of (4.37) is the limit of the relativistic version, and that (7.3) follows in the limit from (4.37) and (5.33) together.

Of more interest is the derivation of (7.10) and (7.23) by limiting processes. To achieve this, we first write the left-hand side of (7.37) in the form

$$-c^3 \rho v^\alpha (1 + u^\beta v_\beta) + c^3 u^\beta (\rho v_\beta v^\alpha - T_{\beta}{}^\alpha). \tag{7.38}$$

By (2-8.21) and (2-8.32) the first term has the limit

$$\bar{\rho} v^\alpha \bar{g}_{\beta\delta\epsilon\gamma} \bar{u}^\beta \bar{u}^\gamma \bar{v}^\delta \bar{v}^\epsilon.$$

The limit of the second term of (7.38) must thus exist for all u^α. Since $c u^\alpha \to \bar{u}^\alpha$, this can only happen if

$$c^2(\rho v_\beta v^\alpha - T_{\beta}{}^\alpha) \tag{7.39}$$

also has a well-defined limit, say $\bar{\sigma}_\beta{}^\alpha$. Substitution into (7.37) now shows that

$$\bar{T}_{\beta\gamma;}{}^\alpha \bar{u}^\beta \bar{u}^\gamma = (\bar{\rho} \bar{g}_{\beta\delta\epsilon\gamma} \bar{v}^\alpha \bar{v}^\delta \bar{v}^\epsilon + \bar{t}_{(\beta} \bar{\sigma}_{\gamma)}{}^\alpha) \bar{u}^\beta \bar{u}^\gamma. \tag{7.40}$$

Since $\bar{u}^\beta$ is arbitrary and both the coefficients of the product $\bar{u}^\beta \bar{u}^\gamma$ in (7.40) are symmetric in β and γ, it follows that $\bar{T}_{\beta\gamma;}{}^\alpha$ equals the bracket on the right-hand side. This agrees with (7.10) as required.

To recover (7.23), note first that in virtue of (5.19),

$$q_\alpha = (g_{\alpha\beta} + v_\alpha v_\beta) q^\beta \quad \text{and} \quad \sigma_\beta{}^\alpha = (g_{\beta\gamma} + v_\beta v_\gamma) \sigma^{\gamma\alpha}. \tag{7.41}$$

When taken together with (2-8.28) and (7.34), these show that

$$c q_\alpha \to 0, \quad c^2 \sigma_\beta{}^\alpha \to 2 \bar{g}_{\beta\delta\epsilon\gamma} \bar{v}^\delta \bar{v}^\epsilon \bar{\sigma}^{\gamma\alpha}. \tag{7.42}$$

But by (5.16), the tensor (7.39) may be written as

$$-(c^{-1}v_\beta)\{c^2(\mu-\rho)(cv^\alpha)+c^3q^\alpha\}-(cv^\alpha)(cq_\beta)+c^2\sigma_\beta{}^\alpha. \quad (7.43)$$

If the limits of the individual terms are substituted into this from
(2-8.21), (7.34), (7.35) and (7.42), it gives the right-hand side of (7.23)
as required. This confirms the correctness of the limit

$$c^2(\rho v_\beta v^\alpha - T_\beta{}^\alpha)\to\bar{\sigma}_\beta{}^\alpha. \quad (7.44)$$

Since it follows from (4.37), (5.33) and the normalization condition
$v_\alpha v^\alpha = -1$ that

$$v^\beta\partial_\alpha(\rho v_\beta v^\alpha - T_\beta{}^\alpha) = 0, \quad (7.45)$$

we also see from (7.44) that (7.27) may be obtained directly as the
limit of (7.45).

4
Relativistic simple fluids

1 The role of thermodynamics in continuum mechanics

The results of the preceding chapter provide a framework within which continuum mechanics can be developed and interpreted. To obtain a complete theory for a particular type of continuous material, these general results must be augmented by constitutive laws which characterize the particular type of material being studied. By their very nature, these constitutive laws are not universal laws of physics. However, experience shows us that most materials fall into one of a small number of distinctive types. The materials within a given type obey common general laws but are distinguished from one another by differing values of a small number of parameters. As far as their mechanical properties are concerned, many materials can be put into the three categories of gases, viscous liquids and elastic solids. Further separation occurs when electromagnetic behaviour is taken into account, starting with a division into conductors and insulators but with further subdivision when more detailed behaviour is considered. But even with such subdivisions, the number of distinct types is very small in comparison with the number of materials which can be so classified.

Without further guidelines it would be necessary to extract the constitutive laws of each type of material separately from experiment. Historically this is what did happen. However, it was found that the laws of the different material types had a common pattern, and that certain general statements could be made which were equally applicable to all types. These general statements form the basic of the subject known as *thermodynamics*. Currently this subject has three distinct levels, which correspond to three distinct stages in its historical development. Attention was originally restricted to homogeneous equilibrium states of matter at rest. The laws thus found appear to have universal validity for all materials. Their study is known as *classical* or *equilibrium* thermodynamics, although the more logical term *thermostatics* is gradually gaining acceptance. Starting with the work of Onsager (1931a, b), near-equilibrium states began to be incorporated systematically into the theory. These developments led to the *thermodynamics of irreversible processes*, a detailed account of which is given in the book of de Groot & Mazur (1962). Its laws do not have the universality of those of classical thermodynamics, due largely to the vagueness of the restriction to near-equilibrium states. However, they do appear to be valid for a large class of materials under a wide range of conditions. The most recent development stems from the work of Coleman (1964) and is aimed at a thermodynamics valid for arbitrary non-equilibrium states. At this level the theory is still in the process of evolution, and it will not here be considered further.

The general rules of the thermodynamics of irreversible processes enable the constitutive laws of the different material types to be recovered from characterizations of these types which are much simpler than these laws themselves. This provides the guideline that we require when we wish to develop a relativistic continuum mechanics of specific simple materials. It suggests that it is the simple thermodynamic characterization of the material that should be retained in the relativistic theory, together with the general statements of the thermodynamics of irreversible processes. The relativistic generalizations of the constitutive laws can then be deduced from these assumptions. The range of validity of the resulting theory is somewhat vague, as is always the case with the thermodynamics of irreversible processes. However, an ideal gas can also be studied using relativistic kinetic theory. The calculations of Israel & Stewart (1976) and Stewart (1977) agree with the results of the thermodynamic theory even in extreme relativistic situations. This shows that the rules of the thermodynamic theory do not rely on non-relativistic conditions for their validity, and it suggests that the thermodynamic theory has as wide a range of applicability in relativistic physics as it has been confirmed experimentally to have in Newtonian physics.

In the present chapter the relativistic theory of a simple fluid will be developed as an example of the general procedure discussed above. Although it is the simplest possible example, it illustrates all the main features of the method. In the next chapter the same techniques will be applied to study the interaction of such a fluid with an electromagnetic field. In contrast to Chapter 3, only the relativistic theory will be developed in detail. The corresponding Newtonian results will be obtained by taking the Newtonian limit.

2 The entropy law

From now on, the term 'thermodynamics' will be used to mean the thermodynamics of irreversible processes and classical (equilibrium) thermodynamics will be referred to as thermostatics. The primary law of thermodynamics is the entropy law. Roughly speaking, this asserts that there exists an additive function S of the state of a material system, known as its *entropy*, which for an isolated system is a non-decreasing function of time. In conformity with the macroscopic approach adopted throughout this book, we shall accept this law without attempting to justify it from a microscopic viewpoint. The reader interested in the microscopic interpretation of entropy will find a stimulating discussion in the article of Prigogine (1974).

Our first task is to make the entropy law precise and to give it a spacetime formulation. An additive function of state is a function S which assigns a value $S(V, t)$ to the material in a region V at time t in such a way that

$$S(V_1 \cup V_2, t) = S(V_1, t) + S(V_2, t) \tag{2.1}$$

for any two disjoint regions V_1 and V_2. For a homogeneous state $S(V, t)$ will thus be proportional to the volume of V. It is assumed in continuum mechanics that around any point there exists a volume element which is both large enough in comparison with the microscopic structure of the matter for the continuum approximation to be valid, and at the same time small enough for the state of the material to be considered as uniform throughout it. This assumption allows every additive function of state S to be described by a corresponding density $s(\mathbf{x}, t)$, which is a function of position $\mathbf{x}$ and time t such that

$$S(V, t) = \int_V s(\mathbf{x}, t) \, dV. \tag{2.2}$$

Since the entropy law must hold in all inertial frames, the entropy density s must be defined in all such frames. However, at a given spacetime point z the value of s may depend on the four-velocity u^α of the frame in which it is evaluated. We must thus consider s as a function $s(z, u_\alpha)$ of both z and u_α. The argument vector u_α is taken in its covariant form for later convenience With this notation , (2-7.8) can be used to write (2.2) in the spacetime form

$$S(\Sigma) = \int_\Sigma s(z, n_\beta) \, n^\alpha \, dS_\alpha. \tag{2.3}$$

Here Σ is the spacelike hyperplane region which corresponds to V at time t, n^{α} is the future-pointing unit normal to Σ and Σ has been given the outer orientation determined by n^{α} in accordance with § 2-6a.

The right-hand side of (2.3) is meaningful when Σ is any spacelike hypersurface. The only change is that n^{α} becomes a function of z when Σ does not lie in a hyperplane. It is convenient to allow Σ to have an arbitrary outer orientation while restricting n^{α} to be future-pointing. This causes $S(\Sigma)$ to change sign when the outer orientation of Σ is reversed. As a recognition of this fact we shall call $S(\Sigma)$ the *entropy flux* across Σ. This brings the terminology for entropy into line with that used for momentum in § 3-1a and for inert mass in § 3-3b, except that entropy flux is so far defined only across a spacelike hypersurface while the fluxes of momentum and inert mass are defined across an arbitrary hypersurface.

Now let Σ_1 and Σ_2 be spacelike sections of the isolated system under consideration, with their natural outer orientations towards the future and with Σ_1 to the future of Σ_2 as illustrated in figure 3 of § 3-1a. To give the entropy law a spacetime formulation we assume that

$$S(\Sigma_1) \geqslant S(\Sigma_2) \tag{2.4}$$

for all such sections Σ_1 and Σ_2. This is a generalization of the version given at the beginning of this section, but it is a very natural one.

Since entropy is a property of matter, the entropy density is zero outside the system. There is thus no loss of generality in supposing that Σ_1 and Σ_2 meet outside the matter and thus enclose some region R of spacetime. Give the piecewise smooth boundary ∂R of R the outer orientation determined by the outward direction from R. Then $\partial R = \Sigma_1 \cup (-\Sigma_2)$, where $(-\Sigma_2)$ denotes Σ_2 with its orientation reversed, so that (2.4) can be written equivalently as

$$S(\partial R) \geqslant 0. \tag{2.5}$$

Suppose now that R is continuously deformed by squeezing Σ_1 and Σ_2 into contact over an increasing area. By continuity, (2.5) must remain valid throughout. But such squeezing can bring ∂R to lie wholly within the matter. Consequently (2.5) is valid for *any* spacetime region R whose boundary ∂R is piecewise smooth and is composed of spacelike hypersurfaces.

The increase of entropy between the future and past portions of ∂R must be due to physical processes which occur within R. If $V(R)$ is the volume of R as defined by (2-6.22), it follows that

$$S(\partial R) = O(V(R)) \quad \text{as} \quad V(R) \to 0. \tag{2.6}$$

Note that this is not a consequence of (2.5) even when smoothness assumptions are added. It is an additional postulate concerning the entropy concept, albeit one that is implicit in the physical interpretation of (2.5).

This postulate enables the treatment which was applied in §3-4a to the momentum flux to be applied with a few modifications also to the entropy flux. First extend the definition of $s(z, n_\alpha)$ to timelike vectors n_α which are not necessarily of unit length by requiring

$$s(z, \lambda n_\alpha) = \lambda s(z, n_\alpha) \tag{2.7}$$

for arbitrary scalar λ. Next adopt the notation of the paragraph containing (3-4.4) but take u_α and $e_\alpha^{(\gamma)}$ to be timelike. It then follows from (2.3) that

$$\lim_{\lambda \to 0} S(\Sigma(\lambda))/\lambda = -s(z, u_\alpha) \tag{2.8}$$

and

$$\lim_{\lambda \to 0} S(\Sigma^\gamma(\lambda))/\lambda = u_\gamma s(z, e_\alpha^{(\gamma)}), \tag{2.9}$$

and from (2.6) that

$$S(\Sigma(\lambda)) + \sum_\gamma S(\Sigma^\gamma(\lambda)) = O(\lambda^{\frac{3}{2}}) \quad \text{as} \quad \lambda \to 0. \tag{2.10}$$

On dividing (2.10) by λ and taking the limit $\lambda \to 0$, we find that

$$s(z, u_\alpha) = -\sum_\gamma s^\gamma(z) u_\gamma \tag{2.11}$$

where

$$s^\gamma(z) \equiv -s(z, e_\alpha^{(\gamma)}). \tag{2.12}$$

Since $s(z, u_\alpha)$ is a scalar for arbitrary timelike vectors u_α, the quotient rule of §2-2a enables us to deduce from (2.11) that the $s^\gamma(z)$ are components of a vector. It is known as the *entropy flux density* at z, and gives the primary description of entropy in the relativistic theory.

In the notation of (2.3) we have

$$n_\beta n^\alpha dS_\alpha = -dS_\beta. \tag{2.13}$$

Substitution from (2.11) into (2.3) thus enables the total entropy flux across Σ to be expressed as

$$S(\Sigma) = \int_\Sigma s^\alpha dS_\alpha. \tag{2.14}$$

Use of Stokes' Theorem in the form (2-6.26) now gives

$$S(\partial R) = \int_R \partial_\alpha s^\alpha dV. \tag{2.15}$$

In virtue of the arbitrariness in the region R occurring in (2.5), it follows that (2.5) is equivalent to

$$\partial_\alpha s^\alpha \geqslant 0. \qquad (2.16)$$

This is known as the *Clausius–Duhem inequality*, and it is the differential form of the entropy law. The positive quantity

$$\sigma \equiv \partial_\alpha s^\alpha \qquad (2.17)$$

is known as the *rate of production of entropy*, or as the *entropy source strength*. By adopting (2.14) as the definition of the entropy flux across Σ even when Σ is not spacelike, we see from (2.15) and (2.16) that (2.5) holds even without the restriction on the nature of the boundary ∂R.

The three-dimensional form of (2.16) is obtained by setting

$$s^\alpha = (c^{-1}\mathbf{s}, c^{-1}s) \qquad (2.18)$$

so that

$$\frac{\partial s}{\partial t} + \nabla \cdot \mathbf{s} \geqslant 0. \qquad (2.19)$$

The vector $\mathbf{s}$ is known as the entropy flux three-vector. That the s of (2.18) is the entropy density of (2.2) may be seen by putting $u_\alpha = -c\delta_\alpha^4$ in (2.11). This choice of u_α makes u^α a future-pointing unit vector as required.

The form (2.16) of the entropy law makes no explicit mention of the isolation of the system under consideration. The question thus arises of whether it should nevertheless be restricted to isolated systems, or whether it is in fact valid for all systems. Now many interacting physical systems can be considered as part of some larger isolated system. As a simple example, a liquid being heated in a vessel by a battery-powered electric heater can be considered as part of an isolated system comprising the liquid, its containing vessel, and the heater with its battery. The law (2.16) will hold for every part of the system, including the liquid of interest. This shows that the isolation requirement is essentially irrelevant to the validity of (2.16). The entropy law in its primary form (2.16) can thus be assumed valid for all systems.

3   The nature of equilibrium

Since thermodynamics is concerned with near-equilibrium situations, it is important to know what characterizes the equilibrium state. Intuitively, an equilibrium situation is one which maintains a steady state without external aid. Such a system need not be at rest. If a system in rotation is isolated, it will eventually settle into a steady state but will still be in rotation due to conservation of angular momentum. A system in equilibrium also need not be isolated. One is frequently interested in the equilibrium state of a system subject to given external constraints. Nevertheless, not all steady states are equilibrium states. A steady flow of heat along a metal bar can be maintained by an imposed temperature difference between its ends, but this is not considered to be a state of equilibrium. So somehow we must distinguish between applied constraints, which are acceptable in equilibrium, and external aid, which is not.

To make this distinction we look at the apparatus maintaining the constraint or aid. If this is changing, e.g. a fuel is being consumed, then we have an external aid. If it too is steady, as for a liquid in a thermos flask, then we have an allowed constraint. To formalize this we can say that a system is in equilibrium if it forms part of a larger isolated system, the whole of which is in a steady state. As a consequence of this steadiness, the total entropy of the overall system (as measured in an arbitrary inertial frame) must be independent of time. Since this overall system is isolated, it follows that the entropy source strength must vanish everywhere. In particular, it vanishes in that part which forms the system of interest.

It is this vanishing of the entropy source strength which is the essential distinction between equilibrium and non-equilibrium steady states. In a non-equilibrium steady state, entropy can be produced within the system provided that it flows out through the boundary into the apparatus maintaining the state at the same rate that it is being produced. We can thus dispense with any mention of the maintaining apparatus by characterizing an equilibrium state as a steady state in which the entropy source strength σ is zero. In most materials we can go even further, for dissipative processes often ensure that σ is non-zero in every non-steady state. In such materials, $\sigma = 0$ is alone enough to characterize equilibrium.

4   Thermostatics of a relativistic simple fluid

4a   Detailed characterization of a fluid in equilibrium

The entropy flux density s^α introduced in § 2 is a function of the local state of the material under consideration. This is true also of the energy–momentum tensor $T^{\alpha\beta}$ and the flux density ρ^α of inert mass which were studied in Chapter 3. From a thermodynamic viewpoint, the simplest material would thus be one in which $T^{\alpha\beta}$ and ρ^{α} characterized the local material state sufficiently to determine s^{α}. We would then have a functional dependence

$$s^{\alpha} = s^{\alpha}(T^{\beta\gamma}, \rho^{\delta}). \tag{4.1}$$

Such an ideal material will be described as a *relativistic simple fluid*.

Specification of the amount of information needed to determine s^{α} is the general way which thermodynamics adopts for characterizing the various basic types of material. For more complex materials, internal variables describing deformation are needed in addition to $T^{\alpha\beta}$ and ρ^{α}. When an electromagnetic field is present, its field variables must also be included. But implicit in the use of all such functional forms for s^{α} is a restriction to near-equilibrium states. The kinetic theory of gases gives good reasons for believing that no such relationship as (4.1) can hold for *arbitrary* states of even an ideal gas. The restriction of such thermodynamic theories to near-equilibrium conditions is thus present even at this early stage. However, since it is implicit throughout the following development, (4.1) and other such relationships will be used without further mention of this inherent restriction on their validity.

Consider now an equilibrium state of a simple fluid. Let Γ_0 denote symbolically the local state of the fluid at some point, remembering that the equilibrium state may in be rotation and thus not necessarily homogeneous. The entropy source strength σ is a function of the local state, it satisfies $\sigma(\Gamma_0) = 0$ from §3, and we know from §2 that $\sigma \geqslant 0$ for any other local state Γ. It follows that $\Gamma = \Gamma_0$ is a stationary point of σ in the manifold of local states close to Γ_0.

To extract quantitative information from this result, we note from (2.17) and (4.1) that

$$\sigma = \frac{\partial s^{\alpha}}{\partial T^{\beta\gamma}} \partial_{\alpha} T^{\beta\gamma} + \frac{\partial s^{\alpha}}{\partial \rho^{\beta}} \partial_{\alpha} \rho^{\beta}. \tag{4.2}$$

Consider first a homogeneous equilibrium state, so that

$$\partial_{\alpha} T^{\beta\gamma} = 0, \quad \partial_{\alpha} \rho^{\beta} = 0. \tag{4.3}$$

Then trivially $\sigma = 0$ as required, and a small departure from this state produces an increment $\delta\sigma$ in σ given by

$$\delta\sigma = \frac{\partial s^{\alpha}}{\partial T^{\beta\gamma}} \delta(\partial_{\alpha} T^{\beta\gamma}) + \frac{\partial s^{\alpha}}{\partial \rho^{\beta}} \delta(\partial_{\alpha} \rho^{\beta}). \tag{4.4}$$

Now $T^{\alpha\beta}$ and ρ^α satisfy the conservation equations (3-4.37) and (3-5.29), namely

$$\partial_\alpha T^{\alpha\beta} = 0, \quad \partial_\alpha \rho^\alpha = 0. \tag{4.5}$$

Since we have no other *a priori* restrictions on $T^{\alpha\beta}$ and ρ^α, and we know that $\delta\sigma$ must vanish for all first-order departures from the equilibrium state, it seems reasonable to suppose that $\delta\sigma$ must vanish *as a consequence of the restrictions* (4.5). We shall accept this hypothesis. It requires the derivatives of s^α which occur in (4.4) to have the forms

$$\frac{\partial s^\alpha}{\partial T^{\beta\gamma}} = -A^\alpha_{(\beta}\Theta_{\gamma)}, \quad \frac{\partial s^\alpha}{\partial \rho^\beta} = -\Phi A^\alpha_\beta, \tag{4.6}$$

where A^α_β is as usual the unit tensor and Θ_α and Φ are respectively vector and scalar functions of the equilibrium state. The minus signs are purely for later convenience.

We shall see below that (4.6) is sufficient to completely determine the possible homogeneous equilibrium states. To treat the inhomogeneous (rotating) states of equilibrium, note that every small volume element can separately be considered as a system in equilibrium. For the continuum approximation to be valid, the length scale of the inhomogeneities in the system must be large in comparison with the minimum size of a macroscopic volume element. Each such element can thus be considered on its own as a homogeneous state, and so will satisfy (4.6). The conditions (4.6) must thus hold throughout the original inhomogeneous equilibrium state, with Θ_α and Φ now being dependent on position.

As a check on the consistency of this argument, we see that (4.2) gives $\sigma = 0$ throughout this state, as required, as a consequence of (4.6) and (4.5). The condition $\delta\sigma = 0$ for σ to be stationary now becomes

$$\delta\left(\frac{\partial s^\alpha}{\partial T^{\beta\gamma}}\right)\partial_\alpha T^{\beta\gamma} + \delta\left(\frac{\partial s^\alpha}{\partial \rho^\beta}\right)\partial_\alpha \rho^\beta = 0. \tag{4.7}$$

It yields the further restrictions on $\partial_\alpha T^{\beta\gamma}$ and $\partial_\alpha \rho^\beta$ which govern the inhomogeneities that are possible in equilibrium.

The conclusions of this discussion can be formalized into a postulate which is also applicable to more complex situations than that considered above. This is that *in a deviation from equilibrium, the contributions to $\delta\sigma$ which arise from variations in the derivatives of the arguments of s^α must be such as to be eliminable by use of the differential equations satisfied by these quantities.* In the present case this elimination requires these contributions to vanish separately, but when the differential equations are inhomogeneous they may not do so. The electromagnetic theory of § 5-4a will provide an example of this latter situation. Equations (4.6) follow immediately from this postulate for any equilibrium state, without the necessity of considering the homogeneous and inhomogeneous cases separately.

4b The local state in equilibrium

It will be shown next that Lorentz invariance of (4.6) places a severe restriction on the form of $T^{\alpha\beta}$ in equilibrium. Note first that it is implicit in (4.1) that the components of s^α are functions of the metric components $g_{\alpha\beta}$ as well as of the components of $T^{\alpha\beta}$ and ρ^α. However, the discussion in § 2-3 shows that this is the only hidden dependence that s^α can have. A coordinate transformation which leaves the $g_{\alpha\beta}$ unchanged will thus also leave unchanged the functional dependence of s^α on $T^{\alpha\beta}$ and ρ^α.

Consider an infinitesimal coordinate transformation $(\alpha) \to (a)$ whose coefficients A_α^a have the form

$$A_\alpha^a = \delta_\beta^a(A_\alpha^\beta + \epsilon^\beta_{.\alpha}), \tag{4.8}$$

where δ_β^a is the Kronecker symbol defined by (2-3.8). If we retain only terms linear in $\epsilon^\beta_{.\alpha}$, its effect on the components of a tensor $t^{a\cdots}{}_{\beta\cdots}$ is given by

$$t^{a\cdots}{}_{b\cdots} = \delta_\alpha^a \dots \delta_b^\beta \dots (t^{a\cdots}{}_{\beta\cdots} + \delta t^{a\cdots}{}_{\beta\cdots}) \tag{4.9}$$

where

$$\delta t^{\alpha\cdots}{}_{\beta\cdots} = \epsilon^\alpha_{.\gamma}\, t^{\gamma\cdots}{}_{\beta\cdots} + \dots - \epsilon^\gamma_{.\beta}\, t^{\alpha\cdots}{}_{\gamma\cdots} - \dots . \tag{4.10}$$

It follows that the metric components are invariant if and only if

$$\epsilon_{(\alpha\beta)} = 0 \quad \text{where} \quad \epsilon_{\alpha\beta} \equiv g_{\alpha\gamma}\epsilon^\gamma_{.\beta}. \tag{4.11}$$

The conditions (4.6) require that any infinitesimal change in $T^{\alpha\beta}$ and ρ^α produces in an equilibrium state an increment in s^α given by

$$\delta s^\alpha = -\Theta_\beta \delta T^{\alpha\beta} - \Phi\, \delta\rho^\alpha. \tag{4.12}$$

In particular, this must hold when δs^α, $\delta T^{\alpha\beta}$ and $\delta\rho^\alpha$ are generated by a Lorentz transformation and are thus given by (4.10) in terms of an $\epsilon^\alpha_{.\beta}$ satisfying (4.11). Substitution from (4.10) into (4.12) gives

$$\epsilon_{\beta\gamma}\{g^{\alpha\beta}(s^\gamma + T^{\gamma\delta}\Theta_\delta + \rho^\gamma\Phi) + T^{\alpha\gamma}\Theta^\beta\} = 0, \tag{4.13}$$

which holds for all $\epsilon_{\alpha\beta}$ satisfying (4.11) if and only if

$$g^{\alpha[\beta}(s^{\gamma]} + T^{\gamma]\delta}\Theta_\delta + \rho^{\gamma]}\Phi) + T^{\alpha[\gamma}\Theta^{\beta]} = 0. \tag{4.14}$$

Lower the index α and multiply this by $\eta_{\beta\gamma\epsilon\zeta}\Theta^\zeta$, where $\eta_{\beta\gamma\epsilon\zeta}$ is the fundamental alternating tensor given by (2-5.23). We obtain

$$\eta_{\alpha\gamma\epsilon\zeta}\,\Theta^{\zeta}(s^{\gamma}+T^{\gamma\delta}\Theta_{\delta}+\rho^{\gamma}\Phi)=0, \tag{4.15}$$

which is equivalent to the existence of a scalar p such that

$$s^{\alpha}+T^{\alpha\beta}\Theta_{\beta}+\rho^{\alpha}\Phi=p\Theta^{\alpha}. \tag{4.16}$$

Equation (4.14) now simplifies to the form

$$(T^{\alpha[\beta}-pg^{\alpha[\beta})\,\Theta^{\gamma]}=0. \tag{4.17}$$

This in turn is equivalent to the existence of a vector κ^{α} such that

$$T^{\alpha\beta}-pg^{\alpha\beta}=\kappa^{\alpha}\Theta^{\beta}. \tag{4.18}$$

But since the left-hand side is symmetric, κ^{α} must be parallel to Θ^{α}. Set

$$\kappa^{\alpha}=\kappa\Theta^{\alpha}. \tag{4.19}$$

Then
$$T^{\alpha\beta}=\kappa\Theta^{\alpha}\Theta^{\beta}+pg^{\alpha\beta} \tag{4.20}$$

which may be used in (4.16) to show that

$$s^{\alpha}=-\Phi\rho^{\alpha}-\kappa\Theta_{\beta}\Theta^{\beta}\Theta^{\alpha}. \tag{4.21}$$

Since (4.14) is satisfied identically by (4.20) and (4.21), we have now extracted the full implications of the Lorentz invariance of (4.6).

Next consider (4.12) for an arbitrary variation which maintains equilibrium. Such a variation must preserve the general forms (4.20) and (4.21) but it will change the values of Θ^{α}, Φ, p, κ. Substitution from (4.20) and (4.21) into (4.12) gives

$$\rho^{\alpha}\delta\Phi=\Theta^{\alpha}(\delta p-\kappa\Theta^{\beta}\delta\Theta_{\beta}), \tag{4.22}$$

which gives rise to two distinct possibilities. If ρ^{α} and Θ^{α} are not parallel, (4.22) requires both

$$\delta\Phi=0 \quad\text{and}\quad \delta p=\kappa\Theta^{\beta}\delta\Theta_{\beta}. \tag{4.23}$$

For this case set
$$\mu=\Theta^{\alpha}\Theta_{\alpha}. \tag{4.24}$$

Then (4.23) requires
$$\delta\Phi=0,\quad \delta p=\tfrac{1}{2}\kappa\,\delta\mu. \tag{4.25}$$

It follows that on the manifold of equilibrium states close to the originally chosen one, Φ must be constant and both p and κ are functions of μ related by
$$\kappa=2dp/d\mu. \tag{4.26}$$

Such a system behaves as a superposition of two independent materials, one with

$$\left.\begin{aligned}
&T^{\alpha\beta}=\kappa\Theta^{\alpha}\Theta^{\beta}+pg^{\alpha\beta},\quad s^{\alpha}=-\kappa\mu\Theta^{\alpha},\quad \rho^{\alpha}=0,\\
&\mu=\Theta^{\alpha}\Theta_{\alpha},\quad p=p(\mu),\quad \kappa=2dp/d\mu,\quad \Theta^{\alpha}\text{ arbitrary,}
\end{aligned}\right\} \tag{4.27}$$

and the other with

$$T^{\alpha\beta} = 0, \quad s^\alpha = -\Phi\rho^\alpha, \quad \Phi \text{ constant}, \quad \rho^\alpha \text{ arbitrary.} \quad (4.28)$$

We shall reject this case as physically unrealistic.

The remaining possibility is that ρ^α and Θ^α are parallel. Since $\rho^\alpha = \rho v^\alpha$ by (3-5.32), where v^α is the material velocity, we may set

$$\Theta^\alpha = \Theta v^\alpha, \quad s^\alpha = s\rho^\alpha. \quad (4.29)$$

The scalar s is known as the *specific entropy* of the fluid. Then (4.21) and (4.22) reduce respectively to the scalar equations

$$\rho s = \kappa\Theta^3 - \rho\Phi \quad (4.30)$$

and

$$\rho \, \delta\Phi = \Theta \, \delta p + \kappa\Theta^2 \, \delta\Theta \quad (4.31)$$

while (4.20) gives

$$T^{\alpha\beta} = \kappa\Theta^2 v^\alpha v^\beta + p g^{\alpha\beta}. \quad (4.32)$$

If we put (4.32) into (3-5.14) and use (3-5.10), we find that the energy flux vector q^α and the stress tensor $\sigma^{\alpha\beta}$ in equilibrium are given by

$$q^\alpha = 0, \quad \sigma^{\alpha\beta} = -p(g^{\alpha\beta} + v^\alpha v^\beta). \quad (4.33)$$

In virtue of its relationship to the stress tensor, the scalar p is known as the *pressure* of the fluid.

The corresponding three-dimensional variables $\mathbf{q}$ and $\boldsymbol{\sigma}$ defined by (3-6.11) and (3-6.12) now take the forms

$$\mathbf{q} = 0, \quad \boldsymbol{\sigma} = -pc^2\mathbf{I}, \quad (4.34)$$

where $\mathbf{I}$ is the unit dyadic. This confirms that the mathematically defined 'relativistic simple fluid' has equilibrium properties which agree with one's intuitive notions of the properties of a fluid, namely the stress is isotropic and the energy (heat) flux vanishes. It is the coefficient pc^2 in (4.34) which is known in the three-dimensional theory as the pressure of the fluid. We shall apply this name instead to p itself as it is inconvenient to have factors of c^2 occurring in the space-time formulation. When there is a danger of confusion, the prefixes three- and four- may be used for clarity.

The specific internal energy u is defined by (3-5.35) with (3-5.17). For the $T^{\alpha\beta}$ of (4.32) these yield

$$\rho(1+u) = \kappa\Theta^2 - p. \quad (4.35)$$

This may be used to eliminate κ from (4.30) to give

$$\Phi = \Theta(1+u) + \rho^{-1}\Theta p - s \quad (4.36)$$

and hence κ and Φ from (4.31) to give

$$\delta u = \Theta^{-1}\delta s - p\,\delta(\rho^{-1}). \qquad (4.37)$$

It is conventional to set

$$v = \rho^{-1}, \quad T = \Theta^{-1}, \qquad (4.38)$$

and to call v the *specific volume* and T the *absolute temperature* of the fluid.

If we now write differentials du, etc. in place of the increments δu, etc. then (4.37) becomes

$$du = T\,ds - p\,dv. \qquad (4.39)$$

This is known as the *Gibbs relation*, and it implies that u, T and p can be considered as functions of s and v. Its equivalent form

$$ds = (1/T)\,du + (p/T)\,dv \qquad (4.40)$$

similarly implies that s, T and p can be considered as functions of u and v. The two mutually equivalent equations which give the functional relations

$$u = u(s, v), \quad s = s(u, v) \qquad (4.41)$$

are known as *fundamental equations* for the fluid, as the other thermostatic functions p, T can be determined from them. We see from (4.39) and (4.40) that

$$\left. \begin{array}{ll} T = \partial u/\partial s, & p = -\partial u/\partial v \\ 1/T = \partial s/\partial u, & p/T = \partial s/\partial v. \end{array} \right\} \qquad (4.42)$$

and

In the deduction of (4.42) from (4.39) and (4.40) there is an implicit assumption that v can be varied independently of s or u. This is false for an ideal incompressible fluid, for which v is unchangeable. However, incompressibility is inconsistent with the basic principles of special relativity as it implies an infinite speed of sound. It is thus only a mathematical idealization that is convenient for certain nonrelativistic purposes, and its exclusion is of no practical significance.

Following Israel (1972) we shall call Φ the *thermal potential* of the fluid. Equation (4.36) can now be written as

$$\Phi = T^{-1}(1 + u + pv - Ts), \qquad (4.43)$$

which expresses Φ in terms of the more common thermostatic variables u, p, v, T, s. The variables Θ and Θ^α also deserve names as they arise naturally in the spacetime formulation of thermostatics. According to (4.38) and (4.29) they are given by

$$\Theta = T^{-1}, \quad \Theta^\alpha = \Theta v^\alpha. \qquad (4.44)$$

It thus seems appropriate to call Θ the *inverse temperature* and Θ^α the *inverse temperature vector*.

The Gibbs relation (4.39) with its corresponding fundamental equation $u = u(s, v)$ is the starting point of classical equilibrium thermodynamics. It forms the closest point of contact between the classical theory and the present approach to the thermodynamics of fluids. The alternative form (4.40) based on the fundamental equation $s = s(u, v)$ is important in the non-equilibrium theory as the independent variables u and v are well defined even for non-equilibrium states. It will be used in this context in § 5. In the relativistic theory of equilibrium states a third version is often the most convenient. This is

$$\Theta \, dp = \rho \, d\Phi - [\rho(1+u)+p] \, d\Theta, \qquad (4.45)$$

which corresponds to a fundamental equation $p = p(\Theta, \Phi)$. It is the version which remains closest to the form (4.31) from which (4.39) and (4.40) were derived, and it is obtained from (4.31) simply by substituting for $\kappa\Theta^2$ from (4.35). Its advantages are that ρ^α and $T^{\alpha\beta}$ can be expressed as

$$\rho^\alpha = \Theta \, \frac{\partial p}{\partial \Phi} \, v^\alpha, \quad T^{\alpha\beta} = p g^{\alpha\beta} - \Theta \, \frac{\partial p}{\partial \Theta} \, v^\alpha v^\beta. \qquad (4.46)$$

These are simple expressions, and they involve only the independent variables Θ, Φ and v^α in addition to the fundamental equation $p = p(\Theta, \Phi)$.

4c The global state in equilibrium

The basic equation which governs the position dependence of the local state in an equilibrium situation is (4.7). The variation symbol δ there denotes the difference between the given equilibrium situation and an *arbitrary* neighbouring state which is not necessarily in equilibrium. This is in contrast to the variations considered in §4b, which did preserve equilibrium.

To extract the full content of (4.7) we need to know the most general increments $\delta T^{\alpha\beta}$ and $\delta\rho^\alpha$ which are possible now that the comparison state is not necessarily in equilibrium. We shall in fact assume that $\delta T^{\alpha\beta}$ and $\delta\rho^\alpha$ can be given arbitrary values at a single point, subject only to preservation of the symmetry of $T^{\alpha\beta}$. This assumption of arbitrariness for $\delta\rho^\alpha$ excludes incompressible fluids, as discussed above. That for $\delta T^{\alpha\beta}$ also excludes certain ideal materials, such as inviscid fluids and fluids with zero thermal conductivity.

In the notation of (4.7) we have

$$\delta\left(\frac{\partial s^\alpha}{\partial T^{\beta\gamma}}\right) = \frac{\partial^2 s^\alpha}{\partial T^{\delta\epsilon}\partial T^{\beta\gamma}}\,\delta T^{\delta\epsilon} + \frac{\partial^2 s^\alpha}{\partial\rho^\delta\partial T^{\beta\gamma}}\,\delta\rho^\delta. \tag{4.47}$$

If this and the corresponding result for $\delta(\partial s^\alpha/\partial\rho^\beta)$ are substituted into (4.7), exchange of the order of differentiation in the partial derivatives enables the result to be written as

$$\frac{\partial}{\partial x^\alpha}\left(\frac{\partial s^\alpha}{\partial T^{\delta\epsilon}}\right)\delta T^{\delta\epsilon} + \frac{\partial}{\partial x^\alpha}\left(\frac{\partial s^\alpha}{\partial\rho^\delta}\right)\delta\rho^\delta = 0. \tag{4.48}$$

Since the derivatives are evaluated in the equilibrium state, their values are given by (4.6). This enables (4.48) to be simplified to the form

$$(\partial_\alpha\Theta_\beta)\,\delta T^{\alpha\beta} + (\partial_\alpha\Phi)\,\delta\rho^\alpha = 0. \tag{4.49}$$

But $\delta T^{\alpha\beta}$ and $\delta\rho^\alpha$ are arbitrary, subject only to $\delta T^{\alpha\beta}$ being symmetric. Equation (4.49) can thus hold for all possible variations only if

$$\partial_{(\alpha}\Theta_{\beta)} = 0 \tag{4.50}$$

and

$$\partial_\alpha\Phi = 0. \tag{4.51}$$

These are the equations which govern the global behaviour of a simple fluid in equilibrium.

The conservation equations (4.5) must hold also, but they are in fact satisfied identically as a consequence of (4.50) and (4.51). This may be proved as follows. Multiply (4.50) by $v^\alpha v^\beta$ and (4.51) by v^α and use (4.44) on the first result. This gives

$$\dot\Theta = 0, \quad \dot\Phi = 0, \tag{4.52}$$

where the superscript dots denote the operator $v^\alpha\partial_\alpha$ which differentiates along the world line of a material particle. But (4.45) shows that p,ρ and u can be considered as functions of Θ and Φ. Hence (4.52) implies that also

$$\dot p = 0, \quad \dot\rho = 0, \quad \dot u = 0. \tag{4.53}$$

If we now multiply (4.50) by $g^{\alpha\beta}$ and use (4.52), we obtain

$$\partial_\alpha v^\alpha = 0. \tag{4.54}$$

The second of equations (4.5) is satisfied as required as an immediate consequence of (4.53) and (4.54).

To verify the first of equations (4.5) we deduce from (4.45) and (4.51) that

$$\Theta\partial_\alpha p = -[\rho(1+u)+p]\,\partial_\alpha\Theta \tag{4.55}$$

and from (4.50) on multiplication by v^β that

$$\partial_\alpha \Theta = \Theta v^\beta \partial_\beta v_\alpha. \tag{4.56}$$

But in equilibrium

$$T^{\alpha\beta} = [\rho(1+u)+p]\,v^\alpha v^\beta + pg^{\alpha\beta} \tag{4.57}$$

in virtue of (4.33). Use of (4.53) and (4.54) enables us to deduce from (4.57) that

$$\partial_\beta T_\alpha^{\cdot\beta} = [\rho(1+u)+p]\,v^\beta \partial_\beta v_\alpha + \partial_\alpha p, \tag{4.58}$$

which vanishes as required on application of (4.55) and (4.56).

Equation (4.50) can be integrated explicitly to determine the velocity and temperature distribution in a general equilibrium state. Put

$$\Omega_{\alpha\beta} = \partial_{[\alpha} \Theta_{\beta]}. \tag{4.59}$$

This satisfies the identity

$$\partial_\gamma \Omega_{\alpha\beta} = \partial_\alpha \partial_{(\beta} \Theta_{\gamma)} - \partial_\beta \partial_{(\alpha} \Theta_{\gamma)} \tag{4.60}$$

whose right-hand side vanishes by (4.50). The antisymmetric tensor $\Omega_{\alpha\beta}$ is thus constant throughout the fluid. Together, (4.50) and (4.59) show that

$$\partial_\alpha \Theta_\beta = \Omega_{\alpha\beta} \tag{4.61}$$

which integrates to give $\Theta_\beta = \Omega_{\alpha\beta} x^\alpha + b_\beta$ \hfill (4.62)

where b_β is some constant vector.

4d Hyperbolic motion

The result (4.62) is most easily interpreted in its three-dimensional form. In the notation of (2-4.5) we see from (4.44) and (2-8.18) that

$$\Theta_\alpha = (\gamma/cT)\,(\mathbf{v}, -c^2) \tag{4.63}$$

where γ is given by (2-7.10). Set

$$b_\alpha = (1/cT_0)\,(\mathbf{u}, -c^2), \tag{4.64}$$

thus defining T_0 and $\mathbf{u}$, and let $\boldsymbol{\omega}$ and $\mathbf{g}$ be three-vectors such that

$$cT_0\,\Omega_{\alpha\beta} = \left(\begin{array}{c|c} \epsilon_{abc}\,\omega_c & -g_a \\ \hline g_a & 0 \end{array} \right). \tag{4.65}$$

The notation here is that of (2-4.4) and ϵ_{abc} is the three-dimensional Levi–Civita symbol defined by (2-5.12). Then (4.62) gives

$$\mathbf{v} = \frac{\boldsymbol{\omega} \times \mathbf{x} + \mathbf{g}t + \mathbf{u}}{1 + \mathbf{g}\cdot\mathbf{x}/c^2}, \quad T = \frac{T_0}{(1+\mathbf{g}\cdot\mathbf{x}/c^2)\sqrt{(1-\mathbf{v}^2/c^2)}}. \tag{4.66}$$

Consider first the velocity field given by (4.66). In the Newtonian limit $c \to \infty$ it becomes

$$\mathbf{v} = \boldsymbol{\omega} \times \mathbf{x} + \mathbf{g}t + \mathbf{u}. \tag{4.67}$$

This describes a fluid which is simultaneously undergoing a rigid rotation with angular velocity ω and a uniform acceleration of value g. The formula for v given in (4.66) describes the relativistic equivalent of this motion, and is identical with it when $g = 0$. To understand the modification which arises when $g \neq 0$, consider the special case $\omega = 0$, $u = 0$. Choose axes such that $g = (g, 0, 0)$. The path $x(t)$ followed by a particular fluid element then satisfies

$$(1 + gx/c^2)\frac{dx}{dt} = gt. \tag{4.68}$$

This integrates to give

$$x = \frac{c^2}{g}\left\{\left[\left(1 + \frac{gx_0}{c^2}\right)^2 + \left(\frac{gt}{c}\right)^2\right]^{\frac{1}{2}} - 1\right\} \tag{4.69}$$

and

$$v \equiv \frac{dx}{dt} = gt\left[\left(1 + \frac{gx_0}{c^2}\right)^2 + \left(\frac{gt}{c}\right)^2\right]^{-\frac{1}{2}} \tag{4.70}$$

where x_0 is the value of x when $t = 0$. For fixed x_0 such that $gx_0 \ll c^2$, these reduce for small t to the Newtonian results

$$x \simeq x_0 + \tfrac{1}{2}gt^2, \quad v \simeq gt \tag{4.71}$$

but for large t they give asymptotically

$$x \simeq ct, \quad v \simeq c. \tag{4.72}$$

The existence of c as a limiting speed compels special relativity to modify the concept of a uniformly accelerated motion. We see now that the rather complicated motion (4.69) with which it is replaced is indeed consistent with c as a limiting speed. The type of motion described by (4.69) is known as *hyperbolic motion* since the paths of the material particles in the (x, t) plane form, as x_0 varies, a family of similar hyperbolas. This is evident if (4.69) is written in the equivalent form

$$\left(1 + \frac{gx}{c^2}\right)^2 - \left(\frac{gt}{c}\right)^2 = \left(1 + \frac{gx_0}{c^2}\right)^2. \tag{4.73}$$

A corresponding modification is not needed to the concept of a rigidly rotating motion since the speed of each material element in such a motion is constant. However, since it has $v = \omega \times x$ we see that there is a limit on the *size* of body which can rotate with a given angular velocity. This is imposed by the need to have $|v| < c$.

The general motion described by (4.66) is now seen to be a combination of a rigid rotation and a hyperbolic motion. This can reasonably be interpreted as a steady motion, in a sense which will be made precise in §6c. It thus fulfils all the properties laid down in §3 as characteristic of equilibrium. Nevertheless the temperature distribution is not uniform, even though it becomes so in the Newtonian limit. This is another manifestation of the 'inertia of energy' which was seen in § 3-6 to associate a momentum density with every energy flux.

It also serves as a reminder that one must not attribute properties to physical variables just because one has attached a suggestive name to them. It is perfectly acceptable to define temperature mathematically by (4.38) but its physical properties must then also be explored within the framework of a consistent mathematical theory. Linkages between theory and experiment must be made from the predictions of the overall theory, as there may not be a simple one-to-one correspondence between measured physical parameters and mathematically convenient variables. Lack of appreciation of this fact has led to lengthy debates as to the appropriate relativistic measure of temperature in a given inertial reference frame. In addition to T itself, T/γ (Planck 1907) and $T\gamma$ (Ott 1963, Arzelies 1965) also have their supporters. A general discussion of the Lorentz transformation properties of heat and related variables has been given by Landsberg & Johns (1970), who show clearly how the various possibilities fit into different, but equally consistent, overall schemes. Their paper can also serve as a guide to the literature on the subject for the reader who wishes to pursue this topic further.

5   Description of near-equilibrium states

The results of § 4b show that the equilibrium properties of a particular fluid material are completely determined by the fundamental equation $s = s(u, v)$ which expresses its specific entropy s as a function of the specific internal energy u and specific volume v. A local state may then be described completely by giving u, v and the velocity v^α. In particular, this determines the energy–momentum tensor through (4.57) where the pressure p is given by (4.42). But u, v and v^α can all be defined for an arbitrary state of the fluid. They are related to its energy–momentum tensor $T^{\alpha\beta}$ and flux ρ^α of inert mass by

$$\rho^\alpha = \rho v^\alpha, \quad v^\alpha v_\alpha = -1, \quad v = 1/\rho, \quad \rho(1+u) = T^{\alpha\beta}v_\alpha v_\beta. \quad (5.1)$$

These parameters can thus be used to define a *comparison equilibrium state*.

For a fluid near equilibrium it is reasonable to suppose that the other thermostatic variables of this comparison state will have physical significance for the actual state being studied. From the actual values of u, v, v^α given by (5.1) we thus *define*

$$
\begin{aligned}
s &= s(u, v) && \text{to be the } \textit{specific entropy} \\
p_0 &= -\partial u(s, v)/\partial v && \text{to be the } \textit{thermostatic pressure} \\
T &= \partial u(s, v)/\partial s && \text{to be the } \textit{absolute temperature}
\end{aligned}
\Bigg\} \quad (5.2)
$$

of the nonequilibrium state. Its *stress tensor* $\sigma^{\alpha\beta}$ and *energy flux vector* q^α are defined in terms of $T^{\alpha\beta}$ and v^α by (3-5.14). In equilibrium (4.33) can provide an alternative expression for the pressure, in terms of $\sigma^{\alpha\beta}$. It shows that

$$p = -\tfrac{1}{3} g_{\alpha\beta} \, \sigma^{\alpha\beta}. \tag{5.3}$$

For a nonequilibrium state this quantity is known as the *kinetic pressure*. It can differ from the thermostatic pressure p_0 defined by (5.2), which shows the need for the distinct notations and for the qualifying adjectives in the names.

The energy–momentum tensor of the comparison equilibrium state is

$$T_0^{\alpha\beta}(u, \rho^\gamma) \equiv \rho(1 + u) \, v^\alpha v^\beta + p_0 \, h^{\alpha\beta} \tag{5.4}$$

where

$$h^{\alpha\beta} \equiv g^{\alpha\beta} + v^\alpha v^\beta. \tag{5.5}$$

Let us set

$$\tau^{\alpha\beta} \equiv T^{\alpha\beta} - T_0^{\alpha\beta}(u, \rho^\gamma). \tag{5.6}$$

This satisfies

$$\tau^{\alpha\beta} v_\alpha \, v_\beta = 0 \tag{5.7}$$

in virtue of the definition (5.1) of u. If it is decomposed in the manner of (3-5.16) it thus has $\mu = 0$ and so takes the simplified form

$$\tau^{\alpha\beta} \equiv q_{\mathrm{h}}^\alpha \, v^\beta + v^\alpha q_{\mathrm{h}}^\beta - \sigma_{\mathrm{v}}^{\alpha\beta} \tag{5.8}$$

where

$$v_\alpha q_{\mathrm{h}}^\alpha = 0, \quad v_\beta \sigma_{\mathrm{v}}^{\alpha\beta} = 0, \quad \sigma_{\mathrm{v}}^{\alpha\beta} = \sigma_{\mathrm{v}}^{\beta\alpha}. \tag{5.9}$$

The vector q_{h}^α is known as the *heat flux vector* and the symmetric tensor $\sigma_{\mathrm{v}}^{\alpha\beta}$ as the *viscous stress tensor* of the state. It follows from (5.4), (5.6) and (3-5.16) that

$$q^\alpha = q_{\mathrm{h}}^\alpha, \quad \sigma^{\alpha\beta} = \sigma_{\mathrm{v}}^{\alpha\beta} - p_0 h^{\alpha\beta}. \tag{5.10}$$

In the present situation there is thus no distinction between the energy flux and heat flux vectors. It will however be seen in the next chapter that such a distinction does arise in the presence of an electromagnetic field.

It follows from (5.3) and (5.10) that

$$p = p_0 + p_{\mathrm{v}} \tag{5.11}$$

where

$$p_{\mathrm{v}} \equiv -\tfrac{1}{3} g_{\alpha\beta} \sigma_{\mathrm{v}}^{\alpha\beta}. \tag{5.12}$$

The scalar $p_{\rm v}$ is known as the *viscous pressure*. Its contribution to the viscous stress may be separated off to leave

$$\pi^{\alpha\beta} \equiv \sigma_{\rm v}^{\alpha\beta} + p_{\rm v} h^{\alpha\beta} = \sigma^{\alpha\beta} + p h^{\alpha\beta}. \tag{5.13}$$

This satisfies $v_{\beta}\pi^{\alpha\beta} = 0, \quad \pi^{\alpha\beta} = \pi^{\beta\alpha}, \quad g_{\alpha\beta}\pi^{\alpha\beta} = 0$ (5.14)

and it is known as the *shear stress tensor*.

These definitions give us a considerable amount of language with which to discuss deviations from a local state of equilibrium. We also need some terminology concerned with departure from the conditions (4.50) and (4.51) for global equilibrium. Clearly $\partial_{\alpha}\Phi$ is sufficient to measure departure from (4.51), but (4.50) needs more detailed consideration. This is provided by a decomposition of the velocity gradient tensor $\partial_{\alpha}v_{\beta}$ which enters (4.50) through (4.29).

Since v^{α} is a unit vector, it satisfies

$$v^{\beta}\partial_{\alpha}v_{\beta} = 0 \tag{5.15}$$

identically. Hence if B_{α}^{β} is defined by (3-5.10), the tensor

$$v_{\alpha\beta} \equiv B_{\alpha}^{\gamma}\partial_{\gamma}v_{\beta} \tag{5.16}$$

is orthogonal to v^{α} on both its indices. Its symmetric and antisymmetric parts

$$\theta_{\alpha\beta} \equiv v_{(\alpha\beta)}, \quad \omega_{\alpha\beta} \equiv v_{[\alpha\beta]} \tag{5.17}$$

are known as the *strain rate* and *vorticity* tensors respectively. By substituting from (3-5.10) into (5.16) we find that

$$\partial_{\alpha}v_{\beta} = \theta_{\alpha\beta} + \omega_{\alpha\beta} - v_{\alpha}a_{\beta} \tag{5.18}$$

where $a_{\alpha} \equiv v^{\beta}\partial_{\beta}v_{\alpha}.$ (5.19)

The vector a_{α} is known as the *acceleration* of the motion, and it too is orthogonal to v^{α}. The tensors $\theta_{\alpha\beta}$, $\omega_{\alpha\beta}$ and a_{α} have six, three and three linearly independent components respectively, which together comprise the twelve that $\partial_{\alpha}v_{\beta}$ has when (5.15) is taken into account.

It is possible to decompose $\theta_{\alpha\beta}$ further by setting

$$\theta \equiv g^{\alpha\beta}\theta_{\alpha\beta} = \partial_{\alpha}v^{\alpha} \tag{5.20}$$

and $e_{\alpha\beta} \equiv \theta_{\alpha\beta} - \tfrac{1}{3}\theta h_{\alpha\beta}.$ (5.21)

Then $e_{\alpha\beta}$ satisfies

$$e_{\alpha\beta} = e_{\beta\alpha}, \quad v^{\beta}e_{\alpha\beta} = 0, \quad g^{\alpha\beta}e_{\alpha\beta} = 0 \tag{5.22}$$

which leaves it with five linearly independent components while θ provides the sixth that $\theta_{\alpha\beta}$ possesses. The tensor $e_{\alpha\beta}$ is called the *shear rate* tensor. It is more commonly denoted by $\sigma_{\alpha\beta}$, but also in accordance with common usage we have already used this symbol for the stress tensor. In virtue of (4.5) and (5.1) we find from (5.20) that

$$\theta = v^{-1}\dot{v}. \tag{5.23}$$

The superscript dot denotes the operator $v^\alpha \partial_\alpha$, as in (4.52). From the geometric interpretation of (5.23), θ is known as the *expansion rate* of the motion.

With the definitions given above, it is easily seen that (4.50) and (4.59) together imply

$$\theta_{\alpha\beta} = 0 \tag{5.24}$$

and

$$\Omega_{\alpha\beta} = \Theta(\omega_{\alpha\beta} - 2v_{[\alpha}a_{\beta]}). \tag{5.25}$$

It follows from (5.24) that a fluid in equilibrium is both shearfree $(e_{\alpha\beta} = 0)$ and nonexpanding $(\theta = 0)$. However, at one point the velocity v^α, vorticity $\omega_{\alpha\beta}$ and acceleration a_α can take arbitrary values subject only to the algebraic restrictions

$$\omega_{(\alpha\beta)} = 0, \quad v^\beta\omega_{\alpha\beta} = 0, \quad v^\alpha a_\alpha = 0. \tag{5.26}$$

For suppose that these are given at some point x_0 together with the thermostatic variables Θ, Φ. Construct $\Omega_{\alpha\beta}$ from them by (5.25) and set

$$b_\beta = \Theta v_\beta - \Omega_{\alpha\beta}x_0^\alpha. \tag{5.27}$$

Then (4.62) determines from the constants $\Omega_{\alpha\beta}$ and b_α a vector field Θ_α which satisfies (4.50). Taken together with Φ, which is constant by (4.51), this specifies the unique global state of equilibrium which has at x_0 the required values for Θ, Φ, v^α, $\omega_{\alpha\beta}$ and a_α. The local kinematical variable which measures departure from equilibrium is thus the strain rate tensor $\theta_{\alpha\beta}$.

6 Phenomenological laws – first approximation

6a *The entropy source strength near local equilibrium*

Let Γ denote symbolically a state which is close to local equilibrium. With Γ we can associate a comparison equilibrium state Γ_0 in accordance with §5. The entropy flux density in the state Γ_0 is given by (4.29) and (4.41) to be

$$s^\alpha(\Gamma_0) = s(u, v)\rho^\alpha. \tag{6.1}$$

In virtue of (5.6) we also know that

$$s^\alpha(\Gamma) = s^\alpha(\Gamma_0) + \tau^{\beta\gamma}\frac{\partial s^\alpha}{\partial T^{\beta\gamma}} + O((\tau^{\beta\gamma})^2). \tag{6.2}$$

When (6.2) is taken together with (6.1) and (4.6), it shows that

$$s^\alpha(\Gamma) = s_1^\alpha(\Gamma) + O((\tau^{\beta\gamma})^2) \tag{6.3}$$

where

$$s_1^\alpha(\Gamma) \equiv s\rho^\alpha - \tau^{\alpha\beta}\Theta_\beta. \tag{6.4}$$

Here Θ_α is the inverse temperature vector of the comparison equilibrium state Γ_0. In keeping with the conventions of §5 we shall define this to be the inverse temperature vector also of Γ, so that (4.44) now holds in all states. Substitution from (4.44) and (5.8) into (6.4) gives that equation the alternative form

$$s_1^\alpha = s\rho^\alpha + q_{\mathrm{h}}^\alpha/T. \tag{6.5}$$

This shows clearly the origin of the special role taken in thermodynamics by the heat flux vector q_{h}^α.

The thermostatic theory gives no information about the $O((\tau^{\beta\gamma})^2)$ terms in (6.3). The simplest approximation that is consistent with thermostatics is thus to neglect these terms completely. This is the approximation adopted in the standard development of thermodynamics. Let us accept it for the time being and investigate its implications. It follows from (5.2) and (4.44) that

$$\partial_\alpha s = \Theta(\partial_\alpha u + p_0 \partial_\alpha v). \tag{6.6}$$

Differentiation of (6.4) with use of (6.6) and of the conservation law $\partial_\alpha \rho^\alpha = 0$ gives

$$\partial_\alpha s_1^\alpha = \Theta(\rho^\alpha \partial_\alpha u + p_0 \partial_\alpha v^\alpha - v_\beta \partial_\alpha \tau^{\alpha\beta}) - \tau^{\alpha\beta}\partial_\alpha \Theta_\beta. \tag{6.7}$$

But the expression in brackets is zero in virtue of (5.6), (5.4) and (4.5). In this approximation the entropy source strength σ of (2.17) is thus given by

$$\sigma = -\tau^{\alpha\beta}\partial_{(\alpha}\Theta_{\beta)}. \tag{6.8}$$

The symmetrizing brackets are redundant since $\tau^{\alpha\beta}$ is symmetric. They have been included merely to emphasize that it is only the symmetric part of $\partial_\alpha \Theta_\beta$ which contributes to σ.

If the tensor fields $\tau^{\alpha\beta}$ and $\partial_{(\alpha}\Theta_{\beta)}$ were independent of one another in a general near-equilibrium state, it would be possible to violate the requirement $\sigma \geqslant 0$. This follows immediately from (6.8). Since such a violation is forbidden, these two tensor fields must be related to one another by some law that we have so far neglected. To see why we should expect there to be such a law, consider the roles of these two fields in the theory so far. The field $\tau^{\alpha\beta}$, defined by (5.6), measures the extent to which the conditions of §4b for local equilibrium are violated.

On the other hand, $\partial_{(\alpha}\Theta_{\beta)}$ measures the violation of the condition (4.50) for global equilibrium. The immediate effect of disturbing a fluid in equilibrium, say by stirring and heating it, is to violate this global condition on the macroscopic velocity and temperature distribution. On a microscopic scale it is a disturbance with a large length-scale of the equilibrium velocity distribution of the molecules. It will be followed by an evolution towards a new equilibrium state caused by the randomizing effect of molecular collisions. This will rapidly generate a small-scale disturbance of the molecular velocity distribution which is perceived macroscopically as a deviation from the local equilibrium condition $\tau^{\alpha\beta} = 0$. The generation of entropy during the process of evolution is the macroscopic manifestation of the process of randomization itself.

6b The Lie derivative – an aid to interpretation

The macroscopic laws which describe the causal link between violations of local and global equilibrium are known as the *phenomenological laws* of the system. These laws form part of the physical characterization of any material. They must be consistent with the entropy law, but they are not determined by it. However, some useful guidelines as to the possible form of these laws can be obtained from an interpretation of the expression (6.8) for σ. This interpretation requires a new mathematical construction, the Lie derivative, which we now develop. We shall use language associated with the motion of material systems, but this is purely for convenience. The construction could be formulated in an entirely abstract manner if so desired.

The material orbits of a continuous medium are defined by (3-5.1). This definition parametrizes them by the proper time τ along each orbit. If we use instead an arbitrary parameter u, a material orbit $x^{\alpha}(u)$ will satisfy

$$dx^{\alpha}/du = U^{\alpha}(x) \qquad (6.9)$$

where

$$U^{\alpha} \equiv (d\tau/du)\,v^{\alpha}. \qquad (6.10)$$

Consider now a one-parameter family $x^{\alpha}(u, w)$ of such orbits. By this we mean that the point $x^{\alpha}(u, w)$ depends smoothly on both u and w, and that for each fixed value of w, $x^{\alpha}(u, w)$ is a material orbit parametrized by u. Then

$$\partial x^{\alpha}/\partial u = U^{\alpha} \qquad (6.11)$$

by (6.9). Let us also set $W^{\alpha} \equiv \partial x^{\alpha}/\partial w. \qquad (6.12)$

The identity $\qquad \partial^2 x^\alpha / \partial u \partial w = \partial^2 x^\alpha / \partial w \partial u$ $\qquad$ (6.13)

then gives $\qquad \partial W^\alpha / \partial u = W^\beta \partial_\beta U^\alpha.$ $\qquad$ (6.14)

This is a first-order ordinary differential equation for $W^\alpha(u)$ which governs the dependence of W^α on u along each material orbit. A contravariant vector field W^α which satisfies (6.14) is said to be *dragged along* by U^α. Geometrically $W^\alpha(x)\,\delta w$ can be considered as an infinitesimal position vector relative to x. Dragging along by U^α corresponds to both ends of this position vector being displaced by equal increments in u along the material orbits on which they lie.

A covariant vector field P_α is said to be dragged along by U^α if its scalar product with any dragged-along contravariant vector field W^α is constant along each material orbit. It must thus satisfy

$$W^\alpha dP_\alpha / du = -P_\alpha dW^\alpha / du = -W^\alpha P_\beta \partial_\alpha U^\beta. \qquad (6\ 15)$$

But at any point W^α can be chosen arbitrarily. It follows that

$$dP_\alpha / du = -P_\beta \partial_\alpha U^\beta. \qquad (6.16)$$

Consider now an arbitrary tensor field $t^{\alpha\beta\cdots}{}_{\gamma\delta\ldots}$. Let P_α, Q_α, ... and W^α, X^α, ... be any vector fields which are dragged along by U^α. Use of (6.14) and (6.16) then shows that

$$\frac{d}{du}\,(t^{\alpha\beta\cdots}{}_{\gamma\delta\ldots}\,P_\alpha Q_\beta \ldots W^\gamma X^\delta \ldots) = (L_U\,t^{\alpha\beta\cdots}{}_{\gamma\delta\ldots})\,P_\alpha Q_\beta \ldots W^\gamma X^\delta \ldots$$

$$(6.17)$$

where

$$L_U\,t^{\alpha\cdots}{}_{\gamma\ldots} \equiv U^\epsilon \partial_\epsilon t^{\alpha\cdots}{}_{\gamma\ldots} - t^{\epsilon\cdots}{}_{\gamma\ldots}\,\partial_\epsilon U^\alpha - \ldots + t^{\alpha\cdots}{}_{\epsilon\ldots}\,\partial_\gamma U^\epsilon + \ldots .$$

$$(6.18)$$

In this expression there is one term involving a derivative of U^α for each index of the tensor $t^{\alpha\cdots}{}_{\gamma\ldots}$. The tensor $L_U\,t^{\alpha\cdots}{}_{\gamma\ldots}$ is known as the *Lie derivative* of $t^{\alpha\cdots}{}_{\gamma\ldots}$ with respect to U^α.

If this Lie derivative vanishes, the tensor field $t^{\alpha\cdots}{}_{\gamma\ldots}$ is said to be dragged along by U^α. This is consistent with the separate definitions made for contravariant and covariant vectors in (6.14) and (6.16). Dragging along by U^α corresponds in a natural sense to the physical concept of a quantity being carried along with the motion described by U^α. When the Lie derivative of a quantity is nonzero it provides a measure of the rate at which that quantity is changing relative to this motion.

It is easily seen from (6.18) that the Lie derivative of outer and inner products of tensors, and of sums of tensors with the same valence, may be expanded by the usual rules. Examples are

$$L_U(a^{\alpha\beta}b_\beta) = a^{\alpha\beta}L_U\,b_\beta + b_\beta\,L_U\,a^{\alpha\beta},\left.\right\}$$
$$L_U(a^\alpha + b^\alpha) = L_U\,a^\alpha + L_U\,b^\alpha. \qquad (6.19)$$

This operation also commutes with contraction, e.g. the Lie derivative of the vector $t^\alpha_{.\alpha\gamma}$ is also the contraction on α and β of the tensor $L_U\,t^\alpha_{.\beta\gamma}$ of valence $(1,2)$. However, the Lie derivative operation in general does *not* commute with the raising and lowering of tensor indices. This is because the Lie derivative of the metric tensor is not necessarily zero. Indeed, use of (6.18) shows that

$$L_U\,g_{\alpha\beta} = 2\partial_{(\alpha}U_{\beta)}. \qquad (6.20)$$

6c *General form of the phenomenological laws*

It follows from (6.20) that the condition (4.50) for global equilibrium can be expressed as

$$L_\Theta\,g_{\alpha\beta} = 0. \qquad (6.21)$$

The metric is thus dragged along by the inverse temperature vector Θ^α. This justifies the description of the motion as 'steady', as already discussed in §4d. In combination with (6.10) it shows also that the steadiness determines a natural parametrization of the material orbits, proportional to the product $T\tau$. Except for the special case of the homogeneous equilibrium states, this differs from parametrization by the proper time τ itself. The approximation (6.8) for σ can be similarly re-expressed, as

$$\sigma = -\tfrac{1}{2}\tau^{\alpha\beta}L_\Theta\,g_{\alpha\beta}. \qquad (6.22)$$

This shows that $T\tau$ remains the natural parametrization of the orbits even for non-equilibrium states.

The discussion of §6a shows that the phenomenological laws describe a relaxation phenomenon, the microscopic adjustment of the system to a new equilibrium state after a macroscopic disturbance. This suggests that these laws should concern the variables describing the local state of the system, together with their rates of change. These local variables are those upon which the s^α of (4.1) depends, namely $T^{\alpha\beta}$, ρ^α and (implicitly) $g_{\alpha\beta}$. The most natural rate-of-change operator has been seen to be L_Θ. It is necessary to remember the usually implicit variable $g_{\alpha\beta}$ as its rate of change in this sense is nonzero. However, we shall leave $g_{\alpha\beta}$ itself implicit as usual. The above suggestion thus corresponds to phenomenological laws with the general form

$$f^A(T^{\alpha\beta}, \rho^\alpha, L_\Theta\,g_{\alpha\beta}, L_\Theta\,T^{\alpha\beta}, L_\Theta\,\rho^\alpha) = 0 \qquad (6.23)$$

where $A = 1, 2, \ldots N$ labels the N equations of the system. This is the simplest reasonable possibility and we shall now add it as an additional hypothesis to the characterization of a 'relativistic simple fluid' given in §4a. We shall also assume that these laws are non-degenerate near equilibrium in a sense which will be made precise below.

An equivalent set of argument variables for the phenomenological laws is

$$u, \rho^\alpha, \tau^{\alpha\beta}, X_{\alpha\beta}, \dot{u}, \dot{\rho}, L_\Theta \tau_{\alpha\beta} \qquad (6.24)$$

where

$$X_{\alpha\beta} \equiv v_\alpha v_\beta v^\gamma v^\delta \partial_\gamma \Theta_\delta - \partial_{(\alpha} \Theta_{\beta)} \qquad (6.25)$$

and the superscript dots denote the operator $v^\alpha \partial_\alpha$ as in (4.52). To see the equivalence, note first that $T^{\alpha\beta}$ and ρ^α are together equivalent to u, ρ^α and $\tau^{\alpha\beta}$ in virtue of (5.1) and (5.6). Next, (6.20) shows that $L_\Theta g_{\alpha\beta}$ may be replaced by both $X_{\alpha\beta}$ and $\dot{\Theta}$. The identity

$$L_\Theta \rho^\alpha = \Theta^\alpha L_\Theta(\rho/\Theta) = \Theta^\alpha(\dot{\rho} - \rho\dot{\Theta}/\Theta) \qquad (6.26)$$

now enables $L_\Theta \rho^\alpha$ to be replaced in this modified set by $\dot{\rho}$. These changes convert the argument set of (6.23) into the equivalent set

$$u, \rho^\alpha, \tau^{\alpha\beta}, X_{\alpha\beta}, \dot{\Theta}, \dot{\rho}, L_\Theta T^{\alpha\beta}.$$

Since each of the pairs (Θ, ρ) and (u, ρ) can be considered as a function of the other, we may further replace $\dot{\Theta}$ by $\dot{u}$. Finally the Lie derivatives of (5.4) and (5.6) enable $L_\Theta T^{\alpha\beta}$ to be replaced by $L_\Theta \tau_{\alpha\beta}$ to give (6.24) as required. We choose the covariant form of $\tau_{\alpha\beta}$ in $L_\Theta \tau_{\alpha\beta}$ as (5.7) implies

$$v^\alpha v^\beta L_\Theta \tau_{\alpha\beta} = \Theta^{-2} L_\Theta (\Theta^\alpha \Theta^\beta \tau_{\alpha\beta}) = 0. \qquad (6.27)$$

The corresponding quantity $v_\alpha v_\beta L_\Theta \tau^{\alpha\beta}$ for the contravariant form will not vanish in general since L_Θ does not commute with the raising of indices. For all the other variables it is immaterial whether the covariant or contravariant form is chosen.

One advantage of the variables (6.24) is that they include those which are independently variable in equilibrium, namely u and ρ^α, while the remainder all vanish in equilibrium. Another advantage is that (6.25) separates off from $\partial_{(\alpha} \Theta_{\beta)}$ that part which cannot contribute to the σ of (6.8) due to the constraint (5.7) on $\tau^{\alpha\beta}$. The tensor $X_{\alpha\beta}$ obeys the corresponding constraint

$$v^\alpha v^\beta X_{\alpha\beta} = 0 \qquad (6.28)$$

and (6.8) is equivalent to

$$\sigma = \tau^{\alpha\beta} X_{\alpha\beta}. \qquad (6.29)$$

Denote the variables $\tau^{\alpha\beta}$, $X_{\alpha\beta}$, $\dot{u}$, $\dot{\rho}$ and $L_\Theta \tau_{\alpha\beta}$ generically by y^Γ, where the index Γ runs from 1 to 29 to label the 29 linearly independent components of these tensors. The laws (6.23) can then be expressed in terms of N new functions F^A as

$$F^A(u, \rho^\alpha, y^\Gamma) = 0. \tag{6.30}$$

The y^Γ are the arguments which vanish in equilibrium. Since all equilibrium states are physically possible, the F^A must satisfy

$$F^A(u, \rho^\alpha, 0) = 0 \tag{6.31}$$

identically. We shall assume also that the $(N \times 29)$ matrix

$$[\partial F^A / \partial y^\Gamma]_0$$

has rank N, where the subscript 0 denotes evaluation at $y^\Gamma = 0$. This is the assumption of non-degeneracy referred to above. By a suitable numbering of the ys it enables us to assume without loss of generality that the $(N \times N)$ submatrix

$$[\partial F^A / \partial y^\Gamma]_0, \quad (A = 1, 2, ..., N; \quad \Gamma = 1, 2, ..., N) \tag{6.32}$$

is nonsingular. It then follows by the implicit function theorem of calculus that, close enough to equilibrium, the equations (6.30) can be solved for $y^1, ..., y^N$ in terms of u, ρ^α and $y^{N+1}, ..., y^{29}$. When this is done the functions F^A take the special form

$$F^A(u, \rho^\alpha, y^\Gamma) \equiv y^A - G^A(u, \rho^\alpha, y^{N+1}, ..., y^{29}) = 0 \quad (1 \leqslant A \leqslant N). \tag{6.33}$$

6d Implications of the entropy law

Consider fixed values of u and ρ^α and let $\{\bar{y}^\Gamma\}$ be a set of constants such that

$$\sum_\Gamma [\partial F^A / \partial y^\Gamma]_0 \, \bar{y}^\Gamma = 0. \tag{6.34}$$

By taking $y^\Gamma(\lambda) = \lambda \bar{y}^\Gamma$ for $(N+1) \leqslant \Gamma \leqslant 29$ and defining $y^\Gamma(\lambda)$ for $1 \leqslant \Gamma \leqslant N$ by (6.33) we obtain a parametrized family of solutions of (6.30) such that

$$y^\Gamma = 0, \quad dy^\Gamma / d\lambda = \bar{y}^\Gamma \quad \text{when} \quad \lambda = 0. \tag{6.35}$$

Let $\tau^{\alpha\beta}(\lambda)$ and $X_{\alpha\beta}(\lambda)$ be the values of $\tau^{\alpha\beta}$ and $X_{\alpha\beta}$ in these solutions and let $\bar{\tau}^{\alpha\beta}$ and $\bar{X}_{\alpha\beta}$ be the corresponding constants of the set $\{\bar{y}^\Gamma\}$. In virtue of (6.29) the entropy law $\sigma \geqslant 0$ requires that $\tau^{\alpha\beta}(\lambda) X_{\alpha\beta}(\lambda)$ has a minimum at $\lambda = 0$. Since

$$[d^2(\tau^{\alpha\beta} X_{\alpha\beta}) / d\lambda^2]_0 = 2\bar{\tau}^{\alpha\beta} \bar{X}_{\alpha\beta} \tag{6.36}$$

by (6.35), it follows that every solution of (6.34) for $\bar{y}^\Gamma$ must satisfy

$$\bar{\tau}^{\alpha\beta}\bar{X}_{\alpha\beta} \geq 0. \tag{6.37}$$

In §4c we introduced the assumption that there are no algebraic constraints on $\delta T^{\alpha\beta}$ in a small deviation from equilibrium. By (6.35) this implies that a solution of (6.34) must exist corresponding to any value of $\bar{\tau}^{\alpha\beta}$ compatible with the constraint (5.7). Suppose now that $\bar{y}^\Gamma$ is *any* solution of (6.34), as above, and that $\bar{\bar{y}}^\Gamma$ is a solution with $\bar{\bar{\tau}}^{\alpha\beta} = 0$. Since (6.34) is linear, $\bar{y}^\Gamma + k\bar{\bar{y}}^\Gamma$ is also a solution of (6.34) for all values of the scalar k. Application of (6.37) to this solution gives

$$\bar{\tau}^{\alpha\beta}(\bar{X}_{\alpha\beta} + k\bar{\bar{X}}_{\alpha\beta}) \geq 0 \tag{6.38}$$

for all k, which is possible only if

$$\bar{\tau}^{\alpha\beta}\bar{\bar{X}}_{\alpha\beta} = 0. \tag{6.39}$$

But by hypothesis $\bar{\tau}^{\alpha\beta}$ can take any value consistent with (5.7). It follows that $\bar{\bar{X}}_{\alpha\beta}$ must have the form $X v_\alpha v_\beta$ for some scalar X, which is consistent with (6.28) only when

$$\bar{\bar{X}}_{\alpha\beta} = 0. \tag{6.40}$$

The result (6.40) shows that (6.34) must imply $\bar{X}_{\alpha\beta} = 0$ whenever $\bar{\tau}^{\alpha\beta} = 0$. This imposes severe restrictions on the coefficient matrix in (6.34). In particular it requires the submatrix $[\partial F^A/\partial X_{\alpha\beta}]_0$ to have rank 9, since 9 is the number of linearly independent components of $X_{\alpha\beta}$. This implies that $N \geq 9$ and it enables $y^1, ..., y^9$ in (6.32) and (6.33) to be chosen as components of $X_{\alpha\beta}$. The phenomenological laws thus include equations of the form

$$X_{\alpha\beta} = X_{\alpha\beta}(u, \rho^\alpha, \tau^{\alpha\beta}, z^a) \tag{6.41}$$

where $\{z^a; 1 \leq a \leq 11\}$ is a set of linearly independent components of the variables $\dot{u}, \dot{\rho}$ and $L_\Theta \tau_{\alpha\beta}$.

The nine independent equations of (6.41) are all that are needed to complete a deterministic set of equations governing the time evolution of the physical system. We already have the five equations (4.5), and together with these nine phenomenological equations this gives us a total of fourteen equations. There are also just fourteen linearly independent variables amongst the components of $T^{\alpha\beta}$ and ρ^α. The thermostatic variables such as Θ become known functions of these fourteen variables once the fundamental equation $s = s(u, v)$ is given. We thus have as many equations as we have variables, which will be adequate to completely determine the evolution of the system from given initial data. Further phenomenological equations are unnecessary and they could at most be constraints on the allowed initial conditions. We shall suppose that no such constraints exist. This gives us precisely $N = 9$ and shows the equations (6.41) to be the complete set of phenomenological laws for the system. In order for $\bar{\tau}^{\alpha\beta} = 0$ to imply $\bar{X}_{\alpha\beta} = 0$ we see that these equations must satisfy

$$[\partial X_{\alpha\beta}/\partial z^a]_0 = 0. \tag{6.42}$$

6e *Linearization*

For a system near equilibrium the laws (6.41) can be approximated by the first few terms of their Taylor series expansion in the variables $\tau^{\alpha\beta}$ and z^a. Close enough to equilibrium it will be necessary to retain only the linear terms in these variables. But the terms linear in z^a are absent by (6.42). It follows that the linear approximation has the simple form

$$X_{\alpha\beta} = c_{\alpha\beta\gamma\delta}\tau^{\gamma\delta} \tag{6.43}$$

where the coefficient tensor $c_{\alpha\beta\gamma\delta}$ depends only on u, ρ^a and the metric tensor $g_{\alpha\beta}$. In virtue of (5.7), (6.28) and the symmetry of both $X_{\alpha\beta}$ and $\tau^{\alpha\beta}$, this tensor can be chosen to satisfy

$$c_{\alpha\beta\gamma\delta} = c_{(\alpha\beta)(\gamma\delta)}, \quad v^\alpha v^\beta c_{\alpha\beta\gamma\delta} = 0, \quad v^\gamma v^\delta c_{\alpha\beta\gamma\delta} = 0. \tag{6.44}$$

It can be shown that any tensor function of u, ρ^a and $g_{\alpha\beta}$ must be a linear combination of outer products of A^α_β, $g_{\alpha\beta}$, $g^{\alpha\beta}$, v_α and v^α whose coefficients are scalar functions of u and ρ. This seems plausible but a formal proof requires techniques from the theory of group representations. This is the second illustration that we have come across of the use of these powerful techniques in tensor algebra. The previous illustration was in §2-5a and concerned tensor symmetries. The interested reader should consult Weyl (1946) for further details. To apply this theorem systematically to $c_{\alpha\beta\gamma\delta}$ one should write down the most general such linear combination with valence $(0, 4)$ and then apply the restrictions (6.44). In practice it is easier to apply (6.44) if one uses $h_{\alpha\beta}$ in place of $g_{\alpha\beta}$. This is defined by (5.5) and is clearly equivalent in this context to $g_{\alpha\beta}$ since v_α may also be used. One finds that the most general tensor consistent with (6.44) has the form

$$c_{\alpha\beta\gamma\delta} = c_1 h_{\alpha\beta} h_{\gamma\delta} + c_2 h_{\alpha(\gamma} h_{\delta)\beta} + c_3 v_{(\alpha} h_{\beta)(\gamma} v_{\delta)} \tag{6.45}$$

where c_1, c_2. c_3 are arbitrary scalar functions of u and ρ. The additional symmetry

$$c_{\alpha\beta\gamma\delta} = c_{\gamma\delta\alpha\beta} \tag{6.46}$$

possessed by (6.45) has not been imposed. It is a consequence of the other hypotheses as to the form of this tensor.

It follows from (6.45) and (5.8) with the aid of (5.10) and (5.13) that

$$c_{\alpha\beta\gamma\delta}\tau^{\gamma\delta} = (3c_1+c_2)p_{\rm v}h_{\alpha\beta}-c_2\pi_{\alpha\beta}-c_3v_{(\alpha}q_{\beta)}. \qquad (6.47)$$

Similarly (6.25) can be combined with (5.18) and (5.21) to give

$$X_{\alpha\beta} = -\Theta(\tfrac{1}{3}\theta h_{\alpha\beta}+e_{\alpha\beta})-v_{(\alpha}(B^{\gamma}_{\beta)}\partial_\gamma\Theta-\Theta a_{\beta)}) \qquad (6.48)$$

where we have again set $B^{\alpha}_{\beta} \equiv A^{\alpha}_{\beta}+v^{\alpha}v_{\beta}.$ (6.49)

In virtue of (5.9), (5.14) and (5.22) these expressions enable (6.43) to be decomposed into the three equations

$$p_{\rm v} = -\zeta\theta, \quad \pi_{\alpha\beta} = 2\eta e_{\alpha\beta} \qquad (6.50)$$

and $$q_\alpha = -\kappa(B^{\beta}_{\alpha}\partial_\beta T+Ta_\alpha) \qquad (6.51)$$

where $\zeta = 1/[3T(3c_1+c_2)], \quad \eta = 1/(2Tc_2), \quad \kappa = 1/(T^2c_3).$ (6.52)

These are the linearized phenomenological laws in their standard forms. The notation in (6.50) and (6.51) is as follows:

$p_{\rm v}$ is the viscous pressure
$\pi_{\alpha\beta}$ is the shear stress } of the local state,
q_α is the heat flux
T is the absolute temperature

θ is the expansion rate
$e_{\alpha\beta}$ is the shear rate } of the fluid flow.
a^α is the acceleration

The parameters ζ, η, κ are functions of u and ρ and are named thus:

ζ is the bulk viscosity
η is the shear viscosity } of the material.
κ is the thermal conductivity

The simple forms (6.50) and (6.51) of the linearized laws give a direct physical significance to these variables which were introduced formally in §5 to describe near-equilibrium states.

For $\tau^{\alpha\beta}$ to be free from algebraic constraints, as we have supposed, we clearly need ζ, η and κ all to be nonzero. We show now that the entropy law requires them all to be positive. For when (6.43), (6.47) and (6.52) are used to simplify (6.29) it is found that

$$\sigma = p_{\rm v}^2/(\zeta T)+\pi_{\alpha\beta}\pi^{\alpha\beta}/(2\eta T)+q_\alpha q^\alpha/(\kappa T^2). \qquad (6.53)$$

It is an empirical fact, which can be formalized as a law of thermo-statics, that absolute temperatures are always positive. The tensor squares $\pi_{\alpha\beta}\pi^{\alpha\beta}$ and $q_{\alpha}q^{\alpha}$ are both non-negative since $\pi_{\alpha\beta}$ and q^{α} are orthogonal to v^{α}. The entropy law is thus satisfied for all states close to equilibrium if and only if

$$\zeta > 0, \quad \eta > 0, \quad \kappa > 0. \tag{6.54}$$

7 Phenomenological laws – second approximation

7a Forces, fluxes, impedance and relaxation

Since the tensor $X_{\alpha\beta}$ of (6.25) is the part of $\partial_{(\alpha}\Theta_{\beta)}$ which contributes to the entropy source strength σ, it is an effective measure of the departure of the state from global equilibrium. The discussion in § 6a shows that departures from local equilibrium, measured by $\tau^{\alpha\beta}$, are driven by such departures from global equilibrium. In recognition of this fact the components of $X_{\alpha\beta}$ are said to be *thermodynamic forces* and the components of $\tau^{\alpha\beta}$, which describe the response to these forces, are called *thermodynamic fluxes*. According to (6.43) there is a linear relationship between these forces and fluxes. The coefficient tensor $c_{\alpha\beta\gamma\delta}$ of this relationship will be called the *impedance tensor*, by ana-logy with electric circuit theory in which the driving force (voltage) is expressed as a multiple (impedance) of the response (current).

A direct linear relationship between the forces and fluxes implies an instantaneous response of the local state to departures from global equilibrium. This is physically unrealistic and conflicts with the con-clusions of our general discussions in §§ 6a and c. We should expect the microscopic processes of adjustment described by the phenomeno-logical laws to take a short but nonzero time. If this time is very small in comparison with the time-scale of the macroscopic disturbance of global equilibrium then it can be neglected, but it should be present in the basic laws which describe these processes. This should be true even in their linear approximation, as this approximation is based on the amplitude of the disturbance and not on its time-scale.

It was pointed out by Israel (1976) that to put right this omission we must improve on the approximation adopted for s^{α} in § 6a. We saw there that the standard thermodynamic theory neglects completely the terms in (6.3) which are quadratic in $\tau^{\alpha\beta}$. Retention of these terms but neglect of the cubic terms would give

$$s^{\alpha} = s_1^{\alpha} - \tfrac{1}{2}a^{\alpha}{}_{.\beta\gamma\delta\epsilon}\tau^{\beta\gamma}\tau^{\delta\epsilon} \tag{7.1}$$

where $a^\alpha_{.\beta\gamma\delta\epsilon}$ is some tensor function of u, ρ^α and $g_{\alpha\beta}$. The minus sign is for later convenience. These quadratic terms will contribute to σ some terms which involve $\partial_\alpha \tau^{\beta\gamma}$, which will force the phenomenological laws to contain similar terms even in their linear approximation. If we are to maintain the general form (6.23) of the phenomenological laws, these derivatives must occur as a rate of change $v^\alpha \partial_\alpha \tau^{\beta\gamma}$. This requires $a^\alpha_{.\beta\gamma\delta\epsilon}$ to have the form

$$a^\alpha_{.\beta\gamma\delta\epsilon} = \rho^\alpha a_{\beta\gamma\delta\epsilon}. \tag{7.2}$$

As the next approximation to s^α we shall thus take

$$s^\alpha = s_1^\alpha + s_2^\alpha \tag{7.3}$$

where s_1^α is given by (6.4) and

$$s_2^\alpha \equiv -\tfrac{1}{2}\rho^\alpha a_{\beta\gamma\delta\epsilon}\tau^{\beta\gamma}\tau^{\delta\epsilon}. \tag{7.4}$$

The tensor $a_{\alpha\beta\gamma\delta}$ will be called the *relaxation tensor*. It is a function of u, ρ^α and $g_{\alpha\beta}$, and in virtue of (5.7) and the symmetry of $\tau^{\alpha\beta}$ it can be chosen to satisfy

$$a_{\alpha\beta\gamma\delta} = a_{(\alpha\beta)(\gamma\delta)}, \quad a_{\alpha\beta\gamma\delta} = a_{\gamma\delta\alpha\beta}, \quad v^\gamma v^\delta a_{\alpha\beta\gamma\delta} = 0. \tag{7.5}$$

These are precisely the same conditions as given by (6.44) and (6.46) for the impedance tensor. It follows that the relaxation tensor can be expressed in terms of three new scalar functions a_1, a_2, a_3 of u and ρ as

$$a_{\alpha\beta\gamma\delta} = a_1 h_{\alpha\beta} h_{\gamma\delta} + a_2 h_{\alpha(\gamma} h_{\delta)\beta} + a_3 v_{(\alpha} h_{\beta)(\gamma} v_{\delta)}. \tag{7.6}$$

7b The rate-of-change operator $D/D\tau$

The mathematical development of the second approximation is simplified considerably if rates of change are described by a new operator $D/D\tau$ instead of by the Lie derivative L_Θ. Although this Lie derivative operation has a fundamental physical significance, its failure to commute with the raising and lowering of indices leads to a number of algebraic difficulties. For example, the Lie derivative of a tracefree tensor such as the shear stress $\pi^{\alpha\beta}$ is no longer tracefree. Again, although $L_\Theta \pi_{\alpha\beta}$ shares with $\pi_{\alpha\beta}$ the property of being orthogonal to v^α, this is not true of the contravariant form $L_\Theta \pi^{\alpha\beta}$. The decomposition of (6.43) into the three separate laws (6.50) and (6.51) was dependent on such properties. If the rate-of-change operator fails to preserve them, it will complicate the corresponding decomposition when the relaxation tensor is included. We would thus prefer to use a new operator which satisfies

$$Dg_{\alpha\beta}/D\tau = 0, \quad Dv^\alpha/D\tau = 0. \tag{7.7}$$

Any tensor $Q^\alpha_{\cdot\beta}$ of valence (1, 1) can be used to define a rate of-change operator Δ_Q with many properties in common with L_Θ. We set

$$\Delta_Q t^{\alpha\cdots}{}_{\beta\ldots} = \Theta^\gamma \partial_\gamma t^{\alpha\cdots}{}_{\beta\ldots} + Q^\alpha_{\cdot\gamma} t^{\gamma\cdots}{}_{\beta\ldots} + \cdots - Q^\gamma_{\cdot\beta} t^{\alpha\cdots}{}_{\gamma\ldots} - \cdots, \quad (7.8)$$

where there is one term involving $Q^\alpha_{\cdot\beta}$ for each index of the tensor $t^{\alpha\cdots}{}_{\beta\ldots}$. Comparison with (6.18) shows that L_Θ itself corresponds to the case

$$Q^\alpha_{\cdot\beta} = -\partial_\beta \Theta^\alpha. \quad (7.9)$$

All such operators commute with the process of contraction. They also act on outer and inner products of tensors, and on sums of tensors with the same valence, according to the usual rules for derivatives. In particular, the examples (6.19) remain valid if L_U is replaced by Δ_Q.

As special cases of (7.8) we have

$$\Delta_Q g_{\alpha\beta} = -2Q_{(\alpha\beta)}, \quad \Delta_Q v_\alpha = \Theta a_\alpha - Q_{\beta\alpha} v^\beta, \quad (7.10)$$

where a_α is the acceleration vector defined by (5.19). We require both of these to vanish identically. Another essential feature of the new operator is that it must be possible to use $DT^{\alpha\beta}/D\tau$ and $D\rho^\alpha/D\tau$ in the phenomenological laws (6.23) as equivalents of $L_\Theta T^{\alpha\beta}$ and $L_\Theta \rho^\alpha$. This will be possible if $Q^\alpha_{\cdot\beta}$ differs from the value (7.9) by some tensor function of the remaining arguments of (6.23), namely $T^{\alpha\beta}$, ρ^α, $g_{\alpha\beta}$ and $L_\Theta g_{\alpha\beta}$. The choice

$$Q_{\alpha\beta} + \partial_\beta \Theta_\alpha = \tfrac{1}{2} L_\Theta g_{\alpha\beta} \quad (7.11)$$

gives

$$Q_{\alpha\beta} = \partial_{[\alpha} \Theta_{\beta]} \quad (7.12)$$

by (6.20), which makes $\Delta_Q g_{\alpha\beta}$ vanish as required. However, it also gives

$$\Delta_Q v_\alpha = \tfrac{1}{2}(\Theta a_\alpha - \partial_\alpha \Theta - v_\alpha v^\beta \partial_\beta \Theta) \quad (7.13)$$

which generally is nonzero. But we may rewrite (7.13) as

$$\Delta_Q v_\alpha = -v^\beta v^\gamma (L_\Theta g_{\gamma[\alpha}) v_{\beta]}. \quad (7.14)$$

It follows that if we replace (7.11) by

$$Q_{\alpha\beta} + \partial_\beta \Theta_\alpha = \tfrac{1}{2} L_\Theta g_{\alpha\beta} + v^\gamma (L_\Theta g_{\gamma[\alpha}) v_{\beta]}, \quad (7.15)$$

the right-hand side of which is a tensor function of ρ^α, $g_{\alpha\beta}$ and $L_\Theta g_{\alpha\beta}$ as required, then Δ_Q has both the desired properties

$$\Delta_Q g_{\alpha\beta} = 0, \quad \Delta_Q v_\alpha = 0. \quad (7.16)$$

Use of (6.20) and (5.18) enables us to simplify (7.15) to the form

$$Q_{\alpha\beta} = \Theta(\omega_{\alpha\beta} - 2v_{[\alpha} a_{\beta]}), \quad (7.17)$$

where $\omega_{\alpha\beta}$ and a_α are respectively the vorticity and acceleration of the fluid. Since this involves Θ only as a scalar factor, it is convenient to set

$$D/D\tau \equiv \Theta^{-1}\Delta_Q. \qquad (7.18)$$

The operator $D/D\tau$ is then independent of Θ. It is determined purely by the kinematics of the fluid flow and is given by

$$Dt^{\alpha\cdots}{}_{\beta\ldots}/D\tau = v^\gamma \partial_\gamma t^{\alpha\cdots}{}_{\beta\ldots} + S^\alpha_{\cdot\gamma} t^{\gamma\cdots}{}_{\beta\ldots} + \cdots - S^\gamma_{\cdot\beta} t^{\alpha\cdots}{}_{\gamma\ldots} - \cdots \qquad (7.19)$$

where

$$S_{\alpha\beta} \equiv \omega_{\alpha\beta} - 2v_{[\alpha}a_{\beta]}. \qquad (7.20)$$

Its construction shows that it satisfies (7.7) as required, but these are also easily verified directly. It measures the rate of change of the components of $t^{\alpha\cdots}{}_{\beta\ldots}$ with respect to axes which have the same local acceleration and angular velocity as the fluid. For an equilibrium motion satisfying (4.50), $Q_{\alpha\beta} \equiv \Theta S_{\alpha\beta}$ reduces to the constant tensor $\Omega_{\alpha\beta}$ of (4.61).

7c The entropy source strength and its implications

For a scalar field, $D/D\tau$ reduces simply to $v^\gamma \partial_\gamma$. It thus follows from (4.5) and (7.4) that

$$\partial_\alpha s_2^\alpha = -\tfrac{1}{2}\rho D(a_{\alpha\beta\gamma\delta}\tau^{\alpha\beta}\tau^{\gamma\delta})/D\tau$$

$$= -\rho\tau^{\alpha\beta}[a_{\alpha\beta\gamma\delta}D\tau^{\gamma\delta}/D\tau + \tfrac{1}{2}\tau^{\gamma\delta}Da_{\alpha\beta\gamma\delta}/D\tau]. \qquad (7.21)$$

The entropy source strength in the second approximation is obtained by adding this correction to the first approximation (6.29). This gives,

$$\sigma = \tau^{\alpha\beta}Y_{\alpha\beta} \qquad (7.22)$$

where

$$Y_{\alpha\beta} \equiv X_{\alpha\beta} - \rho a_{\alpha\beta\gamma\delta}D\tau^{\gamma\delta}/D\tau - \tfrac{1}{2}\rho\tau^{\gamma\delta}Da_{\alpha\beta\gamma\delta}/D\tau. \qquad (7.23)$$

In virtue of (6.28), (7.5) and (7.7) the tensor $Y_{\alpha\beta}$ satisfies

$$Y_{\alpha\beta} = Y_{\beta\alpha}, \quad v^\alpha v^\beta Y_{\alpha\beta} = 0. \qquad (7.24)$$

The results of §7b show that $L_\Theta \tau_{\alpha\beta}$ may be replaced by $D\tau^{\alpha\beta}/D\tau$ in the list (6.24) of acceptable arguments for the phenomenological laws. Since $Da_{\alpha\beta\gamma\delta}/D\tau$ is a function of $u, \rho^\alpha, g_{\alpha\beta}, \dot{u}$ and $\dot{\rho}$ we may then also replace $X_{\alpha\beta}$ by $Y_{\alpha\beta}$. Let the variables y^Γ of (6.30) now represent $\tau^{\alpha\beta}$, $Y_{\alpha\beta}, \dot{u}, \dot{\rho}$ and $D\tau^{\alpha\beta}/D\tau$. Because of the correspondence between (6.28) and (7.24), the results of §6d then all remain valid provided $X_{\alpha\beta}$ is replaced throughout by $Y_{\alpha\beta}$. In this second approximation the phenomenological laws thus take the form

$$Y_{\alpha\beta} = Y_{\alpha\beta}(u, \rho^\alpha, \tau^{\alpha\beta}, z^a) \qquad (7.25)$$

where $\{z^a; 1 \leqslant a \leqslant 11\}$ is now a set of linearly independent components of the rates of change

$$\dot{u} \equiv Du/D\tau, \quad \dot{\rho} \equiv D\rho/D\tau \quad \text{and} \quad D\tau^{\alpha\beta}/D\tau. \qquad (7.26)$$

By analogy with (6.42) the entropy law requires

$$\partial Y_{\alpha\beta}/\partial z^a = 0 \quad \text{when} \quad z^a = 0. \qquad (7.27)$$

Close to equilibrium we may approximate the laws (7.25) by the first few terms of their Taylor series expansion in the variables $\tau^{\alpha\beta}$ and z^a. By (7.27) this expansion cannot contain linear terms involving the z^a, but it can contain quadratic terms which involve a product $\tau^{\alpha\beta}z^a$. Since it follows from (5.5), (7.6) and (7.7) that

$$Da_{\alpha\beta\gamma\delta}/D\tau = \dot{a}_1 h_{\alpha\beta} h_{\gamma\delta} + \dot{a}_2 h_{\alpha(\gamma} h_{\delta)\beta} + \dot{a}_3 v_{(\alpha} h_{\beta)(\gamma} v_{\delta)} \qquad (7.28)$$

with

$$\dot{a}_i = (\partial a_i/\partial u)\dot{u} + (\partial a_i/\partial \rho)\dot{\rho}, \quad (i = 1, 2, 3) \qquad (7.29)$$

we see that the final term in (7.23) is also of this type. A true linearization of the phenomenological laws (7.25) should thus omit this contribution to $Y_{\alpha\beta}$, or more strictly, it should cancel it with a suitable quadratic term of the Taylor expansion. The linearized form of (7.25) which corresponds to the linearization (6.43) of (6.41) is thus seen from (7.25) and (7.23) to be

$$\rho a_{\alpha\beta\gamma\delta} D\tau^{\gamma\delta}/D\tau + c_{\alpha\beta\gamma\delta}\tau^{\gamma\delta} = X_{\alpha\beta}. \qquad (7.30)$$

In this approximation the thermodynamic force $X_{\alpha\beta}$ is related to the thermodynamic flux $\tau^{\alpha\beta}$ by a first-order differential equation in the proper time τ. Such a relationship describes a response process which takes a nonzero time to occur. This confirms that the retention of the quadratic terms (7.4) in s^α does indeed have the hoped-for effect used to motivate their retention. The coefficient tensor $c_{\alpha\beta\gamma\delta}$ in (7.30) again has all the properties assigned to it in §6e. We shall still call it the impedance tensor even though it is not now solely responsible for determining the response of the system to a given thermodynamic force.

A decomposition of (7.30) into three simpler equations can be performed in close analogy with the decomposition of (6.43) into (6.50) and (6.51). We find that

$$\tau_{\text{b}}\frac{D}{D\tau}p_{\text{v}} + p_{\text{v}} = -\zeta\theta, \qquad (7.31)$$

$$\tau_{\text{s}}\frac{D}{D\tau}\pi_{\alpha\beta} + \pi_{\alpha\beta} = 2\eta e_{\alpha\beta}, \qquad (7.32)$$

$$\tau_{\text{t}}\frac{D}{D\tau}q_\alpha + q_\alpha = -\kappa(B_\alpha^\beta \partial_\beta T + Ta_\alpha). \qquad (7.33)$$

The notation is the same as in (6.50) and (6.51), but now there are three additional parameters τ_b, τ_s, τ_t. These are defined and named as follows:

$$\begin{aligned}
\tau_b &\equiv \rho(3a_1 + a_2)/(3c_1 + c_2) \text{ is the bulk relaxation time} \\
\tau_s &\equiv \rho a_2/c_2 \qquad\qquad\qquad \text{is the shear relaxation time} \\
\tau_t &\equiv \rho a_3/c_3 \qquad\qquad\qquad \text{is the thermal relaxation time.}
\end{aligned}\right\}$$

$$(7.34)$$

In the approximation (6.43) the flux $\tau^{\alpha\beta}$ must vanish when the force $X_{\alpha\beta}$ is zero. This is not necessarily true in the improved approximation (7.30). Consider the situation of a fluid at rest, when $D/D\tau$ reduces by (7.19) and (7.20) simply to ordinary differentiation $d/d\tau$ with respect to proper time τ. When $X_{\alpha\beta}$ is zero so also are the right-hand sides of (7.31) to (7.33). The general solution of (7.31) under these conditions is

$$p_v = A \exp(-\tau/\tau_b) \qquad (7.35)$$

where A is an arbitrary constant. This decays exponentially when $\tau_b > 0$ with a characteristic proper time of τ_b, but when $\tau_b < 0$ it grows exponentially. Since equilibrium states are stable with respect to small disturbances, only the case $\tau_b > 0$ is physically possible. The same argument can be applied also to (7.32) and (7.33). It shows clearly the origin of the name 'relaxation time' and it proves that the relaxation times (7.34) are all positive.

When (7.30) holds, it follows with the aid of (7.22) and (7.23) that the entropy source strength σ is given by

$$\sigma = (c_{\alpha\beta\gamma\delta} - \tfrac{1}{2}\rho Da_{\alpha\beta\gamma\delta}/D\tau)\tau^{\alpha\beta}\tau^{\gamma\delta}. \qquad (7.36)$$

Since (6.45) and (7.28) show that $c_{\alpha\beta\gamma\delta}$ and $Da_{\alpha\beta\gamma\delta}/D\tau$ have the same general forms, it is easily seen that the effect of the term in $Da_{\alpha\beta\gamma\delta}/D\tau$ is to replace the coefficients $1/(\zeta T)$, etc. in (6.53) by

$$\left[\frac{1}{\zeta T} - \tfrac{1}{2}\rho \frac{d}{d\tau}\left(\frac{\tau_b}{\rho\zeta T}\right)\right] \quad \text{etc.}$$

If the time-scale of macroscopic changes within the fluid is large in comparison with the relaxation times, the second terms in these modified coefficients will be small in comparison with the first terms. In such circumstances the positivity of σ will be ensured by the positivity (6.54) of ζ, η, κ. When the time-scale of macroscopic changes approaches that of the relaxation times, linearization of (7.25) in the rates of change z^a will no longer be reasonable. The linear phenomenological laws (7.31) to (7.33) are thus consistent with the entropy law in virtue of (6.54) whenever this linearization is a valid approximation.

8 Thermal diffusion

In Newtonian physics the heat flux in a fluid is proportional to the temperature gradient. Even when the relaxation term is neglected, this is no longer true in the relativistic theory. Equation (7.33) shows that an acceleration can also generate a flux of heat. However, this should not be surprising since we saw in §4d that even in equilibrium the temperature distribution within a rotating fluid is nonuniform. The acceleration field present in a rotating fluid of uniform temperature must thus be able to generate a heat flow to create the required nonuniformity of the equilibrium situation.

Let us investigate next the effect of a nonzero relaxation time on the propagation of heat. Consider a fluid which is initially in a homogeneous state of equilibrium, at rest in some inertial frame. For simplicity suppose it to have a negligible coefficient of thermal expansion, so that the mechanical equilibrium is not upset by expansion when the fluid is heated. We shall also neglect the mechanical recoil produced in reaction to the 'inertia of heat', as we are not interested here in tiny relativistic corrections. With these assumptions the fluid will remain at rest when heated. The energy equation (3-5.37) and thermal conduction equation (7.33) then simplify to give

$$\rho \dot{u} = -\partial_\alpha q^\alpha \qquad (8.1)$$

and

$$\tau_t \dot{q}_\alpha + q_\alpha = -\kappa B_\alpha^\beta \partial_\beta T. \qquad (8.2)$$

But by the Gibbs relation (4.39) we have

$$\dot{u} = C\dot{T} \qquad (8.3)$$

since $\dot{\rho} = 0$, where

$$C \equiv T \partial s(v, T)/\partial T \qquad (8.4)$$

is the specific heat at constant volume (and hence also at constant pressure as thermal expansion is being neglected). Let us eliminate u and q^α from these equations under the assumption that the variation in temperature is small enough for C to be treated as constant. If we use a Minkowskian coordinate system in which the fluid is at rest, so that $v = c^{-1}\delta_4^\alpha$, we obtain

$$\nabla^2 T = \frac{1}{\lambda} \frac{\partial T}{\partial t} + \frac{1}{c_t^2} \frac{\partial^2 T}{\partial t^2} \qquad (8.5)$$

where

$$c_t^2 = \kappa c^2/(\rho C \tau_t), \quad \lambda = \kappa c/(\rho C). \qquad (8.6)$$

Since specific heats are always positive, as also are κ and τ_t, we see that $\lambda > 0$ and that c_t is real and may be taken as positive.

Equation (8.6) is the three-dimensional form of the heat diffusion equation for the idealized situation described above. It is a second-order partial differential equation of hyperbolic type, and the well-known theory of such equations shows that disturbances in T cannot be propagated with a speed greater than c_t. We may thus consider c_t to be the 'speed of heat'. The existence of an upper limit to the speed of heat diffusion is essential for the consistency of a relativistic theory as nothing should propagate faster than the speed of light, but in Newtonian physics c_t is usually taken to be infinite. This is because the relaxation time τ_t is usually neglected in the Newtonian theory. It does not mean that a finite c_t is a relativistic effect. The presence of the speed of light c in the expression (8.6) for c_t is due to differences between the three- and four-dimensional definitions of the scalar parameters κ, τ_t etc. and is not particularly significant. Calculations in the kinetic theory of gases (Kranyš 1972) give a value for c_t of the same order as the speed of sound. But in practice, under laboratory conditions the effects of a finite c_t are swamped by the diffusion effects which are governed by the parameter λ in (8.5). The observed phenomena can thus be described adequately with the final term of (8.5) omitted.

Nonzero values of the other two relaxation times τ_b and τ_s lead similarly to finite speeds of propagation of viscous effects. Again such finite speeds are essential for a consistent relativistic theory. As none of these speeds can exceed c, we see that the relativistic theory imposes lower bounds on all the relaxation times. It thus compels us to include the quadratic terms (7.4) in the entropy flux s^α, terms which are present but are usually and consistently neglected in Newtonian physics.

9   The Newtonian limit

The Newtonian phenomenological laws may be obtained from the relativistic ones by taking the limit $c \to \infty$. We know from §2-8b how the metric and velocity fields behave in this limit. If we again use an overbar to denote Newtonian spacetime variables, we have from that section that

$$cv^\alpha \to \bar{v}^\alpha, \quad c^{-1}v_\alpha \to -\bar{t}_\alpha, \left.\begin{array}{r}\\ \\\end{array}\right\}$$
$$g^{\alpha\beta} \to \bar{g}^{\alpha\beta}, \quad c^{-2}g_{\alpha\beta} \to -\bar{t}_\alpha \bar{t}_\beta, \quad h_{\alpha\beta} \to \bar{h}_{\alpha\beta} \tag{9.1}$$

where
$$h_{\alpha\beta} \equiv g_{\alpha\beta} + v_\alpha v_\beta, \quad \bar{h}_{\alpha\beta} \equiv 2\bar{g}_{\alpha\gamma\delta\beta}\bar{v}^\gamma \bar{v}^\delta. \tag{9.2}$$

We also know from §3-7b how the basic dynamical variables behave.

We saw there that $\quad \rho \to \bar{\rho}, \quad c^2 u \to \bar{u}, \quad c^2 p \to \bar{p},$
$$c^3 q^\alpha \to \bar{q}^\alpha, \quad c^2 \sigma^{\alpha\beta} \to \bar{\sigma}^{\alpha\beta}. \tag{9.3}$$

The viscous and shear stress tensors $\sigma_v^{\alpha\beta}$ and $\pi^{\alpha\beta}$ will behave like $\sigma^{\alpha\beta}$ while the viscous and thermostatic pressures p_v and p_0 will behave like p. By a suitable choice of the units of entropy in the two theories we can arrange that
$$s \to \bar{s}. \tag{9.4}$$

For the Gibbs relation (4.39) to keep the same form in the Newtonian theory, this requires that $\quad c^2 T \to \bar{T}. \tag{9.5}$

The Newtonian acceleration $\bar{a}^\alpha$, vorticity $\bar{\omega}^{\alpha\beta}$, strain rate $\bar{\theta}^{\alpha\beta}$, shear rate $\bar{e}^{\alpha\beta}$ and expansion rate $\bar{\theta}$ are defined by

$$\bar{a}^\alpha = \bar{v}^\beta \partial_\beta \bar{v}^\alpha, \quad \bar{\theta} = \partial_\alpha \bar{v}^\alpha, \quad \bar{g}^{\alpha\gamma}\partial_\gamma \bar{v}^\beta = \bar{\theta}^{\alpha\beta} + \bar{\omega}^{\alpha\beta},$$
$$\bar{\theta}^{[\alpha\beta]} = 0, \quad \bar{\omega}^{(\alpha\beta)} = 0, \quad \bar{e}^{\alpha\beta} = \bar{\theta}^{\alpha\beta} - \tfrac{1}{3}\bar{\theta}\bar{g}^{\alpha\beta}. \tag{9.6}$$

As with $\bar{q}^\alpha$, $\bar{\sigma}^{\alpha\beta}$ etc., those that are tensors are orthogonal to $\bar{t}_\alpha$ on all indices. The three-dimensional counterparts of such four-dimensional contravariant Newtonian tensors are simply the Cartesian tensors formed by their totally spacelike components. Specific examples are given by (3-6.29) and (3-7.29). These counterparts will be denoted by the same kernel symbol but with the overbars omitted. To prevent confusion with the relativistic variables, latin letters will be used exclusively for the indices of such Cartesian tensors and they will all be written as subscripts in accordance with the discussion of §2-4a. Since we also have $\bar{v}^\alpha = (v_a, 1)$ it is easily seen that the definitions (9.6) are equivalent to their more familiar three-dimensional forms such as

$$a_a = \partial v_a/\partial t + v_b\,\partial_b v_a, \quad \omega_{ab} = \partial_{[a} v_{b]}. \tag{9.7}$$

The definitions (9.6) may be combined with (9.1) and those of §5 to show that $\quad c^2 a^\alpha \to \bar{a}^\alpha, \quad c\theta \to \bar{\theta}, \quad c\theta^{\alpha\beta} \to \bar{\theta}^{\alpha\beta},$
$$ce^{\alpha\beta} \to \bar{e}^{\alpha\beta}, \quad c\omega^{\alpha\beta} \to \bar{\omega}^{\alpha\beta}. \tag{9.8}$$

These enable us also to find the Newtonian limit of the operator $D/D\tau$ of §7b. If we rewrite (7.20) as

$$S^\alpha_{.\beta} = (\omega^{\alpha\gamma} - v^\alpha a^\gamma)\,h_{\gamma\beta} + a^\alpha v_\beta \tag{9.9}$$

we find from (9.1) and (9.8) that

$$cS^\alpha_{.\beta} \to \bar{S}^\alpha_{.\beta} \equiv \bar{\omega}^{\alpha\gamma}\bar{h}_{\gamma\beta} - \bar{a}^\alpha \bar{t}_\beta. \tag{9.10}$$

Hence $\quad\quad\quad\quad\quad cD/D\tau \to D/Dt \tag{9.11}$

where D/Dt is given by the right-hand side of (7.19) with overbars added throughout. When D/Dt is applied to a contravariant tensor $\bar{T}^{\alpha\beta\cdots}$ that is orthogonal to $\bar{t}_\alpha$ on each index, the contributions from the final term in the definition (9.10) of $\bar{S}^\alpha{}_\beta$ will all vanish. We have already mentioned the correspondence between such tensors and three-dimensional Cartesian tensors. We know from (2-8.13) that the contraction of two Cartesian tensor indices is represented in spacetime form by taking the inner product with $\bar{h}_{\alpha\beta}$ of the corresponding contravariant tensor indices. It follows that the action of D/Dt on the three-dimensional Cartesian tensor $T_{ab\ldots}$ corresponding to $\bar{T}^{\alpha\beta\cdots}$ is given by

$$\frac{D}{Dt}T_{ab\ldots} = \frac{\partial}{\partial t}T_{ab\ldots} + v_c\partial_c T_{ab\ldots} + \omega_{ac}T_{cb\ldots} + \omega_{bc}T_{ac\ldots} + \ldots. \quad (9.12)$$

This measures the rate of change in a Newtonian reference frame that is comoving and corotating with the fluid.

We are at last in a position to take the Newtonian limit of the linear phenomenological laws (7.31) to (7.33). Raise all the free indices and multiply these equations by c^2, c^2 and c^3 respectively so that the second term on each left-hand side has a finite but nontrivial limit. To keep the other terms finite but nontrivial we need the relativistic and Newtonian scalar parameters to be related by

$$\left.\begin{array}{l} c\zeta\to\bar{\zeta}, \quad c\eta\to\bar{\eta}, \quad c\kappa\to\bar{\kappa}, \\ c^{-1}\tau_{\rm b}\to\bar{\tau}_{\rm b}, \quad c^{-1}\tau_{\rm s}\to\bar{\tau}_{\rm s}, \quad c^{-1}\tau_{\rm t}\to\bar{\tau}_{\rm t}. \end{array}\right\} \quad (9.13)$$

Now take the limit $c\to\infty$ using the above results and express the results so obtained in three-dimensional form. We find that

$$\tau_{\rm b}\frac{D}{Dt}p_{\rm v} + p_{\rm v} = -\zeta\theta, \quad (9.14)$$

$$\tau_{\rm s}\frac{D}{Dt}\pi_{ab} + \pi_{ab} = 2\eta e_{ab} \quad (9.15)$$

and
$$\tau_{\rm t}\frac{D}{Dt}q_a + q_a = -\kappa\partial_a T, \quad (9.16)$$

where for simplicity the overbars have been omitted throughout. Note that the acceleration disappears completely from the equations in this limit, both from the right-hand side of (7.33) and from its involvement in the operator $D/D\tau$.

Relaxation terms equivalent to those appearing in (9.14) and (9.15) are known in nonrelativistic fluid dynamics in the theory of so-called non-Newtonian fluids. They are significant for such fluids only under conditions in which terms quadratic in $\tau^{\alpha\beta}$ are also significant. These quadratic terms were neglected in the derivation of (7.30) from (7.25) but they are easily included if required. A very general theory of higher order approximations in nonrelativistic fluid dynamics has been developed by Coleman & Noll (1960). As was remarked in §8, the relaxation term in the thermal equation (9.16) is not usually considered in Newtonian physics.

Classical fluid dynamics restricts itself to so-called Newtonian fluids, which include all common real fluids such as air, water, etc. For such fluids the linear approximation is valid under all normal physical conditions and the relaxation times are all negligible. Equations (9.14) to (9.16) then reduce to the standard phenomenological equations of classical fluid dynamics such as may be found in Batchelor (1967). For most work in classical fluid dynamics even the effect of bulk viscosity is insignificant, so that ζ is usually taken to be zero. In a simple model of a gas based on structureless molecules, nonrelativistic kinetic theory calculations do indeed give $\zeta = 0$. However relativistic kinetic theory gives nonzero values of ζ even for such idealized gases, with a temperature dependence of the form $\zeta/\eta \propto T^2$ for moderate temperatures as shown for example by Stewart (1971). It thus becomes increasingly significant as the temperature increases, when the thermal velocities of the molecules become large and relativistic effects can be expected to make an appearance.

This completes the programme laid down in §1. We have developed the theory of the simplest type of material that is compatible with the basic laws governing momentum, angular momentum, inert mass and entropy. Moreover, we have linearized the equations in all variables which describe departures from equilibrium. It is perhaps surprising that such simple hypotheses should produce a theory with any significant domain of validity. But the theory in fact has very wide validity. We have seen that it includes all the basic laws of classical fluid dynamics. It is in agreement with relativistic calculations of the kinetic theory of gases (Israel & Stewart 1976, Stewart 1977). Since it takes relaxation processes into consideration it is also consistent with relativistic causality, in that thermal and viscous diffusion processes propagate with a finite speed. These same relaxation terms occur in the non-relativistic theory of non-Newtonian fluids. This all suggests that the equations of the theory can form the basis of a meaningful relativistic generalization of Newtonian fluid dynamics. The laying of such foundations was the aim of this chapter, and so we leave the subject at this point.

10 Perfect fluids

Much of this chapter has been concerned with dissipative phenomena in fluids, i.e. phenomena which generate entropy. However, there is one ideal situation for which all this can be by-passed. This is the case of a so-called *perfect fluid*, which is defined to be a fluid in which the local state never deviates from equilibrium. Before we conclude this chapter, let us look briefly at this case.

In a local state of equilibrium both (4.33) and (4.39) hold. Since a perfect fluid is always in local equilibrium, these results must thus hold always. With their use the energy equation (3-5.37) reduces simply to $ds/d\tau = 0$, so that the specific entropy s is constant along each material orbit. It is possible in principle for a perfect fluid of homogeneous chemical composition to have s varying from one material orbit to another, but it is natural to consider a perfect fluid as homogeneous only if s is constant throughout. The value of s then becomes a characteristic of the fluid material, the fluid loses one of its two internal degrees of freedom and the Gibbs relation (4.39) reduces to

$$du = -p\,d(1/\rho).\tag{10.1}$$

Beyond this point thermodynamics essentially disappears from the theory of the perfect fluid.

It is convenient to describe the remaining internal degree of freedom by the pressure p. If the density ρ is given as a function of p, (10.1) can be integrated to give $u(p)$ up to an additive constant of integration. But the binding energy when $p = 0$ may be considered as inert and incorporated into ρ. When this is done we have $u = 0$ when $p = 0$ and so obtain

$$u(p) = \int_0 \frac{dp}{\rho} - \frac{p}{\rho}.\tag{10.2}$$

The energy–momentum tensor can now be expressed as

$$T^{\alpha\beta} = \rho^* v^\alpha v^\beta + p g^{\alpha\beta}\tag{10.3}$$

where
$$\rho^* = \mu + p = \rho(1 + u + p/\rho) = \rho\left(1 + \int_0 dp/\rho\right)\tag{10.4}$$

and μ is the rest-mass density in accordance with (3-5.35). It follows that the equation of state $\rho = \rho(p)$ gives a complete dynamical description of a perfect fluid material.

Use of these results and the notations (5.5) and (5.19) enables the momentum conservation equation (3-4.37) to be expressed as

$$\rho^* a^\alpha = -h^{\alpha\beta}\partial_\beta p.\tag{10.5}$$

This is the equation of motion for a perfect fluid. The coefficient ρ^* multiplying the acceleration may be interpreted as the density of inertial mass. By (10.4) it includes a contribution from the pressure p and so is not identical with the rest-mass density μ. If we recall from §4b that the usual three-dimensional pressure is pc^2 we see that under normal conditions $\mu \gg p$. The pressure contribution to the inertial mass density is thus a relativistic correction that is not in conflict with observation.

An especially simple example of a perfect fluid is given by a coherently moving cloud of particles, generally known as a dust cloud. The assumptions here are that all the particles which lie on any sufficiently small hypersurface element dS_α have the same velocity v^α, and that they are sufficiently dense to be considered as a continuum. If ρ is their density of inert mass, the total mass on that element dS_α is $\rho v^\beta dS_\beta$. Due to the coherence of the motion the momentum will thus be $(\rho v^\beta dS_\beta) v^\alpha$, which corresponds to an energy–momentum tensor

$$T^{\alpha\beta} = \rho v^\alpha v^\beta. \tag{10.6}$$

This corresponds to a perfect fluid at zero pressure, as might have been anticipated from the microscopic origin of pressure in kinetic theory. It is a trivial material in the absence of body forces. Equation (10.5) shows that the acceleration vanishes, so that the particle paths are all straight lines. It is introduced here as it will be of interest in the next chapter, where the interaction between a charged dust cloud and an electromagnetic field will be considered.

5

Electrodynamics of polarizable fluids

1 The role of thermodynamics in electrodynamics

The matter distributions considered so far have been assumed free from body forces. In the present chapter this restriction will be removed to allow a study of one of the two possible types of body force, namely those of electromagnetic origin. The only other long-range field capable of exerting a body force is gravitation, which was seen in Chapter 1 to be outside the scope of the special theory of relativity. In our considerations of point particles, general forces were defined and analysed since point particles can be subjected to a great variety of forces by direct physical contact. In continuum mechanics, such forces of contact become surface forces which are included in the theory through the boundary conditions of a problem. It may have seemed anomalous that body forces were not also included in the general continuum theory. The reason was this severe limitation on the types of possible body force which makes it unnecessary to do so since electromagnetism has to be treated at some stage as a topic in its own right.

The aim of this chapter is to investigate the interaction between the electromagnetic field and ordinary substantial matter by use of the thermodynamic techniques developed in Chapter 4. As in that chapter, attention will be restricted to the system that is simplest from a thermodynamic viewpoint. The substantial matter will thus be a simple fluid. Thermodynamics has nothing to say about the universal laws governing the electromagnetic field itself. Maxwell's equations for the electromagnetic field will thus be assumed known in their most general three-dimensional form, namely

$$\nabla \cdot \mathbf{B} = 0, \quad \nabla \times \mathbf{E} = -\frac{1}{c}\frac{\partial \mathbf{B}}{\partial t} \tag{1.1}$$

and

$$\nabla \cdot \mathbf{D} = 4\pi\rho, \quad \nabla \times \mathbf{H} = \frac{4\pi}{c}\mathbf{j} + \frac{1}{c}\frac{\partial \mathbf{D}}{\partial t} \tag{1.2}$$

where
$\mathbf{E}$ is the electric field,
$\mathbf{D}$ is the electric displacement,

B is the magnetic induction,
H is the magnetic field,
ρ is the electric charge density,
j is the electric current.

The **D** and **H** fields are distinct from **E** and **B** only within matter whose internal structure is affected by an electromagnetic field. In vacuo, or within matter that is electromagnetically inert,

$$\mathbf{D} = \mathbf{E} \quad \text{and} \quad \mathbf{H} = \mathbf{B}. \tag{1.3}$$

In keeping with our general attitudes to physical variables, no *a priori* interpretation will be given to the fields **E, D, B, H**. In particular, nothing will be assumed about the relationships (if any) between these four fields that replace (1.3) within matter that is not electromagnetically inert. Such relationships are not universal. They belong to the constitutive laws for the material being investigated, and as such they come within the scope of thermodynamics. It will be seen that for a simple fluid both the existence and the form of these relationships can be deduced thermodynamically.

This aspect of the constitutive laws concerns the effect that the fluid has on the electromagnetic field. We are also interested in the effect of the field on the motion of the fluid. This motion is governed by the conservation laws (4-4.5), but to deduce physical consequences from these laws $T^{\alpha\beta}$ must be known as a function of more primitive variables such as the density, pressure and the electromagnetic field vectors. This functional form is also a constitutive law of the system, provided that the system is taken to include the electromagnetic field which itself carries energy and momentum. It too can be deduced by thermodynamic considerations. In this way thermodynamics can be used to determine the body force exerted on the fluid by the electromagnetic field.

Such deductions cannot quite be made on the basis of (1.1) to (1.3) alone. These equations govern the generation of electromagnetic fields, but they do not describe any physical effect by which such fields may be detected. Clearly some such effect must be put into the theory before any physical deductions can be made from it. An electromagnetic field in vacuo can be recognized by its effect on the motion of a charged particle. It exerts on a particle of charge e and three-velocity **v** a pure force **F** given by

$$\mathbf{F} = e(\mathbf{E} + c^{-1}\mathbf{v} \times \mathbf{B}). \tag{1.4}$$

This is known as the *Lorentz force law*. Provided that the particle is free from other forces, its motion is thus governed by (3-3.30) and (3-3.31) with $\mathbf{F}$ given by (1.4) and with $G = 0$. The detection and measurement of electromagnetic fields within a medium can be reduced to that in vacuo. This will be shown in §2c. Equation (1.4) thus provides the only theoretical handle on the physical effects of an electromagnetic field that will be needed.

2 Maxwell's equations in spacetime form

2a *The electromagnetic field tensor in vacuo*

The first steps in the incorporation of electromagnetism into special relativity are to confirm that (1.4) is Lorentz invariant and to find its four-dimensional formulation. It follows from (3-3.25), (3-3.27) and (3-3.29) that the four-force F^α corresponding to (1.4) is

$$F^\alpha = e(c^{-2}\gamma(\mathbf{E} + c^{-1}\mathbf{v} \times \mathbf{B}), c^{-4}\gamma\mathbf{v}\cdot\mathbf{E}). \qquad (2.1)$$

If the vector product is written out explicitly and (3-3.1) is used, it can be seen that F^α has a linear dependence on the four-velocity v^α of the particle. This enables a factorization

$$F^\alpha = (e/c)\,F^\alpha_{\cdot\beta}\,v^\beta \qquad (2.2)$$

to be made, where

$$F^\alpha_{\cdot\beta} = \begin{pmatrix} 0 & c^{-1}B_3 & -c^{-1}B_2 & E_1 \\ -c^{-1}B_3 & 0 & c^{-1}B_1 & E_2 \\ c^{-1}B_2 & -c^{-1}B_1 & 0 & E_3 \\ c^{-2}E_1 & c^{-2}E_2 & c^{-2}E_3 & 0 \end{pmatrix}. \qquad (2.3)$$

Since F^α has to be a four-vector for all possible four-velocities v^α, it follows from (2.2) by the quotient rule of §2-2a that $(e/c)\,F^\alpha_{\cdot\beta}$ must be a tensor of valence $(1, 1)$.

This does not tell us anything about the Lorentz transformation properties of e and $F^\alpha_{\cdot\beta}$ separately, but that is not surprising since (1.4) also only involves the products $e\mathbf{E}$ and $e\mathbf{B}$. However, it is implicit in the interpretation of (1.4) that e is a characteristic of the particle while $\mathbf{E}$ and $\mathbf{B}$ describe the field in which it moves. It would be inconsistent with this to have, say, $e\gamma$ being a Lorentz scalar and $\gamma^{-1}F^\alpha_{\cdot\beta}$ a tensor since this introduces the particle velocity into the transformation of the field variables. The only consistent splitting is for e and $F^\alpha_{\cdot\beta}$ to be a scalar and tensor respectively.

Since F^α is a pure force, it must satisfy

$$F^\alpha v_\alpha \equiv (e/c) F_{\alpha\beta} v^\alpha v^\beta = 0 \qquad (2.4)$$

for all velocities v^α. This is possible only if

$$F_{(\alpha\beta)} = 0, \qquad (2.5)$$

so that $F_{\alpha\beta}$ is antisymmetric. This may be verified by lowering the index α in (2.3). Since the vectors $\mathbf{E}$ and $\mathbf{B}$ are algebraically independent, (2.5) is the only algebraic restriction on the field tensor $F_{\alpha\beta}$. In virtue of (1.3) it provides a complete description of the electromagnetic field in vacuo.

2b *The charge–current vector*

The usefulness of the Lorentz force law is considerably enhanced by the fact that charge is conserved. This conservation law is implicit in Maxwell's equations in virtue of the vector identity

$$\mathbf{\nabla} \cdot (\mathbf{\nabla} \times \mathbf{H}) \equiv 0. \qquad (2.6)$$

When combined with (1.2) this shows that

$$\frac{\partial \rho}{\partial t} + \mathbf{\nabla} \cdot \mathbf{j} = 0. \qquad (2.7)$$

Since $\mathbf{j}$ must vanish outside the body, (2.7) may be integrated over the volume V of an isolated body to give

$$dQ/dt = 0, \quad Q \equiv \int \rho \, dV. \qquad (2.8)$$

The constant Q is known as the *total charge* of the body. The e appearing in (1.4) is just this value of Q for the particle.

Taken together, the Lorentz force law and the law of conservation of charge serve both as an implicit definition of e, $\mathbf{E}$ and $\mathbf{B}$ and as a statement of a fundamental law of nature. This dual role seems to be a characteristic of all the really fundamental physical laws. We have seen it previously in the principle of inertia, which defined the inertial reference frames, and in the law of conservation of momentum, which defined rest mass. To complete the definitions of e, $\mathbf{E}$ and $\mathbf{B}$ it is only necessary to choose a reference particle which has unit charge by definition.

We have already seen that the charge of an isolated particle is a Lorentz scalar, but we also need to know the behaviour of the charge variables ρ and $\mathbf{j}$ for a continuum under a Lorentz transformation. This is not yet determined by the assumptions that we have made so far. First we need a clarification of the nature of $\mathbf{j}$. At a microscopic level all charge occurs in the form of discrete charged particles. The charge density ρ measures their total charge per unit volume while the current $\mathbf{j}$ describes the flow of charge produced by the motion of these particles. This may be made precise by the methods used in §§ 3-3b and 3-5 to treat inert mass. The constant value of the charge of an isolated particle is a valid spacetime variable since it is a Lorentz scalar. The flux $Q(\Sigma)$ of charge through any outer oriented hypersurface Σ may thus be defined by

$$Q(\Sigma) = \Sigma(\pm e). \tag{2.9}$$

The summation is over all particles whose world lines intersect Σ, and the sign convention is as in (3-1.3) and (3-3.11). When Σ is taken as the boundary ∂R of a bounded region R of spacetime, we have

$$Q(\partial R) = 0 \tag{2.10}$$

as an expression of charge conservation. The flux $Q(\Sigma)$ and its conservation law (2.10) must remain meaningful in the continuum description of matter, although we lose its microscopic definition (2.9). The techniques of § 3-4a, which were applied in § 3-5 to inert mass, now lead to the existence of a *charge-current vector* J^α such that

$$Q(\Sigma) = c \int_\Sigma J^\alpha dS_\alpha. \tag{2.11}$$

The factor c on the right-hand side is included purely for later convenience. The law (2.10) gives

$$\partial_\alpha J^\alpha = 0 \tag{2.12}$$

as the differential equation representing charge conservation in a continuum.

The values of ρ and $\mathbf{j}$ in a particular inertial frame may be *defined* in terms of J^α by setting

$$J^\alpha = c^{-2}(\mathbf{j}, \rho). \tag{2.13}$$

When Σ is taken to be a region of a $t = $ constant hyperplane for this frame, use of (2.13) and (3-6.2) enables (2.11) to be put in the form

$$Q(\Sigma) = \int \rho \, dV. \tag{2.14}$$

This agrees with (2.8) while (2.7) is the three-dimensional form taken by (2.12). These agreements confirm the consistency of the definitions and we have in (2.13) the required Lorentz transformation properties of ρ and $\mathbf{j}$.

The simplest material which can interact with an electromagnetic field is a dust cloud of particles which are charged but whose internal structure is not affected by the field. The technical meaning of a dust cloud was given in §4-10. The field will then satisfy (1.3) as in vacuo, and can be described by the tensor $F^{\alpha}{}_{\beta}$ of (2.3). For such a material the first pair (1.1) of Maxwell's equations may be combined into the spacetime form

$$\partial_{\alpha} F_{\beta\gamma} + \partial_{\beta} F_{\gamma\alpha} + \partial_{\gamma} F_{\alpha\beta} = 0. \tag{2.15}$$

In virtue of (2.5) and (2-5.7) this can be written more concisely as

$$\partial_{[\alpha} F_{\beta\gamma]} = 0. \tag{2.16}$$

With the aid of (1.3) and (2.13) the second pair (1.2) may also be combined, to give

$$\partial_{\beta} F^{\alpha\beta} = 4\pi J^{\alpha}. \tag{2.17}$$

The consistency of (2.13) as the definition of ρ and $\mathbf{j}$ is supported by this possibility of combining the equations (1.2) into the four-vector form (2.17). However (2.13) has a universal validity which extends beyond mere dust clouds, so that the argument cannot be inverted to act as a *proof* of the four-vector character of $(\mathbf{j}, \rho)$.

The particles of a homogeneous charged dust cloud will all be identical, and so in particular will all have the same charge e and the same inert mass m. Temporarily let ρ_{m} denote the density of inert mass for the cloud, to distinguish it from the charge density ρ used above. Then the fluxes $cJ^{\alpha} dS_{\alpha}$ of charge and $\rho_{\mathrm{m}} v^{\alpha} dS_{\alpha}$ of inert mass through a hypersurface element with vector area dS_{α} must always be in the same ratio as e and m. It follows that

$$J^{\alpha} = \rho_{\mathrm{m}} z v^{\alpha}, \tag{2.18}$$

where

$$z \equiv e/mc \tag{2.19}$$

is a constant determined by the material of which the cloud is composed As should be expected, the conservation equations (2.12) of charge and (3-5.33) of inert mass become consequences of one another for such a material in virtue of this constancy of z.

To complete the equations governing a charged dust cloud, we need an equation of motion. Each particle will satisfy (3-1.2) with F^{α} given by (2.2). Since $dv^{\alpha}/d\tau = v^{\beta} \partial_{\beta} v^{\alpha}$, this can be written as

$$v^{\beta} \partial_{\beta} v^{\alpha} = z F^{\alpha}{}_{.\beta} v^{\beta} \tag{2.20}$$

On multiplication by ρ_{m} and use of (2.18) and (3-5.33), (2.20) gives

$$\partial_\beta T_{\mathrm{d}}^{\alpha\beta} = F^{\alpha\beta} J_\beta \qquad (2.21)$$

where $\qquad\qquad T_{\mathrm{d}}^{\alpha\beta} \equiv \rho_{\mathrm{m}} v^\alpha v^\beta. \qquad (2.22)$

Comparison of (2.22) with (4-10.6) shows that $T_{\mathrm{d}}^{\alpha\beta}$ is the energy-momentum tensor for dust in the absence of external forces. Equation (2.21) thus gives the modification produced by the electromagnetic body forces to the momentum conservation equation for dust.

2c Polarization and magnetization

The electromagnetic variables within a charged dust cloud satisfy two sets of relations which are not valid in a more general material. One set is (1.3). The other is
$$\mathbf{j} = \rho\mathbf{v}, \qquad (2.23)$$

which follows from (2.18) and (2.13). A material in which (1.3) breaks down is said to be *polarizable*. One in which (2.23) fails is said to be *electrically conducting*. The vectors $\mathbf{P}$, $\mathbf{M}$ and $\mathbf{j}_{\mathrm{c}}$ defined by

$$4\pi\mathbf{P} = \mathbf{D} - \mathbf{E}, \quad 4\pi\mathbf{M} = \mathbf{B} - \mathbf{H}, \quad \mathbf{j}_{\mathrm{c}} = \mathbf{j} - \rho\mathbf{v} \qquad (2.24)$$

are known respectively as the *polarization*, the *magnetization* and the *conduction current* of the material. They vanish in vacuo and in materials whose internal structure is not affected by an electromagnetic field.

Electrical conductors present us with no new physical variables as we have already found the physical significance of J^α for a general material. In contrast, the polarization and magnetization vectors of a polarizable material do represent new physical phenomena. Before we study such materials theoretically, let us give some physical significance to $\mathbf{P}$ and $\mathbf{M}$ by considering, at least in principle, how they could be measured. This is a larger problem than it might at first appear to be. We are in fact lacking four physical characterizations, not two, since the Lorentz force law cannot even in principle be retained to define $\mathbf{E}$ and $\mathbf{B}$. This is easily seen if we consider what would happen if an attempt were made to determine $\mathbf{E}$ and $\mathbf{B}$ in this way. Suppose first that a microscopic particle is used as a test probe, such as is provided by an energetic beam of electrons. Then it will interact in a complex way with the individual atoms of the material and its loss of energy due to the excitation or ionization of such atoms cannot be separated from the effect of the Lorentz force at this stage in the logical development of the theory. In a fluid material it would be possible to use instead a macroscopic particle, in the hope that the mechanical forces exerted on it by the fluid could be more easily treated than these excitation and ionization losses of a microscopic particle. However, such a particle would be acted upon not by the fields truly present within the fluid but by the fields present in the hole that it would make and occupy. Not all the fields **E, D, B, H** can be continuous at the boundary of the hole since the difference fields **P** and **M** depend on the nature of the material and this is discontinuous at the boundary. Use of the Lorentz force law to define **E** and **B** within the fluid would be equivalent to an assumption that **E** and **B** are continuous at this boundary, and such an assumption should not be made *a priori*.

These criticisms of the direct use of the Lorentz force law suggest one possible method of attack on the problem. We can use the field equations (1.1) and (1.2) to investigate the relationship between the fields within the material and those within a small hole present in the material. The fields in the hole can be measured by use of the Lorentz force law, as discussed above. These theoretical relationships can then be used to deduce information about the fields within the material adjacent to the hole. For this to be a useful technique, the creation of the hole must have a negligible effect on the field in the surrounding material. Only then can the field at any point within the material be studied in this way by making an appropriate small hole. Despite this possible obstacle to success, let us see what can be done in this direction.

The required boundary conditions are obtained by first writing (1.1) and (1.2) in an equivalent integral form. Let $V(t)$ be an arbitrary time-dependent volume that is moving with the matter. Similarly let $S(t)$ be any comoving surface. Then use of the theorems of Gauss and Stokes, together with standard techniques of three-dimensional vector analysis, enables us to deduce from (1.1) and (1.2) that

$$\int_{\partial V} \mathbf{B} \cdot \mathbf{dS} = 0, \quad \int_{\partial V} \mathbf{D} \cdot \mathbf{dS} = 4\pi \int_V \rho \, dV, \qquad (2.25)$$

$$\left. \begin{aligned} \int_{\partial S} \mathbf{E}^* \cdot \mathbf{dx} &= -\frac{1}{c} \frac{d}{dt} \int_S \mathbf{B} \cdot \mathbf{dS} \\ \text{and} \qquad \int_{\partial S} \mathbf{H}^* \cdot \mathbf{dx} &= \frac{4\pi}{c} \int_S \mathbf{j}_c \cdot \mathbf{dS} + \frac{1}{c} \frac{d}{dt} \int_S \mathbf{D} \cdot \mathbf{dS}. \end{aligned} \right\} \qquad (2.26)$$

Here **v** is the three-velocity of the matter, $\mathbf{j}_c$ is the conduction current defined by (2.24), and the fields **E*** and **H*** are given by

$$\mathbf{E}^* \equiv \mathbf{E} + (\mathbf{v} \times \mathbf{B})/c, \quad \mathbf{H}^* \equiv \mathbf{H} - (\mathbf{v} \times \mathbf{D})/c. \qquad (2.27)$$

The first of the equations (2.26) is simply Faraday's law of electro-magnetic induction expressed in vector notation. For the time being let us restrict attention to materials that are electrically neutral ($\rho = 0$) and non-conducting ($\mathbf{j}_c = 0$). Consider the comoving surface $S_0(t)$ of material discontinuity that separates two different such materials in contact, or one such material from a vacuum. Make the reasonable assumptions that the field vectors $\mathbf{E}$, $\mathbf{D}$, $\mathbf{B}$, $\mathbf{H}$ all remain finite everywhere even though they may be discontinuous at S_0, and that within a given material element they vary smoothly with time. Then by applying (2.25) to a small disc-shaped volume that encloses a portion of S_0, as illustrated in figure 5(a), it may be deduced that the components of $\mathbf{B}$ and $\mathbf{D}$ normal to S_0 are continuous across S_0. Similarly by applying (2.26) to a small strip-shaped surface that intersects S_0, as illustrated in figure 5(b), it may be deduced that the components of $\mathbf{E}^*$ and $\mathbf{H}^*$ tangential to S_0 are continuous across S_0. If $\mathbf{n}$ is the unit normal to S_0, these results may alternatively be expressed as the continuity across S_0 of

$$\mathbf{n}\cdot\mathbf{B}, \quad \mathbf{n}\cdot\mathbf{D}, \quad \mathbf{n}\times\mathbf{E}^*, \quad \mathbf{n}\times\mathbf{H}^*. \tag{2.28}$$

They are simple generalizations of the boundary conditions within a medium at rest that are well known from elementary electromagnetic theory.

In virtue of (1.3), knowledge of the quantities (2.28) in vacuo completely determines the electromagnetic field there. Consider now the field within a small comoving disc-shaped hole in a neutral, non-conducting medium. Let $\mathbf{n}$ be the unit normal to the parallel plane faces of the disc. For a thin disc these faces will have a much larger area than the curved cylindrical surface which joins their edges. The field within the hole will thus be close to the uniform field determined completely by continuity of the quantities (2.28) across these plane faces, and the accuracy of this approximation will increase as the thickness h of the disc is reduced. It follows that measurement of the electromagnetic field within the hole determines the quantities (2.28) within the neighbouring material to an accuracy that increases as h decreases. Let us make the reasonable assumption that the effect of such a hole on the field within the material decreases smoothly with h. Then by extrapolation to the limit $h \to 0$ we can determine from such measurements the values of the quantities (2.28) within the intact material after the hole has been closed up.

Note that the shape of the hole is important. The above reasoning

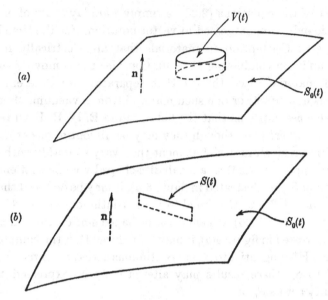

Figure 5. The volume $V(t)$ and surface $S(t)$ of integration used in the derivation of the electromagnetic boundary conditions at a comoving surface $S_0(t)$ of material discontinuity.

depends critically on the fact that the normal direction is constant over almost all the surface of the hole. If this were not so, continuity of (2.28) over the whole surface would require an inhomogeneous field within the hole. This field would have to vary more and more rapidly with position as the size of the hole was reduced. The limit as the hole was shrunk to zero would then not be well behaved and no conclusions could be drawn about the field in the intact material.

The sequential use of disc-shaped holes parallel to each of the three coordinate planes enables every component of **B**, **D**, **E*** and **H*** to be determined by the above method. Use of (2.27) then enables **E** and **H** to be calculated. We thus have in principle a method for measuring all four fields, **E**, **D**, **B** and **H**, from which the polarization **P** and magnetization **M** can be obtained by (2.24) if required. It was seen in §2b that the Lorentz force law contains implicit definitions of the electromagnetic field variables in vacuo. We see now that within a polarizable medium the Maxwell equations themselves contain implicit definitions of the field variables whose behaviour they describe.

2d The field tensors in a polarizable medium

Use of (2.27) shows that the continuity across S_0 of (2.28) is equivalent to the continuity of

$$\mathbf{n} \times \mathbf{E} - \frac{1}{c}(\mathbf{n} \cdot \mathbf{v})\mathbf{B} \quad \text{and} \quad \mathbf{n} \cdot \mathbf{B}, \tag{2.29}$$

and of

$$\mathbf{n} \times \mathbf{H} + \frac{1}{c}(\mathbf{n} \cdot \mathbf{v})\mathbf{D} \quad \text{and} \quad \mathbf{n} \cdot \mathbf{D}. \tag{2.30}$$

Let us now consider this same situation at a surface S_0 of material discontinuity in a neutral, non-conducting medium from a four-dimensional viewpoint. The comoving surface S_0 becomes a hypersurface Σ to which the four-velocity v^α corresponding to the three-velocity $\mathbf{v}$ is necessarily tangent. The covariant unit normal n_α to Σ must thus be parallel to $(\mathbf{n}, -\mathbf{n} \cdot \mathbf{v})$. It follows that the continuity of (2.29) and (2.30) can also be expressed as the continuity of

$$n_{[\alpha} F_{\beta\gamma]} \quad \text{and} \quad n_\beta I^{\alpha\beta}$$

respectively, where

$$F_{\alpha\beta} = \begin{pmatrix} 0 & c^{-1}B_3 & -c^{-1}B_2 & E_1 \\ -c^{-1}B_3 & 0 & c^{-1}B_1 & E_2 \\ c^{-1}B_2 & -c^{-1}B_1 & 0 & E_3 \\ -E_1 & -E_2 & -E_3 & 0 \end{pmatrix}, \tag{2.31}$$

and

$$I^{\alpha\beta} = \begin{pmatrix} 0 & c^{-1}H_3 & -c^{-1}H_2 & -c^{-2}D_1 \\ -c^{-1}H_3 & 0 & c^{-1}H_1 & -c^{-2}D_2 \\ c^{-1}H_2 & -c^{-1}H_1 & 0 & -c^{-2}D_3 \\ c^{-2}D_1 & c^{-2}D_2 & c^{-2}D_3 & 0 \end{pmatrix}. \tag{2.32}$$

The first of these antisymmetric arrays is identical in form to the tensor $\hat{F}_{\alpha\beta}$ defined by (2.3), where the circumflex is temporarily used as a distinguishing mark for the vacuum variables. More significant, however, is the fact that in vacuo they both reduce to the appropriate tensorial forms of $\hat{F}_{\alpha\beta}$. At a boundary between matter and vacuum we thus have

$$n_{[\alpha} F_{\beta\gamma]} = n_{[\alpha} \hat{F}_{\beta\gamma]}, \quad n_\beta I^{\alpha\beta} = n_\beta \hat{F}^{\alpha\beta}, \tag{2.33}$$

where the left- and right-hand sides are evaluated at neighbouring points just within the matter and vacuum respectively. Since $\hat{F}_{\alpha\beta}$ is already known to be a tensor, the right-hand sides of the two equations (2.33) are both tensors.

By the formation of a suitable small comoving disc-shaped hole within the matter the direction of n_α can be chosen arbitrarily, subject only to the orthogonality condition

$$n_\alpha v^\alpha = 0 \qquad (2.34)$$

which expresses the fact that v^α is tangential to a comoving boundary. Now let a^α, b^α be two arbitrary vectors and consider $a_\alpha b_\beta I^{\alpha\beta}$. In virtue of the antisymmetry of $I^{\alpha\beta}$ we have

$$a_\alpha b_\beta I^{\alpha\beta} = (a_\alpha - \lambda b_\alpha) b_\beta I^{\alpha\beta} \qquad (2.35)$$

identically for all scalars λ. If $v^\alpha b_\alpha = 0$ then the left-hand side of (2.35) is a scalar since $I^{\alpha\beta} b_\beta$ is a vector by the above result. If $v^\alpha b_\alpha \neq 0$ then λ can be chosen to make

$$v^\alpha (a_\alpha - \lambda b_\alpha) = 0. \qquad (2.36)$$

In this case $I^{\alpha\beta}(a_\alpha - \lambda b_\alpha)$ is a vector by the above result and so once again the left-hand side of (2.35) is a scalar. A double application of the quotient rule now shows that $I^{\alpha\beta}$ is a tensor. If $\eta^{\alpha\beta\gamma\delta}$ is the fundamental alternating tensor given by (2-5.23), a similar examination of

$$\eta^{\alpha\beta\gamma\delta} a_\alpha b_\beta F_{\gamma\delta}$$

with the aid of the first of the results (2.33) enables us to deduce that $F_{\alpha\beta}$ is also a tensor. Note that both the new tensors $I^{\alpha\beta}$ and $F_{\alpha\beta}$ are antisymmetric.

Historically the vectors **E** and **H** have been considered as analogous, as have **D** and **B**. This is reflected in the convention which calls both **E** and **H** field vectors and which uses the alternative names of displacement and induction for **D** and **B**. We see from (2.31) and (2.32) that this pairing is logically incorrect. It is **B** which unites with **E**, and **H** with **D**, in the tensors of the spacetime description. It is nowadays conventional to regard **B** and **H** as the misnamed fields, and in consequence $F_{\alpha\beta}$ is known as the *electromagnetic field tensor* and $I^{\alpha\beta}$ as the *electromagnetic induction tensor*. The book of de Groot & Suttorp (1972) actually goes against tradition to the extent of calling **B** the magnetic field and **H** the magnetic displacement. However, the traditional names are too well established for this not to cause some confusion and so although logical, this will not be followed here.

Use of the tensors $F_{\alpha\beta}$ and $I^{\alpha\beta}$ enables us to write the Maxwell equations (1.1) and (1.2) for an electrically neutral, non-conducting but polarizable medium in the spacetime form

$$\partial_{[\alpha} F_{\beta\gamma]} = 0, \quad \partial_\beta I^{\alpha\beta} = 0. \qquad (2.37)$$

A tensor $P^{\alpha\beta}$ may also be constructed from $\mathbf{P}$ and $\mathbf{M}$. It is known as the *polarization tensor* and is defined by

$$4\pi P^{\alpha\beta} = F^{\alpha\beta} - I^{\alpha\beta}. \tag{2.38}$$

In any inertial frame it satisfies

$$P^{\alpha\beta} = \begin{pmatrix} 0 & c^{-1}M_3 & -c^{-1}M_2 & c^{-2}P_1 \\ -c^{-1}M_3 & 0 & c^{-1}M_1 & c^{-2}P_2 \\ c^{-1}M_2 & -c^{-1}M_1 & 0 & c^{-2}P_3 \\ -c^{-2}P_1 & -c^{-2}P_2 & -c^{-2}P_3 & 0 \end{pmatrix} \tag{2.39}$$

as may be seen from (2.31), (2.32) and (2.24).

The alterations to the development of §§2c and d that are needed to remove the restriction to an electrically neutral medium are very slight. The boundary conditions are unchanged and they lead as before to the tensorial character of the arrays (2.31) and (2.32). With the aid of (2.13) the spacetime form of Maxwell's equations for a general non-conducting medium is found to be

$$\partial_{[\alpha} F_{\beta\gamma]} = 0, \quad \partial_\beta I^{\alpha\beta} = 4\pi J^\alpha. \tag{2.40}$$

The situation for a conducting medium is somewhat different, however. On a microscopic scale, electrical conduction is a process allied to diffusion. Conduction currents are caused by electrons drifting, relative to the macroscopic velocity of the matter, in response to an applied electric field. The creation of a hole, such as was used in §2c to define the fields within a medium, will interrupt this electron drift and so a layer of charge will build up in the surface of the hole. This will significantly affect the field within the medium even for an arbitrarily thin disc-shaped hole. The whole procedure of §2c for the measurement of the fields within the medium will thus break down. The strictly macroscopic theory can be developed instead by leaving the fields undefined and simply postulating that the arrays (2.31) and (2.32) are tensors even in a conducting medium. The Maxwell equations will then again have the form (2.40). The validity of such a theory based on undefined fields can be judged by experimentally testing the predictions of the theory as a whole, so that this is a perfectly acceptable procedure. Alternatively the fields $\mathbf{E}$, $\mathbf{D}$, $\mathbf{B}$ and $\mathbf{H}$ can all be defined in terms of the microscopic structure of matter. It is then possible to prove the tensorial character of $F_{\alpha\beta}$ and $I^{\alpha\beta}$ even within a conductor. This approach is carried out in great detail in the book of de Groot & Suttorp (1972). Note that the possibility of writing (1.1) and (1.2) formally in the form (2.40) is not in itself a guarantee that $F_{\alpha\beta}$ and $I^{\alpha\beta}$ are tensors. As the fluids studied in the following sections will be assumed to be non-conducting, we shall not need to pursue either of these approaches further.

3   The electromagnetic energy–momentum tensor in vacuo

The presence of an electromagnetic field can only be detected through its interaction with matter. It is thus possible to study theoretically the electromagnetic field even in vacuo only by examining its interaction with some material system. When one's primary interest is with the field itself, it is natural to choose the simplest possible form of matter that can interact with an electromagnetic field. In this way one minimizes the problems that can arise in separating electromagnetic and mechanical effects within the matter. A suitable material system for such purposes is the charged dust cloud introduced in §2b. Let us use it to show that the electromagnetic field in vacuo is a carrier of energy and momentum.

The result of eliminating J_α between (2.17) and (2.21) can be written as

$$\partial_\beta \left(\rho_m v_\alpha v^\beta + \frac{1}{4\pi} F_{\alpha\gamma} F^{\beta\gamma} \right) = -\frac{1}{4\pi} F^{\beta\gamma} \partial_\beta F_{\gamma\alpha}. \qquad (3.1)$$

But (2.5) and (2.15) imply

$$\partial_{[\beta} F_{\gamma]\alpha} = -\tfrac{1}{2}\partial_\alpha F_{\beta\gamma}, \qquad (3.2)$$

from which it follows that

$$F^{\beta\gamma}\partial_\beta F_{\gamma\alpha} = -\tfrac{1}{4}\partial_\alpha(F_{\beta\gamma} F^{\beta\gamma}). \qquad (3.3)$$

If this is substituted into (3.1) it shows that

$$\partial_\beta(T_d^{\alpha\beta} + T_{em}^{\alpha\beta}) = 0 \qquad (3.4)$$

where

$$T_d^{\alpha\beta} = \rho_m v^\alpha v^\beta \qquad (3.5)$$

and

$$T_{em}^{\alpha\beta} = \frac{1}{4\pi}(F^\alpha_{.\gamma} F^{\beta\gamma} - \tfrac{1}{4}g^{\alpha\beta}F^{\gamma\delta}F_{\gamma\delta}). \qquad (3.6)$$

A deliberately suggestive notation has been used here. We see that (3.4) has the form (3-4.37) of the momentum conservation equation for matter free from external forces, with

$$T^{\alpha\beta} = T_d^{\alpha\beta} + T_{em}^{\alpha\beta}. \qquad (3.7)$$

Each of the tensors on the right-hand side of (3.7) is symmetric and the first of them, $T_d^{\alpha\beta}$, is identical in form with the energy–momentum tensor (4-10.6) of a freely moving uncharged dust cloud. It is thus natural to consider $T^{\alpha\beta}$ as the total energy–momentum tensor of the whole system comprising dust and electromagnetic field, and to consider $T_{\mathrm{d}}^{\alpha\beta}$ and $T_{\mathrm{em}}^{\alpha\beta}$ as the contributions to this total from the dust and field respectively. With this viewpoint, electromagnetism ceases to be an external body force acting on the matter and becomes instead a part of the material system. Outside the dust cloud $T^{\alpha\beta}$ consists solely of $T_{\mathrm{em}}^{\alpha\beta}$, which must be regarded as the energy–momentum tensor of the electromagnetic field in vacuo.

The electromagnetic field in vacuo differs from substantial matter by its lack of inert mass and consequent lack of a well-defined velocity field. It is nevertheless possible to interpret $T_{\mathrm{em}}^{\alpha\beta}$ in a particular inertial frame by comparing it with the situation for substantial matter at rest, which is given by (3-6.16) with $\mathbf{v} = 0$. In the notation of (2-4.4) we find from (2.3) and (3.6) that

$$T_{\mathrm{em}}^{\alpha\beta} = \left(\begin{array}{c|c} -c^{-2}\mathbf{T} & c^{-4}\mathbf{S} \\ \hline c^{-4}\mathbf{S} & c^{-4}\epsilon \end{array} \right) \tag{3.8}$$

where
$$\epsilon = \frac{1}{8\pi}\,(\mathbf{E}^2 + \mathbf{B}^2), \quad \mathbf{S} = \frac{c}{4\pi}\,\mathbf{E} \times \mathbf{B},$$

$$\mathbf{T} = \frac{1}{4\pi}\,[\mathbf{E}\mathbf{E} + \mathbf{B}\mathbf{B} - \tfrac{1}{2}\mathbf{I}(\mathbf{E}^2 + \mathbf{B}^2)] \tag{3.9}$$

and $\mathbf{I}$ is the unit dyadic. The substitutions needed in (3-6.16) to bring it to the form (3.8) are

$$\mathbf{v} \to 0, \quad \epsilon \to \epsilon, \quad \mathbf{p} \to \mathbf{S}/c^2, \quad \mathbf{q} \to \mathbf{S}, \quad \boldsymbol{\sigma} \to \mathbf{T}. \tag{3.10}$$

This enables us to interpret

ϵ as the energy density,
$\mathbf{S}$ as the energy flux,
$\mathbf{S}/c^2$ as the momentum density
and $\mathbf{T}$ as the stress tensor

of the electromagnetic field in vacuo. $\mathbf{S}$ is known as the *Poynting vector* and $\mathbf{T}$ as the *Maxwell stress tensor*. These interpretations may also be obtained directly by three-dimensional considerations.

We know now that an electromagnetic field in vacuo and a freely moving material continuum both possess conserved symmetric energy–momentum tensors. It is thus natural to suppose that a system consisting of a material continuum in interaction with an electromagnetic field also has a conserved symmetric energy–momentum tensor $T^{\alpha\beta}$. This assumption will form one of the foundations for the thermodynamic theory of such interacting systems that will be developed in the next section. For charged dust we have seen that this total energy–momentum tensor separates naturally into contributions from the matter and from the field. This is a consequence of the extreme simplicity of the interaction between dust and the field, and is not a general phenomenon. In general, only the total $T^{\alpha\beta}$ is well defined and any interpretation of it as a sum of terms with distinct physical origins is either ambiguous or conventional.

4   Local equilibrium of a polarizable simple fluid

4a   Characterization of the fluid and its equilibrium

Certain tensor fields are defined throughout every material that is in interaction with an electromagnetic field. These are the total energy–momentum tensor $T^{\alpha\beta}$ whose existence was discussed in §3, the flux ρ^{α} of inert mass which is a purely mechanical variable, the charge–current vector J^{α} introduced in §2b, and the electromagnetic field and induction tensors $F_{\alpha\beta}$ and $I^{\alpha\beta}$. The general theory of continuous media given in Chapter 3 was restricted for simplicity to single-component systems, as discussed in §3-5. This excludes diffusion processes, and as mentioned in §2d electrical conduction is essentially a diffusion process. For consistency let us thus now restrict attention to non-conducting media. Most such media are electrically neutral, but it is convenient to retain the possibility of a volume charge distribution so that the results can be checked against the simple case of charged dust. When the conduction current $\mathbf{j}_c$ of (2.24) vanishes, it follows from (2.13) that J^{α} and ρ^{α} are parallel. Let us set

$$J^{\alpha} = z\rho^{\alpha}. \tag{4.1}$$

The conservation laws of charge and inert mass then imply that

$$v^{\alpha}\partial_{\alpha}z = 0, \tag{4.2}$$

so that z is constant for each material element. Since z is thus a characteristic of the material, a homogeneous medium will have z uniform throughout. This was seen in §2b to be true for charged dust. We now see that it is true for any homogeneous non-conducting medium.

In virtue of (4.1) we are left with four independent universal variables for a non-conducting medium, namely $T^{\alpha\beta}$, ρ^{α}, $F_{\alpha\beta}$ and $I^{\alpha\beta}$. The simplest such material from a thermodynamic viewpoint is one in which the entropy flux density s^α is a function of just these four fields,

$$s^\alpha = s^\alpha(T^{\alpha\beta}, \rho^\alpha, F_{\alpha\beta}, I^{\alpha\beta}). \tag{4.3}$$

We shall describe such an ideal material as a *polarizable simple fluid*. It is a natural generalization to include electromagnetism of the ideal relativistic simple fluid defined in §4-4a.

The universal laws satisfied by the argument tensors of s^α are given by (4-4.5) and (2.40) with (4.1), namely

$$\left.\begin{array}{ll} \partial_\beta T^{\alpha\beta} = 0, & \partial_\alpha \rho^\alpha = 0, \\ \partial_{[\alpha} F_{\beta\gamma]} = 0, & \partial_\beta I^{\alpha\beta} = 4\pi z \rho^\alpha. \end{array}\right\} \tag{4.4}$$

Let $\delta\sigma$ be the increment in the entropy source strength $\sigma \equiv \partial_\alpha s^\alpha$ produced by a small departure from equilibrium. The general principles laid down in §4-4a require that the derivatives of $T^{\alpha\beta}$, etc. present in the contribution

$$\frac{\partial s^\alpha}{\partial T^{\beta\gamma}} \delta(\partial_\alpha T^{\beta\gamma}) + \frac{\partial s^\alpha}{\partial \rho^\beta} \delta(\partial_\alpha \rho^\beta) + \frac{\partial s^\alpha}{\partial F_{\beta\gamma}} \delta(\partial_\alpha F_{\beta\gamma}) + \frac{\partial s^\alpha}{\partial I^{\beta\gamma}} \delta(\partial_\alpha I^{\beta\gamma}) \tag{4.5}$$

to $\delta\sigma$ must be eliminable in equilibrium by use of the laws (4.4). This is true if and only if the derivatives of s^α that occur in (4.5) have the forms

$$\left.\begin{array}{ll} \dfrac{\partial s^\alpha}{\partial T^{\beta\gamma}} = -A^\alpha_{(\beta} \Theta_{\gamma)}, & \dfrac{\partial s^\alpha}{\partial \rho^\beta} = -\Phi A^\alpha_\beta, \\[2ex] \dfrac{\partial s^\alpha}{\partial F_{\beta\gamma}} = -\Psi^{\alpha\beta\gamma}, & \dfrac{\partial s^\alpha}{\partial I^{\beta\gamma}} = -A^\alpha_{[\beta} \Xi_{\gamma]} \end{array}\right\} \tag{4.6}$$

where A^α_β is the unit tensor and

$$\Psi^{\alpha\beta\gamma} = \Psi^{[\alpha\beta\gamma]}. \tag{4.7}$$

The minus signs in (4.6) are for later convenience. The variables Θ_α, Φ, $\Psi^{\alpha\beta\gamma}$ and Ξ_α are functions of the equilibrium state whose properties are to be investigated. They are tensors of the valence indicated by their indices.

The conditions (4.6) characterize the local state of the fluid in equilibrium. When they are satisfied, the technique used to derive (4-4.49) can be used to simplify $\delta\sigma$ to the form

$$\delta\sigma = -(\partial_\alpha \Theta_\beta) \delta T^{\alpha\beta} - (\partial_\alpha \Phi - 4\pi z \Xi_\alpha) \delta\rho^\alpha - (\partial_\alpha \Psi^{\alpha\beta\gamma}) \delta F_{\beta\gamma} - (\partial_\alpha \Xi_\beta) \delta I^{\alpha\beta}. \tag{4.8}$$

This too must vanish in equilibrium for all increments $\delta T^{\alpha\beta}$, etc. which connect the equilibrium state with an arbitrary neighbouring state that is not necessarily in equilibrium. As in §4-4c let us exclude materials in which idealized properties such as incompressibility impose restrictions on the possible values of these increments at a single point. Then $\delta\sigma$ vanishes for an arbitrary variation in the state if and only if

$$\left.\begin{array}{ll} \partial_{(\alpha}\Theta_{\beta)} = 0, & \partial_{\alpha}\Phi = 4\pi z\Xi_{\alpha}, \\ \partial_{\alpha}\Psi^{\alpha\beta\gamma} = 0, & \partial_{[\alpha}\Xi_{\beta]} = 0. \end{array}\right\} \tag{4.9}$$

These are the additional conditions required to ensure global equilibrium.

4b Implications of Lorentz invariance

In virtue of (4.6) an infinitesimal change in the arguments of s^{α} must produce in an equilibrium state an increment δs^{α} in s^{α} given by

$$\delta s^{\alpha} = -\Theta_{\beta}\delta T^{\alpha\beta} - \Phi\delta\rho^{\alpha} - \Psi^{\alpha\beta\gamma}\delta F_{\beta\gamma} - \Xi_{\beta}\,\delta I^{\alpha\beta}. \tag{4.10}$$

In particular, this must hold when δs^{α}, $\delta T^{\alpha\beta}$ etc. are generated by an infinitesimal Lorentz transformation in accordance with (4-4.10). The method of derivation of (4-4.14) shows that (4.10) will hold for increments generated by an arbitrary infinitesimal Lorentz transformation if and only if

$$g^{\alpha[\beta}(s^{\gamma]} + T^{\gamma]\delta}\Theta_{\delta} + \rho^{\gamma]}\Phi + I^{\gamma]\delta}\Xi_{\delta}) = T^{\alpha[\beta}\Theta^{\gamma]} - 2\Psi^{\delta\alpha[\beta}F^{\gamma]}_{.\delta} + I^{\alpha[\beta}\Xi^{\gamma]}. \tag{4.11}$$

Antisymmetrization of (4.11) on $\alpha\beta\gamma$ gives

$$2\Psi^{\delta[\alpha\beta}F^{\gamma]}_{.\delta} - I^{[\alpha\beta}\Xi^{\gamma]} = 0 \tag{4.12}$$

which may be used to rewrite (4.11) as

$$g^{\alpha[\beta}(s^{\gamma]} + T^{\gamma]\delta}\Theta_{\delta} + \rho^{\gamma]}\Phi + I^{\gamma]\delta}\Xi_{\delta}) = T^{\alpha[\beta}\Theta^{\gamma]} + \Psi^{\delta\beta\gamma}F^{\alpha}_{.\delta} - \tfrac{1}{2}I^{\beta\gamma}\Xi^{\alpha}. \tag{4.13}$$

This is equivalent to (4.11) since (4.12) can be recovered by antisymmetrization of (4.13) on $\alpha\beta\gamma$.

Set

$$I^{*}_{\alpha\beta} \equiv \tfrac{1}{2}\eta_{\alpha\beta\gamma\delta}I^{\gamma\delta}, \qquad \Psi_{\alpha} \equiv \tfrac{1}{2}\eta_{\alpha\beta\gamma\delta}\Psi^{\beta\gamma\delta}. \tag{4.14}$$

Use of (2-5.33) enables us to deduce from (4.14) first that

$$I^{\gamma\delta} = \tfrac{1}{2}\eta^{\alpha\beta\gamma\delta}I^{*}_{\alpha\beta}, \qquad \Psi^{\beta\gamma\delta} = \eta^{\alpha\beta\gamma\delta}\Psi_{\alpha} \tag{4.15}$$

and then from the second of these results that

$$\eta_{\gamma\delta\varepsilon\zeta}\Psi^{\beta\gamma\delta} = 4\Psi_{[\varepsilon}A^{\beta}_{\zeta]}. \tag{4.16}$$

Now lower the index α in (4.13) and multiply it by $\eta_{\beta\gamma\epsilon\zeta}\Theta^\zeta$. Use of (4.14) and (4.16) and a relabelling of indices enables the answer to be expressed as

$$\eta_{\alpha\beta\gamma\delta}\Theta^\gamma(s^\delta + T^{\delta\epsilon}\Theta_\epsilon + \rho^\delta\Phi + I^{\delta\epsilon}\Xi_\epsilon) + 2F_{\alpha\beta}\Theta^\gamma\Psi_\gamma$$
$$= 2F_{\alpha\gamma}\Theta^\gamma\Psi_\beta - \Xi_\alpha I^*_{\beta\gamma}\Theta^\gamma. \quad (4.17)$$

When (4.17) is symmetrized on α and β it gives

$$\Xi_\alpha I^*_{\beta\gamma}\Theta^\gamma + I^*_{\alpha\gamma}\Theta^\gamma\Xi_\beta = 2\Psi_\alpha F_{\beta\gamma}\Theta^\gamma + 2F_{\alpha\gamma}\Theta^\gamma\Psi_\beta. \quad (4.18)$$

The relation (4.18) is possible only if both Ξ_α and $I^*_{\alpha\gamma}\Theta^\gamma$ are linearly dependent on Ψ_α and $F_{\alpha\gamma}\Theta^\gamma$. There thus exist scalars a, b, c, d such that

$$\left.\begin{array}{l} \Xi_\alpha = a\Psi_\alpha + bF_{\alpha\beta}\Theta^\beta, \\ I^*_{\alpha\beta}\Theta^\beta = c\Psi_\alpha + dF_{\alpha\beta}\Theta^\beta. \end{array}\right\} \quad (4.19)$$

Substitution of these expressions back into (4.18) shows that unless all the four vectors occurring in (4.19) are parallel, we must have

$$ac = bd = 0, \quad ad + bc = 2. \quad (4.20)$$

It follows that either $\quad a = d = 0, \quad bc = 2 \quad (4.21)$

or $\quad\quad\quad\quad\quad b = c = 0, \quad ad = 2. \quad (4.22)$

These two possibilities can be considered as corresponding to two distinct classes of theoretical materials. The degenerate situation in which the vectors are all parallel can occur in both classes.

Materials of the second class have strange electromagnetic properties which enable us to reject them as theoretical models of normal fluids. To see this, consider a weak electromagnetic field in such a fluid at rest. Since the field is weak, Θ^α can be taken as having its value for zero field. This is given by (4-4.29) so that Θ^α and v^α are parallel. In the rest frame of the fluid the second of the relations (4.19) then reduces to

$$\mathbf{H} = d\mathbf{E} \quad (4.23)$$

for a material in which (4.22) holds. The field within a normal non-conducting fluid does not differ as radically as this from the vacuum behaviour given by (1.3). We shall thus develop the theory further only for the first class of materials, in which (4.21) holds. For such materials let us set

$$b = 2\lambda, \quad c = \lambda^{-1} \quad (4.24)$$

which defines λ as a scalar function of the local state in equilibrium. In virtue of (4.14) the relations (4.19) can now be expressed as

$$\Xi_\alpha = 2\lambda F_{\alpha\beta}\Theta^\beta, \quad \Psi^{\alpha\beta\gamma} = 3\lambda\Theta^{[\alpha}I^{\beta\gamma]}. \quad (4.25)$$

Use of (4.25) enables (4.17) to be reduced to the simpler form

$$\eta_{\alpha\beta\gamma\delta}\,\Theta^\gamma(s^\delta + T^{\delta\epsilon}\Theta_\epsilon + \rho^\delta\Phi + I^{\delta\epsilon}\Xi_\epsilon) = 0. \qquad (4.26)$$

This is equivalent to the existence of a scalar p such that

$$s^\alpha + T^{\alpha\beta}\Theta_\beta + \rho^\alpha\Phi + I^{\alpha\beta}\Xi_\beta = p\Theta^\alpha. \qquad (4.27)$$

Substitution from (4.25) and (4.27) into (4.11) gives

$$(T^{\alpha[\beta} - pg^{\alpha[\beta} + 2\lambda F^\alpha_{\cdot\delta}\,I^{\delta[\beta})\,\Theta^{\gamma]} = 0. \qquad (4.28)$$

This in turn is equivalent to the existence of a vector κ^α such that

$$T^{\alpha\beta} - pg^{\alpha\beta} + 2\lambda F^\alpha_{\cdot\gamma}\,I^{\gamma\beta} = \kappa^\alpha\Theta^\beta. \qquad (4.29)$$

The vector κ^α is not arbitrary since $T^{\alpha\beta}$ is symmetric. If we anti-symmetrize (4.29) on α and β we see that it must satisfy

$$\kappa^{[\alpha}\Theta^{\beta]} = 2\lambda F^{\cdot[\alpha}_\gamma I^{\beta]\gamma}. \qquad (4.30)$$

However, this is the only remaining restriction implied by Lorentz invariance. To see this suppose that $F_{\alpha\beta}$, $I^{\alpha\beta}$ and ρ^α are specified. Choose p, λ, Φ, Θ^α and κ^α arbitrarily, subject only to (4.30). From these construct successively $T^{\alpha\beta}$ by (4.29), Ξ_α and $\Psi^{\alpha\beta\gamma}$ by (4.25), and s^α by (4.27). Then $T^{\alpha\beta}$ and $\Psi^{\alpha\beta\gamma}$ are respectively symmetric and totally antisymmetric and (4.11) is satisfied identically, as required.

4c Provisional Gibbs relations

So far we have only considered the implications of (4.10) when the increments are induced by an infinitesimal Lorentz transformation. We now turn to its consequences for an arbitrary variation which preserves equilibrium. For this purpose we shall regard ρ^α, p, $F_{\alpha\beta}$, $I^{\alpha\beta}$, Θ^α, Φ, κ^α and λ as the variables which specify an equilibrium state. In terms of these, $T^{\alpha\beta}$ is given by (4.29) as

$$T^{\alpha\beta} = \kappa^\alpha\Theta^\beta + pg^{\alpha\beta} - 2\lambda F^\alpha_{\cdot\gamma}\,I^{\gamma\beta} \qquad (4.31)$$

which is symmetric in virtue of (4.30). Similarly, use of (4.25) and (4.31) enables us to deduce from (4.27) that

$$s^\alpha = -\,\Phi\rho^\alpha - \kappa_\beta\Theta^\beta\,\Theta^\alpha. \qquad (4.32)$$

First we use (4.25) to express (4.10) in the form

$$ds^\alpha = 2I^{\alpha\beta}F_{\beta\gamma}\Theta^\gamma d\lambda - \Theta_\beta\,d(T^{\beta\alpha} + 2\lambda F^\beta_{\cdot\gamma}I^{\gamma\alpha}) - \Phi\,d\rho^\alpha - \lambda\Theta^\alpha\!I^{\beta\gamma}dF_{\beta\gamma}. \qquad (4.33)$$

We have here reverted to the use of differentials ds^α, etc. in place of the increments δs^α, etc. of (4.10), but this is only a change of notation. Substitution from (4.31) and (4.32) into (4.33) now gives

$$\Theta^\alpha(dp - \kappa_\beta d\Theta^\beta + \lambda I^{\beta\gamma} dF_{\beta\gamma}) = \rho^\alpha d\Phi + 2I^{\alpha\beta} F_{\beta\gamma} \Theta^\gamma d\lambda. \qquad (4.34)$$

This is the analogue for the present situation of the result (4-4.22) which was obtained in the absence of an electromagnetic field. As with that equation, it gives rise to several distinct possibilities. For simplicity, this time we shall develop only the one of most physical interest. We thus reject the possibility of Φ being identically constant for the reasons given in §4-4b.

Another unphysical case corresponds to the vectors

$$\Theta^\alpha \quad \text{and} \quad I^{\alpha\beta} F_{\beta\gamma} \Theta^\gamma \qquad (4.35)$$

being identically parallel. Such parallelism can be expressed as the identity
$$\eta_{\alpha\beta\gamma\delta} \Theta^\gamma I^{\delta\epsilon} F_{\epsilon\zeta} \Theta^\zeta = 0. \qquad (4.36)$$

Use of (4.15) and (2-5.33) enables (4.36) to be put in the form

$$F_{\gamma[\alpha} I^*_{\beta]\delta} \Theta^\gamma \Theta^\delta = 0, \qquad (4.37)$$

which is equivalent to the parallelism of

$$F_{\alpha\beta} \Theta^\beta \quad \text{and} \quad I^*_{\alpha\beta} \Theta^\beta. \qquad (4.38)$$

This has already been rejected as an identity in §4b in connexion with (4.19).

Since $d\Phi$ is not identically zero, it follows from (4.34) that there exist scalars χ, ξ such that

$$\rho^\alpha = \chi \Theta^\alpha - 2\xi I^{\alpha\beta} F_{\beta\gamma} \Theta^\gamma. \qquad (4.39)$$

As the vectors (4.35) are not identically parallel, these two scalars are uniquely defined functions of the local state in equilibrium. Substitution of (4.39) into (4.34) now enables that equation to be separated into the two scalar differential relations

$$d\lambda = \xi d\Phi \qquad (4.40)$$

and
$$dp = \kappa_\alpha d\Theta^\alpha + \chi d\Phi - \lambda I^{\alpha\beta} dF_{\alpha\beta}. \qquad (4.41)$$

These are the Gibbs relations for our present theoretical model.

It will be seen in the next section that ξ is in fact identically zero. This will enable a number of simplifications to be made in the theory, and in this sense the forms (4.40) and (4.41) are only provisional. They will be developed further in §6a. For the time being we shall continue to work with the possibility of a nonzero ξ.

The Gibbs relations (4.40) and (4.41) show that there must exist functional relations

$$\lambda = \lambda(\Phi), \quad p = p(\Theta^\alpha, \Phi, F_{\alpha\beta}). \tag{4.42}$$

These are *fundamental equations* for the fluid in the sense of §4-4b. Once they are known, use of (4.39), (4.40) and (4.41) determines ρ^α, $I^{\alpha\beta}$ and κ^α also as functions of Θ^α, Φ and $F_{\alpha\beta}$. The functions so obtained satisfy the constraint equation (4.30) identically in consequence of the scalar character of p. To see this, it is only necessary to apply (4.41) to increments generated by an infinitesimal Lorentz transformation in accordance with (4-4.10). Equation (4.30) is then seen to be precisely the condition for the function $p(\Theta^\alpha, \Phi, F_{\alpha\beta})$ of (4.42) to be scalar-valued.

5   Interpretation of the conditions for global equilibrium

5a   The stationary nature of the equilibrium state

The results of §4 take the theory of the equilibrium state as far as is possible on the basis of the conditions (4.6) for local equilibrium alone. To continue further we must study the additional conditions (4.9) which characterize the global state. Let us begin by expressing the conditions on Ξ_α and $\Psi^{\alpha\beta\gamma}$ in an alternative form. It follows from (4.40) and (4.9) that

$$\partial_\alpha \lambda = 4\pi z \xi \Xi_\alpha. \tag{5.1}$$

With the aid of (4-6.18), (4.4), (4.39) and (5.1) we can deduce from (4.25) that

$$\partial_{[\alpha} \Xi_{\beta]} = -\lambda L_\Theta F_{\alpha\beta}, \quad \partial_\gamma \Psi^{\alpha\beta\gamma} = \lambda L_\Theta I^{\alpha\beta}. \tag{5.2}$$

Now (4.4) holds for every state of the system, (4.39) is a condition on the local state in equilibrium, and (5.1) follows from the first pair of equations in (4.9). The equations (5.2) thus show that as restrictions on the global state in equilibrium, the second pair of equations in (4.9) is equivalent to

$$L_\Theta F_{\alpha\beta} = 0, \quad L_\Theta I^{\alpha\beta} = 0. \tag{5.3}$$

More simply, we also see from (4.9) and (4.25) that

$$L_\Theta g_{\alpha\beta} = 0, \quad L_\Theta \Phi = 0. \tag{5.4}$$

But the results of §4c show that the entire equilibrium state is determined by a knowledge throughout it of Θ^α, Φ, $F_{\alpha\beta}$ and of course also the metric tensor $g_{\alpha\beta}$. The equations (5.3) and (5.4) show that all these variables are dragged along by Θ^α in the sense of §4-6b. It follows that this must also be true of every other physical variable for the system. Every such variable thus has vanishing Lie derivative with respect to Θ^α. In particular this must hold for the density ρ of inert mass, which must satisfy

$$L_\Theta \rho \equiv \Theta^\alpha \partial_\alpha \rho = 0. \qquad (5.5)$$

Consider now a rigid vessel containing a fluid in equilibrium. When this is considered from a spacetime viewpoint, the fluid possesses a bounding hypersurface Σ which represents the inner surface of the container. Since ρ is discontinuous at Σ, it follows from (5.5) that at Σ, Θ^α must be tangential to Σ. Since the fluid cannot flow through the walls of the container, the flux ρ^α of inert mass must also be tangential to Σ. It thus follows from (4.39) either that $\xi = 0$ or that $I^{\alpha\beta} F_{\beta\gamma} \Theta^\gamma$ is tangential to Σ. We know from experience that containers can be made out of materials that are impervious to matter but not to electromagnetic fields, so that $I^{\alpha\beta} F_{\beta\gamma} \Theta^\gamma$ cannot always be tangential to Σ. Hence

$$\xi \equiv 0. \qquad (5.6)$$

Equations (4.39) and (4.32) now show that both Θ^α and s^α are parallel to ρ^α, and thus also to the velocity v^α. The same situation was found in §4-4b. As in that section, it can be used to define the specific entropy s and absolute temperature T of the fluid by setting

$$\Theta^\alpha = \Theta v^\alpha, \quad s^\alpha = s\rho^\alpha, \quad T = \Theta^{-1}. \qquad (5.7)$$

Equation (4-4.62) is still valid as (4.9) includes (4-4.50) from which it is derived. The velocity and temperature distribution in equilibrium will thus again be given by (4-4.66). Dragging along by Θ^α now displaces each point x along the material orbit through x. The invariance property of the physical variables that was proved above thus shows that every physical variable, including the electromagnetic field tensors, is carried along with the fluid. Equilibrium is thus indeed a steady state of both the fluid and the electromagnetic field, in agreement with the general discussion of §4-3.

5b *The parameter λ as a universal constant*

It follows from (4.40) and (5.6) that λ has the same value for every equilibrium state of a given fluid. This raises the question of whether λ is a characteristic of the fluid material, as is the case with the charge/ mass ratio z defined by (4.1), or whether it is a universal constant that has the same value for every fluid. To investigate this, let us consider an equilibrium state of two different fluids in contact along some hypersurface Σ, such as a bubble of gas within a liquid. Each of the universal laws (4.4) gives rise to a boundary condition at Σ. These conditions may be obtained by the natural generalization to four dimensions of the method used in §2c to prove the continuity of $\mathbf{n} \cdot \mathbf{B}$ and $\mathbf{n} \cdot \mathbf{D}$ across S_0. If n_α is the unit normal to Σ, we find in this way that

$$n_\beta T^{\alpha\beta}, \quad n_\alpha \rho^\alpha, \quad n_{[\alpha} F_{\beta\gamma]}, \quad n_\beta I^{\alpha\beta} \tag{5.8}$$

must all be continuous across Σ.

Since Σ must be comoving, we have

$$n_\alpha v^\alpha = 0 \tag{5.9}$$

as in (2.34). The continuity of $n_\alpha \rho^\alpha$ is thus trivial as it vanishes on both sides of Σ. However, each of the remaining three quantities in (5.8) gives rise to a real physical condition. Two of these have also been obtained in §2d by three-dimensional considerations. It will now be shown that the three of them together imply the continuity of λ across Σ.

In virtue of (5.7) and (5.9), multiplication of (4.31) by n_β and lowering of the index α gives

$$T_{\alpha\beta} n^\beta = p n_\alpha - 2\lambda F_{\alpha\gamma} I^{\gamma\beta} n_\beta \tag{5.10}$$

from which it follows that

$$n_{[\delta} T_{\alpha]\beta} n^\beta = -3\lambda n_{[\delta} F_{\alpha\gamma]} I^{\gamma\beta} n_\beta. \tag{5.11}$$

Both the left-hand side and the coefficient of λ on the right-hand side are continuous across Σ by (5.8), and in general they will not be zero. As (5.11) must hold on both sides of Σ, it follows that λ itself must be continuous across Σ. It thus has the same value for all fluids, and so is a universal constant.

To determine this constant value let us consider the limiting case of an unpolarizable (but possibly charged) fluid. Since $I^{\alpha\beta}$ and $F^{\alpha\beta}$ are then identical, we see from (4.41) that

$$\partial p / \partial F_{\alpha\beta} = -\lambda F^{\alpha\beta}. \tag{5.12}$$

This integrates to give

$$p(\Theta^\alpha, \Phi, F_{\alpha\beta}) = p_0(\Theta^\alpha, \Phi) - \tfrac{1}{2}\lambda F^{\alpha\beta} F_{\alpha\beta} \tag{5.13}$$

for some scalar function $p_0(\Theta^\alpha, \Phi)$. However, a scalar function of Θ^α and Φ must be independent of the direction of Θ^α. Hence p_0 is in fact a function only of the scalar parameters Θ and Φ. It now follows from (4.41) and (5.13) that

$$\kappa_\alpha = -v_\alpha \partial p_0 / \partial \Theta, \quad \chi = \partial p_0 / \partial \Phi. \tag{5.14}$$

Substitution from (5.6), (5.7), (5.13) and (5.14) into (4.39) and (4.31) gives

$$\rho^\alpha = \Theta v^\alpha \partial p_0 / \partial \Phi \tag{5.15}$$

and

$$T^{\alpha\beta} = T_{\mathrm{f}}^{\alpha\beta} + T_{\mathrm{em}}^{\alpha\beta} \tag{5.16}$$

where

$$T_{\mathrm{f}}^{\alpha\beta} = p_0 g^{\alpha\beta} - \Theta v^\alpha v^\beta \partial p_0 / \partial \Theta \tag{5.17}$$

and

$$T_{\mathrm{em}}^{\alpha\beta} = 2\lambda (F^\alpha_{\ \gamma} F^{\beta\gamma} - \tfrac{1}{4} g^{\alpha\beta} F^{\gamma\delta} F_{\gamma\delta}). \tag{5.18}$$

Comparison of (5.15) and (5.17) with (4-4.46) shows that ρ^α and $T_{\mathrm{f}}^{\alpha\beta}$ are identical with the equilibrium values of the flux of inert mass, and of the energy–momentum tensor, for a fluid with pressure p_0, inverse temperature Θ and thermal potential Φ. Equation (5.16) thus shows that the total energy–momentum tensor separates naturally into the sum of two contributions, one $(T_{\mathrm{f}}^{\alpha\beta})$ from the fluid alone and one $(T_{\mathrm{em}}^{\alpha\beta})$ from the electromagnetic field alone. It generalizes the similar separation obtained in §3 for the case of a charged dust cloud. In the limiting case in which the matter is absent, we are left with $T_{\mathrm{em}}^{\alpha\beta}$ as the energy–momentum tensor of the electromagnetic field in vacuo. This tensor was also found in §3 to be given by (3.6). The two expressions agree if and only if

$$\lambda = 1/8\pi. \tag{5.19}$$

This is thus the required value of the universal constant λ.

The form (5.18) for the energy–momentum tensor of the electromagnetic field in vacuo has been derived without any explicit use of the Lorentz force law. This law has been used only to determine the behaviour of the electromagnetic field variables under a Lorentz transformation. Since the Lorentz force law describes the basic interaction between the electromagnetic field and matter, this is a remarkable achievement of the thermodynamic method. We could in fact continue beyond (5.18) and 'derive' the Lorentz force law for a charged but unpolarizable perfect fluid. Recall from §4-10 that in a perfect fluid the local state never departs from equilibrium. Consequently (5.15) to (5.18) hold always. By using (4.4) with $I^{\alpha\beta} = F^{\alpha\beta}$ we can deduce from (5.16) and (5.18) that

$$\partial_\beta T_{\mathrm{f}}^{\alpha\beta} = 8\pi\lambda z F^\alpha_{\ \beta} \rho^\beta. \tag{5.20}$$

Note next that the equations (4.4) are invariant under a simultaneous rescaling of z and $F_{\alpha\beta}$ by the same factor k. To preserve (5.20) such a rescaling must multiply λ by k^{-2}. There are thus only three distinct possibilities for λ, namely $\lambda > 0$, $\lambda = 0$ and $\lambda < 0$. Any positive (negative) value can be converted into any other by such a rescaling, i.e. by a suitable change in the unit of electric field. Experiment easily confirms that $\lambda > 0$. The unit of electric field can then be chosen to give (5.19), which puts (5.20) in the form

$$\partial_\beta T_f^{\alpha\beta} = z F^\alpha{}_{.\beta} \rho^\beta. \tag{5.21}$$

For a charged but unpolarizable perfect fluid, this is the continuum form of the Lorentz force law. The form for a particle can be deduced if desired by integrating (5.21) across a spacelike section of a small body composed of such a fluid material.

6   The Kelvin variables

6a   *The Gibbs relations reconsidered*

To make contact with the classical treatments of the thermodynamics of polarizable media, we need to recast the Gibbs relation (4.41) into a more familiar form. The first step to be taken is the evaluation of κ^α in terms of the specific internal energy u. For this purpose it is convenient to decompose $F^{\alpha\beta}$ and $I^{\alpha\beta}$ with respect to the velocity v^α by setting

$$\left.\begin{array}{ll} E^\alpha \equiv F^{\alpha\beta} v_\beta, & B^{\alpha\beta} \equiv F^{\alpha\beta} - 2v^{[\alpha} E^{\beta]}, \\ D^\alpha \equiv I^{\alpha\beta} v_\beta, & H^{\alpha\beta} \equiv I^{\alpha\beta} - 2v^{[\alpha} D^{\beta]}. \end{array}\right\} \tag{6.1}$$

These new tensors are all orthogonal to v_α, and both $B^{\alpha\beta}$ and $H^{\alpha\beta}$ are antisymmetric. We see from (2.31) and (2.32) that in the instantaneous rest frame of a fluid element, E^α and D^α are given by

$$E^\alpha = (c^{-1}\mathbf{E}, 0) \quad \text{and} \quad D^\alpha = (c^{-1}\mathbf{D}, 0) \tag{6.2}$$

while $B_{\alpha\beta}$ and $H^{\alpha\beta}$ are given by (2.31) and (2.32) respectively with the electric terms omitted.

Multiplication of (4.31) by $v_\alpha v_\beta$ with the use of (4-5.1), (5.7) and (5.19) now gives

$$\Theta \kappa^\alpha v_\alpha = (1/4\pi) E_\alpha D^\alpha - \rho(1+u) - p. \tag{6.3}$$

If we also multiply (4.30) by v_β and use (6.3), we find that

$$\Theta \kappa^\alpha = v^\alpha [\rho(1+u) + p - (1/4\pi) E_\beta D^\beta] + \rho g^\alpha \tag{6.4}$$

where

$$g^\alpha \equiv (1/2\pi\rho) F^{[\alpha}_{.\gamma} I^{\beta]\gamma} v_\beta. \tag{6.5}$$

Note that (6.4) is consistent with (6.3) since

$$g^\alpha v_\alpha = 0. \tag{6.6}$$

Similar expressions for χ and Φ follow immediately from (4.39) and (4.32) when use is made of (5.6), (5.7) and (6.3). We obtain

$$\chi = \rho T \tag{6.7}$$

and
$$\Phi = T^{-1}[1 + u + pv - sT - (v/4\pi) E_\alpha D^\alpha] \tag{6.8}$$

where we have reintroduced the specific volume $v = 1/\rho$ of (4-4.38). It is straightforward to verify from (6.1) and (6.5) that

$$I^{\alpha\beta} dF_{\alpha\beta} = H^{\alpha\beta} dB_{\alpha\beta} - 2D^\alpha dE_\alpha + 8\pi\rho g^\alpha dv_\alpha. \tag{6.9}$$

Substitution of the above expressions into (4.41) yields

$$du = T\,ds - p\,dv + (1/4\pi)[E^\alpha d(vD_\alpha) + \tfrac{1}{2}vH^{\alpha\beta} dB_{\alpha\beta}] \tag{6.10}$$

and into (4.31) it gives

$$T^{\alpha\beta} = [\rho(1 + u) + p - (1/4\pi) E_\gamma D^\gamma] v^\alpha v^\beta + pg^{\alpha\beta} + \rho g^\alpha v^\beta + (1/4\pi) F^\alpha_{\cdot\gamma} I^{\beta\gamma}. \tag{6.11}$$

Most of the variables which occur in the Gibbs relation (6.10) have a simple physical significance. Apart from factors of c, the electromagnetic field variables E^α, D^α, $B^{\alpha\beta}$, $H^{\alpha\beta}$ reduce in the local rest frame of the fluid to the familiar three-dimensional fields **E**, **D**, **B**, **H** of (1.1) and (1.2). The specific volume v and specific entropy s are directly related to the flux ρ^α of inert mass and the entropy flux vector s^α, both of which are primary concepts of the theory, by

$$\rho^\alpha = (1/v) v^\alpha, \quad s^\alpha = s\rho^\alpha. \tag{6.12}$$

The absolute temperature T is significant since it is governed by a global equilibrium condition

$$\partial_{(\alpha} \Theta_{\beta)} = 0, \quad \Theta_\alpha = (1/T) v_\alpha \tag{6.13}$$

which is independent of every other physical variable except the velocity v^α. The remaining two variables are the specific internal energy u and the thermostatic pressure p. These are precisely defined, as they can be considered as given by (6.11) in terms of the energy–momentum tensor $T^{\alpha\beta}$. However, like $T^{\alpha\beta}$ itself they are properties of the system as a whole, i.e. of the fluid and electromagnetic field together. This is true even when the fluid and field do not interact, as with a neutral non-polarizable fluid. In particular, they do not vanish in vacuo. It would be more natural from a physical viewpoint if they could be replaced by alternative variables u_0 and p_0 which reduced, for the non-interacting case, respectively to the specific internal energy and the thermostatic pressure of the fluid alone. Such variables *would* vanish in vacuo.

It follows from (5.13) that p_0 and p must be related by

$$p_0 = p + (1/16\pi) F^{\alpha\beta} F_{\alpha\beta} \tag{6.14}$$

when the fluid and field do not interact. In this case we also know that $F^{\alpha\beta} = I^{\alpha\beta}$. There is thus no unique method of generalizing (6.14) to the interacting situation. The product $F^{\alpha\beta}F_{\alpha\beta}$ in (6.14) could be replaced by $F^{\alpha\beta}I_{\alpha\beta}$ or $I^{\alpha\beta}I_{\alpha\beta}$ in this general case and all three choices super-ficially appear equally natural. But if we want p_0 to be as significant from a thermodynamic viewpoint as is p, we must keep to (6.14) as it stands. This is the only choice for which

$$p_0 = p_0(\Theta^\alpha, \Phi, F_{\alpha\beta}) \qquad (6.15)$$

is a fundamental equation for the system, as it is the only one which preserves the general form of the Gibbs relation (4.41). With p_0 defined by (6.14) for the general case, we find from (4.41) and (2.38) that

$$dp_0 = \kappa_\alpha \, d\Theta^\alpha + \chi \, d\Phi + \tfrac{1}{2}P^{\alpha\beta} \, dF_{\alpha\beta} \qquad (6.16)$$

where $P^{\alpha\beta}$ is the polarization tensor. This shows clearly that p_0 depends on $F_{\alpha\beta}$ only when the fluid is polarizable. Following de Groot & Suttorp (1972, Ch. II, §8a) we shall call p_0 the *Kelvin pressure* of the fluid.

To find a definition for u_0, return to the non-interacting case. We find from (4-5.1), (5.16), (5.18) and (6.1) that in this case u_0 and u must be related by

$$u_0 = u - (v/8\pi)(E^\alpha E_\alpha + \tfrac{1}{2}B^{\alpha\beta}B_{\alpha\beta}). \qquad (6.17)$$

If we adopt this in the general case as the definition of a variable u_1, thus

$$u_1 \equiv u - (v/8\pi)(E^\alpha E_\alpha + \tfrac{1}{2}B^{\alpha\beta}B_{\alpha\beta}), \qquad (6.18)$$

then we find that (6.10) can be put in the form

$$du_1 = T \, ds - p_0 \, dv + E^\alpha \, d(vP_\alpha) - \tfrac{1}{2}vM^{\alpha\beta} \, dB_{\alpha\beta}. \qquad (6.19)$$

Here P^α and $M^{\alpha\beta}$ are defined by setting

$$D^\alpha = E^\alpha + 4\pi P^\alpha, \quad B^{\alpha\beta} = H^{\alpha\beta} + 4\pi M^{\alpha\beta} \qquad (6.20)$$

in analogy with (2.24). These variables are related to the polarization tensor $P^{\alpha\beta}$ of (2.39) by

$$P^\alpha = -P^{\alpha\beta}v_\beta, \quad M^{\alpha\beta} = P^{\alpha\beta} + 2v^{[\alpha}P^{\beta]}. \qquad (6.21)$$

In the instantaneous rest frame of a fluid element,

$$P^\alpha = (c^{-1}\mathbf{P}, 0) \qquad (6.22)$$

and $M^{\alpha\beta}$ is given by (2.39) when the terms in $\mathbf{P}$ are omitted. They thus describe respectively the electric polarization and the magnetization of the element as seen in its local rest frame.

The methods of statistical physics have been applied to a polarizable fluid by de Groot & Suttorp (1972). The techniques currently available do not permit a relativistic treatment of the entropy law to be given within statistical physics, but they do allow a relativistic calculation of the energy–momentum tensor from microscopic considerations. They obtain in this way (*loc. cit.* Ch. V eqn. (249)) an expression for $T^{\alpha\beta}$ which is identical with that obtained when (6.14) and (6.18) are substituted into (6.11). By a non-relativistic treatment (*loc. cit.* Ch. II eqn. (251)) they also obtain the Gibbs relation (6.19). This agreement confirms that the polarizable simple fluid defined in §4*a* is a valid theoretical model of real polarizable fluids in situations of local equilibrium.

Although the calculations of de Groot & Suttorp make u_1 seem the most natural internal energy for the fluid from the viewpoint of statistical physics, it is not also the most natural from a thermodynamic viewpoint. Thermodynamically it seems more natural to use an internal energy variable u_0 which is related to the Kelvin pressure p_0 in such a way that the relationship

$$\Phi = T^{-1}(1 + u_0 + p_0 v - Ts) \qquad (6.23)$$

holds identically. Equation (4-4.43) shows that (6.23) is true when the fluid and electromagnetic field do not interact, and in the interacting case all the other variables that occur in it are naturally associated purely with the fluid. As it is complementary to the Kelvin pressure p_0, it is natural to call the u_0 so defined the *Kelvin internal energy* of the fluid. We find with the aid of (6.8), (6.14) and (6.20) that it is related to the u_1 of (6.18) by

$$u_0 = u_1 - vE^\alpha P_\alpha. \qquad (6.24)$$

This shows that u_0 and u_1 are identical for a non-polarizable fluid, as is required. It enables us to write the Gibbs relation (6.19) in the form

$$du_0 = T\,ds - p_0\,dv - v(P^\alpha\,dE_\alpha + \tfrac{1}{2}M^{\alpha\beta}\,dB_{\alpha\beta}) \qquad (6.25)$$

which treats the electric and magnetic variables in a more symmetrical manner than does (6.19) itself. The expression (6.11) for the energy–momentum tensor can also be written in terms of the Kelvin variables, as

$$T^{\alpha\beta} = [\rho(1 + u_0) + p_0]\,v^\alpha v^\beta + p_0\,g^{\alpha\beta} + \rho g^{\alpha\beta} + \frac{1}{4\pi}\,(F^\alpha_{\cdot\gamma}I^{\beta\gamma} - \tfrac{1}{4}F^{\gamma\delta}F_{\gamma\delta}g^{\alpha\beta}).$$

$$(6.26)$$

Any representation of p, u and $T^{\alpha\beta}$ as the sum of a fluid contribution and an electromagnetic field contribution is purely conventional. From a mathematical viewpoint it is also unnecessary. The Gibbs relations (6.10), (6.19) and (6.25) are all correct and none of them has any absolute claim to be more significant than the others. What is important for theoretical purposes is that any Gibbs relation must be accompanied by a corresponding expression for the total energy-momentum tensor of the fluid and electromagnetic field together. However, if it is desired to adopt some conventional representation of this sort, the discussion given above shows that the Kelvin variables p_0 of (6.14) and u_0 of (6.24) have good reason to be considered as the contributions due to the fluid. The terms

$$T_0^{\alpha\beta} \equiv [\rho(1+u_0)+p_0]\,v^\alpha v^\beta + p_0\,g^{\alpha\beta}$$

in (6.26) can then be considered as the fluid contribution to $T^{\alpha\beta}$. Finally let us observe that (6.4) enables (4.30) to be reduced to the form

$$E^{[\alpha}D^{\beta]} + B_\gamma^{\ [\alpha}H^{\beta]\gamma} = 0. \tag{6.27}$$

By using (6.27) we can check that the expressions (6.11) and (6.26) for $T^{\alpha\beta}$ are symmetric, as required.

6b Global equilibrium reconsidered

It is instructive also to investigate the conditions for global equilibrium in more detail. Both (4.4) and (4.9) must be satisfied in equilibrium, but these conditions are not all independent. Let us start by eliminating those that are redundant.

We know from (4.9), (4.25), (5.3), (5.4) and (5.19) that in equilibrium

$$L_\Theta\,g_{\alpha\beta} = 0, \quad L_\Theta\,F_{\alpha\beta} = 0, \quad \partial_\alpha\Phi = zF_{\alpha\beta}\Theta^\beta. \tag{6.28}$$

The first of these equations is equivalent to (4-4.50), and so (4-4.61) still holds with $\Omega_{\alpha\beta}$ being the constant antisymmetric tensor that was studied in detail in §4-4d. The Maxwell equations

$$\partial_{[\alpha}F_{\beta\gamma]} = 0, \quad \partial_\beta I^{\alpha\beta} = 4\pi z\rho^\alpha \tag{6.29}$$

of (4.4) also hold. It was seen in §5a that in virtue of (6.28) and the Gibbs relation (4.41), all physical variables for the system must have vanishing Lie derivative with respect to Θ^α. In particular this is true for κ_α, so that

$$L_\Theta\,\kappa_\alpha \equiv \Theta^\beta\partial_\beta\kappa_\alpha + \kappa_\beta\,\partial_\alpha\Theta^\beta = 0. \tag{6.30}$$

If we operate on (4.31) with ∂_β and use these equations, we find that

$$\partial_\beta T_\alpha^{\ \beta} = \partial_\alpha p - \kappa_\beta\,\partial_\alpha\Theta^\beta - zF_{\alpha\beta}\rho^\beta + (1/8\pi)\,I^{\beta\gamma}\partial_\alpha F_{\beta\gamma}. \tag{6.31}$$

Since $$\chi\partial_\alpha\Phi = zF_{\alpha\beta}\rho^\beta \tag{6.32}$$

by (6.28) with (6.7) and (6.13), we see that the right-hand side of (6.31) vanishes in virtue of (4.41). By also treating $\partial_\alpha\rho^\alpha$ in a similar but much simpler fashion, we can thus deduce that

$$\partial_\beta T^{\alpha\beta} = 0, \quad \partial_\alpha\rho^\alpha = 0 \tag{6.33}$$

are consequences of (6.28) and (6.29) and the properties of the local state in equilibrium. The equations (6.28) and (6.29) thus either include or imply each of the equations (4.4). They also imply the equations of (4.9) that they do not include, as was shown in §5a. It follows that (6.28) and (6.29) together are sufficient to characterize global equilibrium. It is sometimes convenient to include also the second equation of (5.3), but it is in fact redundant.

We have seen that the local properties of equilibrium are completely determined by the fundamental equation

$$p = p(\Theta^\alpha, \Phi, F_{\alpha\beta}) \tag{6.34}$$

of (4.42) and the corresponding Gibbs relation (4.41). The equations (6.28) govern the equilibrium variation throughout the system of the independent variables in (6.34). There is one equation for each variable, as the first of them is essentially a condition on Θ^α. The combination of (6.28) with (6.34) thus forms a description of the global states of equilibrium which is very natural from a mathematical viewpoint, but the variables that it uses are not very familiar ones. Let us now restrict attention to the most common situation, a static equilibrium state. This corresponds to taking $\Omega_{\alpha\beta} = 0$ in (4-4.61). Such a state is not necessarily homogeneous since the electromagnetic field may not be so. For this case the equilibrium conditions can also be expressed simply in terms of the more intuitive Kelvin variables of §6a.

By using (4-6.18), (4.1), (6.13), (6.23) and (6.25) we find that (6.28) reduces for this case to

$$\partial_\alpha T = 0, \quad v^\gamma\partial_\gamma F^{\alpha\beta} = 0 \tag{6.35}$$

and $$F_{\alpha\beta}J^\beta + P^\beta\partial_\alpha E_\beta + \tfrac{1}{2}M^{\beta\gamma}\partial_\alpha B_{\beta\gamma} - \partial_\alpha p_0 = 0. \tag{6.36}$$

In the rest frame of the fluid these can be written in the three-dimensional forms

$$\partial T/\partial t = 0, \quad \partial p_0/\partial t = 0, \quad \partial \mathbf{E}/\partial t = 0, \quad \partial \mathbf{B}/\partial t = 0, \tag{6.37}$$

$$\nabla T = 0, \tag{6.38}$$

and $$\rho_e\,\mathbf{E} + (\nabla\mathbf{E})\cdot\mathbf{P} + (\nabla\mathbf{B})\cdot\mathbf{M} - \nabla(c^2 p_0) = 0 \tag{6.39}$$

where ρ_e is the electric charge density. The main result of §5a shows that every physical variable is time-independent in the rest frame of a static equilibrium state. The equations (6.37) are just special cases of this. Of the other two conditions (6.38) simply states the expected result that the temperature distribution must be uniform. However, (6.39) is more interesting. To interpret it, recall from §4-4b that the three-pressure $\bar{p}_0$ corresponding to the four-pressure p_0 is just given by $\bar{p}_0 = c^2 p_0$. Equation (6.39) can be considered as expressing mechanical equilibrium between a force density $-\nabla \bar{p}_0$ exerted by the pressure gradient of the fluid, the Lorentz force density $\rho_e \mathbf{E}$ acting on the electric charge distribution, and an additional force density

$$\mathbf{F}_K \equiv \mathbf{P} \cdot \nabla \mathbf{E} + (\nabla \mathbf{B}) \cdot \mathbf{M} \qquad (6.40)$$

exerted on the fluid due to its polarization and magnetization. This force $\mathbf{F}_K$ is known as the *Kelvin force*, although only the electrostatic term $\mathbf{P} \cdot \nabla \mathbf{E}$ is actually due to Kelvin. In writing (6.40) we have used the vector identity

$$\mathbf{P} \times (\nabla \times \mathbf{E}) = (\nabla \mathbf{E}) \cdot \mathbf{P} - \mathbf{P} \cdot \nabla \mathbf{E}, \qquad (6.41)$$

and the Maxwell equation $\nabla \times \mathbf{E} = 0$ \qquad (6.42)

given by (1.1) and (6.37), to replace $(\nabla \mathbf{E}) \cdot \mathbf{P}$ of (6.39) by $\mathbf{P} \cdot \nabla \mathbf{E}$. A similar treatment cannot be given to $(\nabla \mathbf{B}) \cdot \mathbf{M}$ as it is $\nabla \times \mathbf{H}$ rather than $\nabla \times \mathbf{B}$ which vanishes.

The interpretation of (6.39) given above depends critically on the adoption of (6.14) as the definition of the pressure within the fluid. As we have seen, this is purely a matter of convention. Consequently the interpretation of (6.40) as the force exerted on the fluid due to its polarization and magnetization is also purely conventional. In §7b we shall consider a different definition of the fluid pressure. This will lead to an alternative expression for this force, the electrostatic term of which is due to Helmholtz. The Kelvin and Helmholtz expressions are not equivalent and were originally considered as mutual rivals. This conflict was not fully resolved until the early 1950s, when it was first appreciated that the polarization force cannot be considered in isolation. The Kelvin pressure p_0 is so named not because it was introduced by Kelvin but because it is the pressure required to complement his expression for this polarization force. More detailed accounts of this controversy and its resolution may be found in the books of Penfield & Haus (1967, §8.2) and of de Groot & Suttorp (1972, Ch.2 §8).

6c Motion of a perfect polarizable fluid

A perfect fluid was defined in §4-10 as a fluid in which the local state never deviates from equilibrium. When such a fluid is polarizable, the local formulae given in §6a remain valid for all states of the fluid although the global conditions for equilibrium given in §6b are violated. In place of these global conditions we have an equation of motion. This is obtained by substituting from (6.26) into the first equation of (4.4) and using (6.9), (6.20) and (6.25) and the remaining equations of (4.4) to simplify the result. This procedure gives

$$\rho^* dv_\alpha/d\tau = F_{\alpha\beta} J^\beta + B_\alpha^\beta (P^\gamma \partial_\beta E_\gamma + \tfrac{1}{2} M^{\gamma\delta} \partial_\beta B_{\gamma\delta} - \partial_\beta p_0)$$
$$+ 2\rho v^\beta \partial_{[\alpha} g_{\beta]} - \rho T v_\alpha \, ds/d\tau \tag{6.43}$$

where $$\rho^* \equiv \rho(1 + u_0 + p_0/\rho) \tag{6.44}$$

and once again $$d/d\tau \equiv v^\alpha \partial_\alpha, \quad B_\alpha^\beta \equiv A_\alpha^\beta + v_\alpha v^\beta. \tag{6.45}$$

Note that the mixed forms of the magnetic induction tensor $B_{\alpha\beta}$ must be written as $B^\alpha{}_\beta$ and $B_\alpha{}^\beta$, so that the notation B_α^β for the projection tensor is still distinctive.

Multiplication of (6.43) by v^α shows that $ds/d\tau = 0$, so that the specific entropy s is constant along each material orbit. This was also found to be the case in §4-10 for a perfect fluid in the absence of an electromagnetic field. As in that section, we shall consider a perfect fluid to have homogeneous composition only if s is constant throughout. This again eliminates entropy from the theory, but this time the Gibbs relation (6.25) cannot be integrated explicitly as in (4-10.2) to completely remove all thermodynamic results. The coefficient ρ^* of the acceleration in (6.43) may again be interpreted as the density of inertial mass. The expression (6.44) for it is formally identical with the second of the corresponding expressions given by (4-10.4).

In virtue of (6.6) we have

$$2\rho v^\beta \partial_{[\alpha} g_{\beta]} = -\rho g_\beta \partial_\alpha v^\beta - \rho \, dg_\alpha/d\tau. \tag{6.46}$$

This contribution to (6.43) thus vanishes in static equilibrium. It is the only term on the right-hand side of (6.43) whose presence could not have been predicted from considerations of static equilibrium alone. The remaining terms all appear in (6.36), which shows that they mutually cancel in such a state. This confirms that (6.43) is consistent with the equilibrium conditions of §6b, as is necessary. The contribution (6.46) will be considered again in §7b.

7 Linear constitutive relations

7a *Electric and magnetic susceptibilities*

Only three independent scalar fields can be constructed from E_α and $B_{\alpha\beta}$. These may be taken to be

$$\theta \equiv E^\alpha E_\alpha, \quad \phi \equiv B^{\alpha\beta}B_{\alpha\beta}, \quad \psi \equiv B^{\alpha\gamma}B^\beta_{\cdot\gamma}E_\alpha E_\beta. \tag{7.1}$$

Any other scalar function of E_α and $B_{\alpha\beta}$ must be a function of θ, ϕ and ψ. To prove this it is only necessary to show that there exists a Minkowskian coordinate system (dependent on E_α and $B_{\alpha\beta}$) in which the components of E_α and $B_{\alpha\beta}$ are determined completely by these three scalars. This will now be shown.

Since both E_α and $B_{\alpha\beta}$ are orthogonal to v^α, choose the fourth basis vector of the coordinates to be parallel to v^α. The first basis vector can then be chosen so that $B_{12} = B_{13} = 0$ and the third so that $E_3 = 0$. In general this fixes the axes up to reflections. Such reflections of the axes may be used to ensure that B_{23}, E_1, and E_2 are all non-negative. Let

$$a = B_{23} = -B_{32}, \quad b = E_1, \quad c = E_2. \tag{7.2}$$

These four components of $B_{\alpha\beta}$ and E_α are the only ones which can still be nonzero. Hence (7.1) gives

$$\theta = b^2 + c^2, \quad \phi = 2a^2, \quad \psi = a^2 c^2. \tag{7.3}$$

If θ, ϕ and ψ are given, these equations have a unique solution for non-negative a, b, c unless $\phi = 0$. But when $\phi = 0$ the construction given above for the axes is not unique, since then $B_{\alpha\beta} \equiv 0$. It is possible in this case to choose the axes so that $a = c = 0$, and (7.3) then determines the non-negative b uniquely. It follows that in all cases a, b and c are determined completely by θ, ϕ and ψ as required.

The fundamental equation corresponding to (6.25) must express u_0 as a scalar function of s, v, E_α and $B_{\alpha\beta}$. The above result shows that u_0 can be considered equivalently as a function

$$u_0 = u_0(s, v, \theta, \phi, \psi). \tag{7.4}$$

It follows from (6.25) and (7.1) that then

$$\left.\begin{array}{l} P^\alpha = \kappa E^\alpha + \xi B^{\alpha\gamma}B_{\beta\gamma}E^\beta \\[4pt] M^{\alpha\beta} = \chi B^{\alpha\beta} - 2\xi B^{\gamma[\alpha}E^{\beta]}E_\gamma \end{array}\right\} \tag{7.5}$$

where $\quad \kappa = -2\rho\,\partial u_0/\partial\theta, \quad \chi = -4\rho\,\partial u_0/\partial\phi, \quad \xi = -2\rho\,\partial u_0/\partial\psi. \tag{7.6}$

It is easily checked with the aid of (6.20) that the constraint (6.27) is satisfied identically by (7.5). For most fluids, under normal conditions the parameter ξ of (7.5) is negligible while κ and χ are functions only of v and T. Then

$$P^\alpha = \kappa(v, T)\, E^\alpha, \quad M^{\alpha\beta} = \chi(v, T)\, B^{\alpha\beta} \tag{7.7}$$

which are known as the *linear constitutive relations* of the material. The parameters κ and χ will be called respectively its *electric* and *magnetic susceptibility*. This nomenclature agrees with that of de Groot & Suttorp (1972), but often the term 'magnetic susceptibility' is applied instead to $\tilde\chi \equiv \chi/(1-4\pi\chi)$. This gives $M^{\alpha\beta} = \tilde\chi H^{\alpha\beta}$ in place of the second equation of (7.7).

7b The Helmholtz variables

The *specific free energy* f_0 of a fluid is defined by

$$f_0 = u_0 - Ts. \tag{7.8}$$

From (6.25) it satisfies

$$df_0 = -p_0\, dv - s\, dT - v(P^\alpha\, dE_\alpha + \tfrac{1}{2} M^{\alpha\beta}\, dB_{\alpha\beta}), \tag{7.9}$$

so that the corresponding fundamental equation has the form

$$f_0 = f_0(v, T, E_\alpha, B_{\alpha\beta}). \tag{7.10}$$

This contains among its arguments the variables on which the κ and χ of (7.7) depend. For a fluid with linear constitutive relations it is thus a more convenient fundamental equation than that given by u_0.

Let us set

$$f^0(v, T) \equiv f_0(v, T, 0, 0). \tag{7.11}$$

For a state in which an electromagnetic field is present, f^0 is the specific free energy of a comparison field-free state which has the same values of v and T. The pressure, entropy and specific internal energy of this comparison state are defined to be the *Helmholtz variables* of the actual state. They will all be indicated by a superscript 0. We see from (7.8) and (7.9) that they can be obtained as functions of v and T from $f^0(v, T)$ by the relations

$$p^0 = -\partial f^0/\partial v, \quad s^0 = -\partial f^0/\partial T, \quad u^0 = f^0 + Ts^0. \tag{7.12}$$

The Helmholtz pressure, like the Kelvin pressure p_0, clearly reduces to the (unambiguous) pressure of the fluid when the fluid and field do not interact.

The Helmholtz pressure is well defined even in the absence of linear constitutive relations, but it is only for fluids which possess such relations that it becomes a convenient variable from a theoretical viewpoint. For such a fluid, the constitutive relations (7.7) enable (7.9) to be integrated with respect to E_α and $B_{\alpha\beta}$ to give

$$f_0(v, T, E_\alpha, B_{\alpha\beta}) = f^0 - \tfrac{1}{2}v(\kappa E^2 + \chi B^2) \qquad (7.13)$$

where
$$E^2 \equiv E^\alpha E_\alpha, \quad B^2 \equiv \tfrac{1}{2}B^{\alpha\beta}B_{\alpha\beta}. \qquad (7.14)$$

The fundamental equation (7.10) is thus completely determined by the three functions
$$f^0(v, T), \quad \kappa(v, T), \quad \chi(v, T) \qquad (7.15)$$

of the field-free equilibrium states. The relationship between p_0 and p^0 can be obtained by differentiating (7.13) with respect to v. Use of (7.9) and (7.12) shows that this gives

$$p_0 = p^0 + \tfrac{1}{2}(\kappa E^2 + \chi B^2) + \tfrac{1}{2}v\left(\frac{\partial \kappa}{\partial v}E^2 + \frac{\partial \chi}{\partial v}B^2\right). \qquad (7.16)$$

It follows that

$$\partial_\alpha p_0 - P^\beta \partial_\alpha E_\beta - \tfrac{1}{2}M^{\beta\gamma}\partial_\alpha B_{\beta\gamma} = \partial_\alpha p^0 + \tfrac{1}{2}(E^2\partial_\alpha \kappa + B^2\partial_\alpha \chi)$$
$$+ \tfrac{1}{2}\partial_\alpha\left(v\frac{\partial \kappa}{\partial v}E^2 + v\frac{\partial \chi}{\partial v}B^2\right) \qquad (7.17)$$

identically, which enables both the equilibrium condition (6.36) and the equation of motion (6.43) to be expressed in terms of p^0 rather than p_0.

The three-dimensional form (6.39) of the condition for static equilibrium can be written with the aid of (7.17) as

$$\rho_e \mathbf{E} + \mathbf{F}_H - \nabla\bar{p}^0 = 0 \qquad (7.18)$$

where

$$\mathbf{F}_H = -\frac{1}{2}\left[\mathbf{E}^2\nabla\kappa + \nabla\left(\mathbf{E}^2 v\frac{\partial \kappa}{\partial v}\right)\right] - \frac{1}{2}\left[\mathbf{B}^2\nabla\chi + \nabla\left(\mathbf{B}^2 v\frac{\partial \chi}{\partial v}\right)\right] \qquad (7.19)$$

with
$$\bar{p}^0 \equiv c^2 p^0, \quad \mathbf{E}^2 \equiv \mathbf{E}\cdot\mathbf{E}, \quad \mathbf{B}^2 \equiv \mathbf{B}\cdot\mathbf{B}. \qquad (7.20)$$

Note that the notations (7.14) and (7.20) are related by

$$\mathbf{E}^2 = c^2 E^2, \quad \mathbf{B}^2 = c^2 B^2 \qquad (7.21)$$

for a fluid at rest. The vector $\mathbf{F}_H$ is known as the *Helmholtz force*, although as with the Kelvin force (6.40) only the electrostatic term is actually due to Helmholtz. Similarly the Helmholtz pressure is so named as it is the pressure required to complement the Helmholtz force. From an experimental viewpoint (7.18) can be tested more easily than (6.39), as it is easier to measure the density ρ and calculate the Helmholtz pressure $\bar{p}^0$ from it than it is to measure the Kelvin pressure $\bar{p}_0$. The validity of (7.18) for a neutral fluid ($\rho_e = 0$) in an electrostatic field has been confirmed experimentally by Hakim & Higham (1962).

By using (6.1) we can express the definition (6.5) of g^α as

$$g^\alpha = -v(B^{\alpha\beta}P_\beta + M^{\alpha\beta}E_\beta).$$ (7.22)

When (7.7) holds, this simplifies to give

$$g^\alpha = -v(\kappa + \chi)B^{\alpha\beta}E_\beta.$$ (7.23)

Consider now the form taken by the equation of motion (6.43) for a fluid which is instantaneously at rest. With the aid of (6.46), (7.17), (7.19) and (7.23) we find for this special case that it can be put in the three-dimensional form

$$\rho^* \frac{d\mathbf{v}}{dt} = \rho_e\,\mathbf{E} + \mathbf{F}_H - \nabla\bar{p}^0 + \frac{1}{c}\frac{\partial}{\partial t}[(\kappa + \chi)\,\mathbf{E} \times \mathbf{B}].$$ (7.24)

As we have seen, the final term is the only one which cannot be confirmed by measurements in a static field. It is a relativistic effect of very small magnitude. Although it has here been derived for a fluid, a similar term would be expected also in a solid dielectric. Its presence has recently been confirmed by Walker & Lahoz (1976) in an experiment which used the effect to induce torsional oscillations in a suspended disc of barium titanate.

7c The energy–momentum tensors of Abraham and Minkowski

We saw in §6a that there is a considerable element of convention in the representation of $T^{\alpha\beta}$ as the sum of a fluid contribution and an electromagnetic field contribution. The two most well-known expressions for the electromagnetic field contribution are those proposed by Minkowski (1908b) and Abraham (1909). The Minkowski tensor is asymmetric and is given by

$$T_M^{\alpha\beta} = \frac{1}{4\pi}\left(F^\alpha_{.\gamma}I^{\beta\gamma} - \tfrac{1}{4}F_{\gamma\delta}I^{\gamma\delta}g^{\alpha\beta}\right)$$ (7.25)

while the Abraham tensor is explicitly symmetric and is given by

$$T_A^{\alpha\beta} = \frac{1}{4\pi}\left(F^{(\alpha}_{.\gamma}I^{\beta)\gamma} - \tfrac{1}{4}F_{\gamma\delta}I^{\gamma\delta}g^{\alpha\beta}\right) + \rho g^{(\alpha}v^{\beta)}$$ (7.26)

where g^α is defined by (6.5). They both differ from that suggested at the end of §6a which was based on the Kelvin variables. Both Minkowski's and Abraham's proposals were made without any consideration of the material contribution, and were intended to be universally valid for all media. The relative merits of these and of yet other alternatives have been debated ever since. A good account of this debate, together with detailed references, has been given by de Groot & Suttorp (1972, Ch.5 §7).

To understand the relationship between the Minkowski and Abraham tensors and our theory of a polarizable fluid, we need to express the total energy–momentum tensor (6.26) in terms of Helmholtz variables. Being functions of the comparison field-free state introduced in §7b, these variables obey the zero-field Gibbs relation

$$du^0 = T\,ds^0 - p^0\,dv \qquad (7.27)$$

identically, even when an electromagnetic field is present. With the aid of (7.8) and (7.13) we can deduce from (7.12) that

$$u_0 = u^0 - \tfrac{1}{2}v(\kappa E^2 + \chi B^2) + \tfrac{1}{2}vT\left(\frac{\partial \kappa}{\partial T}\,E^2 + \frac{\partial \chi}{\partial T}\,B^2\right) \qquad (7.28)$$

for a fluid with linear constitutive relations. If we substitute from (7.16) and (7.28) into (6.26) and use the identity

$$\kappa E^2 + \chi B^2 = \tfrac{1}{2}P^{\alpha\beta}F_{\alpha\beta} \qquad (7.29)$$

which follows from (6.1), (6.21) and (7.7), we obtain

$$T^{\alpha\beta} = T_1^{\alpha\beta} + T_2^{\alpha\beta} + T_3^{\alpha\beta} + T_4^{\alpha\beta} \qquad (7.30)$$

where

$$\left.\begin{aligned}
T_1^{\alpha\beta} &= \rho(1+u^0)\,v^\alpha v^\beta + p^0(g^{\alpha\beta}+v^\alpha v^\beta)\\[4pt]
T_2^{\alpha\beta} &= \tfrac{1}{2}T\left(\frac{\partial \kappa}{\partial T}E^2 + \frac{\partial \chi}{\partial T}B^2\right)v^\alpha v^\beta + \tfrac{1}{2}v\left(\frac{\partial \kappa}{\partial v}E^2 + \frac{\partial \chi}{\partial v}B^2\right)(g^{\alpha\beta}+v^\alpha v^\beta)\\[4pt]
T_3^{\alpha\beta} &= \rho g^\alpha v^\beta\\[4pt]
T_4^{\alpha\beta} &= \frac{1}{4\pi}(F^\alpha{}_{.\gamma}I^{\beta\gamma} - \tfrac{1}{4}F_{\gamma\delta}I^{\gamma\delta}g^{\alpha\beta}).
\end{aligned}\right\} \qquad (7.31)$$

Successively these contributions have a decreasing dependence on the properties of the fluid and an increasing dependence on the electromagnetic field. The tensor $T_1^{\alpha\beta}$ is a function only of the density $\rho = 1/v$, the temperature T and the velocity v^α of the fluid. As such it is indisputably a property of the fluid alone. The tensor $T_2^{\alpha\beta}$ depends explicitly on all these fluid variables, and in addition it involves the electromagnetic field through the scalars E^2 and B^2. The tensor $T_3^{\alpha\beta}$ depends on both the electromagnetic field and induction tensors $F^{\alpha\beta}$ and $I^{\alpha\beta}$, but the only fluid variable that it involves is the velocity v^α. Its apparent dependence on ρ is spurious as it cancels with a factor $1/\rho$ in the definition (6.5) of g^α. Finally $T_4^{\alpha\beta}$ depends on the two electromagnetic tensors alone.

A separation of $T^{\alpha\beta}$ purely into a fluid contribution and an electro-magnetic field contribution must involve the allocation of each of these tensors to either the fluid or the field. Clearly we must give $T_1^{\alpha\beta}$ to the fluid and $T_4^{\alpha\beta}$ to the field, but the descriptions just given show that there is no logical way of assigning $T_2^{\alpha\beta}$ and $T_3^{\alpha\beta}$ uniquely. Comparison of (7.25) and (7.26) with (7.31) shows that

$$T_{\mathrm{M}}^{\alpha\beta} = T_4^{\alpha\beta}, \quad T_{\mathrm{A}}^{\alpha\beta} = T_3^{\alpha\beta} + T_4^{\alpha\beta}. \qquad (7.32)$$

Note that $T^{\alpha\beta}$, $T_1^{\alpha\beta}$ and $T_2^{\alpha\beta}$ are all symmetric, so that the sum $T_3^{\alpha\beta} + T_4^{\alpha\beta}$ must also be so. The explicit symmetrization occurring in (7.26) is thus redundant in the present situation. We see from (7.32) that the Minkowski and Abraham proposals both correspond to giving $T_2^{\alpha\beta}$ to the fluid. They differ only in their assignment of $T_3^{\alpha\beta}$.

The remaining possibility is to give both $T_2^{\alpha\beta}$ and $T_3^{\alpha\beta}$ to the field to obtain a field contribution of

$$T_{\mathrm{G}}^{\alpha\beta} \equiv T_2^{\alpha\beta} + T_3^{\alpha\beta} + T_4^{\alpha\beta}. \qquad (7.33)$$

De Groot & Suttorp (1972, Ch.5 §6d) have pointed out that of these three schemes of assignment, that given by (7.33) is the only one with a real physical meaning. It is physically possible to switch on the field while keeping v and T constant. The tensor $T_{\mathrm{G}}^{\alpha\beta}$ gives the change produced in $T^{\alpha\beta}$ when this is done. However, the proposal (7.33) is of a somewhat different nature to those of (7.25) and (7.26). It is specific to a fluid and depends for its motivation on a knowledge of the corres-ponding material contribution $T_1^{\alpha\beta}$. The historical controversy over the Minkowski and Abraham tensors could not have arisen within a complete theory of this sort. It arose because of a belief that the energy–momentum tensor of the electromagnetic field must be physically well defined even within a polarizable medium. Once it is realized that this is a somewhat ambiguous quantity, the origin of the controversy disappears.

One final point is worth mentioning concerning the Minkowski tensor. Since it is asymmetric, the ordering of the indices is important if the flux of total four-momentum across a hypersurface is to be cor-respondingly separated into a material and a field contribution. With the momentum flux defined by (3-4.11), Minkowski adopted the ordering of indices given in (7.25). This gives a contribution from $T_3^{\alpha\beta}$ to the material momentum flux across a hypersurface element dS_α of $\rho g^\alpha (v^\beta \, dS_\beta)$, known sometimes as 'hidden momentum' as it is not associated with the motion of the material. The three-momentum density **p** associated with this by (3-6.1) is given in the rest frame of the element by

$$\mathbf{p} = -c^{-1}(\kappa + \chi)\,\mathbf{E} \times \mathbf{B}. \qquad (7.34)$$

The final term in (7.24) can thus be considered as representing a mechanical recoil produced by any change in this hidden momentum.

8   Near-equilibrium states

8a   *The comparison equilibrium state*

It follows from (6.26) and (6.6) that for a system in local equilibrium

$$\rho(1 + u_0) = \left(T^{\alpha\beta} - \frac{1}{4\pi}\,F^\alpha{}_{\cdot\gamma}I^{\beta\gamma}\right)v_\alpha v_\beta - \frac{1}{16\pi}\,F_{\alpha\beta}F^{\alpha\beta}. \qquad (8.1)$$

Every variable in this equation except u_0 is well defined for a general state of the system. It can thus be used to *define* u_0 for non-equilibrium states. The values of u_0, ρ, v^α and $F_{\alpha\beta}$ at a point within a general state then determine a unique local state of equilibrium which we shall call the *comparison equilibrium state*. This generalizes the corresponding definition of §4-5 and reduces to it when $F_{\alpha\beta} = 0$. However, there is a greater element of convention in the present definition than in that of §4-5. A definition based on the u_1 of (6.18) for example, instead of on u_0 would lead to a slightly different comparison state, and thus to slightly different definitions of temperature, etc. away from equilibrium. It is meaningless to ask which of these definitions of temperature, say, is correct. Each can be incorporated into its own theory and all these theories will be equivalent. They will however use slightly different languages to describe the same phenomena, and in consequence will differ slightly in any given order of approximation.

With the aid of (6.1) and (6.21) the Gibbs relation (6.25) can be put in the equivalent form

$$T\,ds = du_0 + p_0\,dv + g_\alpha\,dv^\alpha + \tfrac{1}{2}v P^{\alpha\beta}\,dF_{\alpha\beta} \qquad (8.2)$$

where $v = 1/\rho$ and g^α is defined by (6.5). This form corresponds to a fundamental equation

$$s = s(u_0, v, v^\alpha, F_{\alpha\beta}), \qquad (8.3)$$

whose arguments are those used to define the comparison equilibrium state. We can thus use (8.3) to define the specific entropy s of the general state and can then invert it to give

$$u_0 = u_0(s, v, v^\alpha, F_{\alpha\beta}). \qquad (8.4)$$

From this functional form we define

$$
\begin{aligned}
p_0 &\equiv -\partial u_0/\partial v && \text{to be the \textit{Kelvin pressure}} \\
\tilde{P}^{\alpha\beta} &\equiv -2\rho\,\partial u_0/\partial F_{\alpha\beta} && \text{to be the \textit{thermostatic polarization tensor}} \\
T &\equiv \partial u_0/\partial s && \text{to be the \textit{absolute temperature}}
\end{aligned} \right\} \quad (8.5)
$$

of the state. Since neither p_0 nor T have alternative mechanical or electromagnetic definitions, use of the same notation and terminology as for an equilibrium state cannot cause confusion. However, the true polarization tensor $P^{\alpha\beta}$ is defined by (2.38) and it will in general differ from the thermostatic polarization tensor $\tilde{P}^{\alpha\beta}$ of (8.5). This variable thus does need a distinctive notation and name, as was the case for the thermostatic pressure of (4-5.2). The difference

$$\Pi^{\alpha\beta} \equiv P^{\alpha\beta} - \tilde{P}^{\alpha\beta} \tag{8.6}$$

will be called the *dispersive polarization tensor*.

The one partial derivative of u_0 missing from (8.5) is

$$\tilde{g}_\alpha \equiv -\partial u_0/\partial v^\alpha. \tag{8.7}$$

Since $v_\alpha\,dv^\alpha \equiv 0$, this may be constrained to satisfy

$$v^\alpha \tilde{g}_\alpha = 0. \tag{8.8}$$

With this definition we have as an identity that

$$T\,ds = du_0 + p_0\,dv + \tilde{g}_\alpha\,dv^\alpha + \tfrac{1}{2}v\tilde{P}^{\alpha\beta}\,dF_{\alpha\beta}. \tag{8.9}$$

The method of derivation of (4-4.14) enables us to deduce that

$$\tilde{g}^{[\alpha}v^{\beta]} + v\tilde{P}^{[\alpha}_{\cdot\gamma}F^{\beta]\gamma} = 0 \tag{8.10}$$

from the Lorentz invariance of (8.9). Multiplication of (8.10) by v_β gives

$$\tilde{g}^\alpha = 2v\tilde{P}^{[\alpha}_{\cdot\gamma}F^{\beta]\gamma}v_\beta, \tag{8.11}$$

so that $\tilde{g}^\alpha$ is determined by other known variables. We see that it is equal to the g^α of the comparison equilibrium state as given by (6.5), as is to be expected.

8b *The entropy source strength – first approximation*

The electromagnetic induction tensor $\tilde{I}^{\alpha\beta}$ of the comparison equilibrium state is seen from (2.38) to be

$$\tilde{I}^{\alpha\beta} \equiv F^{\alpha\beta} - 4\pi\tilde{P}^{\alpha\beta}. \tag{8.12}$$

On using (8.6) we thus have

$$I^{\alpha\beta} - \tilde{I}^{\alpha\beta} = -4\pi\Pi^{\alpha\beta}. \tag{8.13}$$

With the notation of (8.12) the energy–momentum tensor $\tilde{T}^{\alpha\beta}$ of the comparison equilibrium state is seen from (6.26) to be

$$\tilde{T}^{\alpha\beta} \equiv \rho(1+u_0)\,v^\alpha v^\beta + p_0(g^{\alpha\beta}+v^\alpha v^\beta) + \rho\tilde{g}^\alpha v^\beta + \frac{1}{4\pi}\,(F^\alpha_{.\gamma}\tilde{I}^{\beta\gamma} - \tfrac{1}{4}F^{\gamma\delta}F_{\gamma\delta}\,g^{\alpha\beta}).$$

$$(8.14)$$

It is symmetric in virtue of (8.10) and (8.12). If we set

$$t^{\alpha\beta} \equiv T^{\alpha\beta} - \tilde{T}^{\alpha\beta} \qquad (8.15)$$

then it follows with the aid of (8.1) and (8.6) that

$$v_\alpha v_\beta(t^{\alpha\beta} + F^\alpha_{.\gamma}\,\Pi^{\beta\gamma}) = 0. \qquad (8.16)$$

This is analogous to the result (4-5.7) which holds in the absence of an electromagnetic field, but it is more complicated due to the extra term which involves the dispersive polarization tensor. The notation $t^{\alpha\beta}$ has been used in (8.15) in preference to the notation $\tau^{\alpha\beta}$ used in (4-5.6) since $\tau^{\alpha\beta}$ will be used later for a slightly different tensor which satisfies (4-5.7) precisely.

The expressions (8.13) and (8.15) for the increments in $I^{\alpha\beta}$ and $T^{\alpha\beta}$ may now be combined with the partial derivatives (4.6) to give the first-order variation in s^α between the actual and comparison states. With the aid of (4.25), (5.7) and (5.19) we find that

$$s^\alpha = s^\alpha_1 + O_2 \qquad (8.17)$$

where

$$s^\alpha_1 \equiv s\rho^\alpha - t^{\alpha\beta}\Theta_\beta - \Pi^\alpha_{.\gamma}\,F^{\beta\gamma}\Theta_\beta \qquad (8.18)$$

and O_2 denotes terms of quadratic and higher orders in $t^{\alpha\beta}$ and $\Pi^{\alpha\beta}$. We also need to evaluate the contribution that s^α_1 makes to the entropy source strength σ. By using (5.7), (8.9) and the conservation law (3-5.29) of inert mass, together with (8.12), (8.13) and the definitions (4-6.18), (8.14) and (8.15), we can deduce from (8.18) that

$$\partial_\alpha s^\alpha_1 + \Theta_\beta\,\partial_\alpha T^{\beta\alpha} = -\tfrac{1}{2}t^{\alpha\beta}L_\Theta\,g_{\alpha\beta} - \tfrac{1}{2}\Pi^{\alpha\beta}L_\Theta\,F_{\alpha\beta} - \frac{1}{4\pi}\,\Theta^\gamma F_{\beta\gamma}\,\partial_\alpha I^{\alpha\beta}$$

$$- \frac{3}{8\pi}\,\Theta^\alpha I^{\beta\gamma}\partial_{[\alpha}F_{\beta\gamma]}. \qquad (8.19)$$

The remaining three universal laws of (4.4) may now be used to simplify this to give

$$\partial_\alpha s^\alpha_1 = -\tfrac{1}{2}t^{\alpha\beta}L_\Theta\,g_{\alpha\beta} - \tfrac{1}{2}\Pi^{\alpha\beta}L_\Theta\,F_{\alpha\beta}. \qquad (8.20)$$

This result is the natural generalization of (4-6.22) to the case when electromagnetic fields are present. However, for application to the phenomenological laws it is preferable to have the rate of change of $F_{\alpha\beta}$ described by $DF_{\alpha\beta}/D\tau$ rather than by $L_\Theta F_{\alpha\beta}$. Here $D/D\tau$ is the operator defined by (4-7.19). With the aid of (4-7.8), (4-7.15) and (4-7.18) we find that (8.20) can also be expressed as

$$\partial_\alpha s_1^\alpha = -\tfrac{1}{2}\tau^{\alpha\beta}L_\Theta\, g_{\alpha\beta} - \tfrac{1}{2}\Theta\Pi^{\alpha\beta}DF_{\alpha\beta}/D\tau \qquad (8.21)$$

where $\tau^{\alpha\beta} \equiv t^{\alpha\beta} + F^{(\alpha}_{\ \cdot\gamma}\Pi^{\beta)\gamma} + v^\alpha F^{[\beta}_{\ \cdot\delta}\Pi^{\gamma]\delta}v_\gamma + v^\beta F^{[\alpha}_{\ \cdot\delta}\Pi^{\gamma]\delta}v_\gamma.$ (8.22)

This is identically symmetric and it satisfies

$$v_\alpha v_\beta \tau^{\alpha\beta} = 0 \qquad (8.23)$$

in virtue of (8.16).

The most natural generalization to non-equilibrium states of the expression (6.26) for $T^{\alpha\beta}$ is given by combining (8.22) with (8.14) and (8.15). We find that

$$T^{\alpha\beta} = [\rho(1+u_0)+p_0]v^\alpha v^\beta + p_0 g^{\alpha\beta} + \rho g^{(\alpha}v^{\beta)}$$

$$+\frac{1}{4\pi}\,(F^{(\alpha}_{\ \cdot\gamma}I^{\beta)\gamma} - \tfrac{1}{4}F^{\gamma\delta}F_{\gamma\delta}g^{\alpha\beta}) + \tau^{\alpha\beta}, \qquad (8.24)$$

where g^α is again defined by (6.5). Note that the $I^{\alpha\beta}$ occurring here is the actual value of the electromagnetic induction tensor and not its value $\overset{*}{I}{}^{\alpha\beta}$ in the comparison equilibrium state. Apart from the final term $\tau^{\alpha\beta}$, this differs from (6.26) only in that it is explicitly symmetrized. This symmetrization is essential for non-equilibrium states as the thermostatic relation (6.27), which ensures the symmetry of (6.26) in equilibrium, is not valid for arbitrary states. The *heat flux vector* q_h^α and *viscous stress tensor* $\rho_v^{\alpha\beta}$ can now be defined by decomposing $\tau^{\alpha\beta}$ in accordance with (4-5.8) and (4-5.9). The energy flux vector q^α and (full) stress tensor $\sigma^{\alpha\beta}$ are then seen from (8.24) and (3-5.16) to be given by

$$q^\alpha = \frac{1}{4\pi}\,H^{\alpha\beta}E_\beta + q_h^\alpha \qquad (8.25)$$

and $$\sigma^{\alpha\beta} = \frac{1}{4\pi}\,(E^{(\alpha}D^{\beta)} - B^{(\alpha}_{\ \cdot\gamma}H^{\beta)\gamma})$$

$$-\left[p_0 + \frac{1}{8\pi}\,(E^\gamma E_\gamma - \tfrac{1}{2}B^{\gamma\delta}B_{\gamma\delta})\right](g^{\alpha\beta} + v^\alpha v^\beta) + \sigma_v^{\alpha\beta}. \qquad (8.26)$$

The notation for the electromagnetic tensors is that of (6.1). If we use

(3-6.5) and (3-6.11) to construct the three-dimensional equivalent of (8.25) we find that

$$q = (c\gamma^2/4\pi)\,E^* \times H^* + q_h \qquad (8.27)$$

where γ is given by (2-7.10) and E^*, H^* by (2.27).

We saw in (5.8) that $n_\beta T^{\alpha\beta}$ is continuous across a hypersurface of material discontinuity. It is easily deduced from this and the definitions (3-6.11), (3-6.5) and (3-5.14) that in any inertial frame, $q \cdot n$ is continuous across a surface S of material discontinuity with unit normal n. In virtue of (3-6.10) this expresses the simple physical fact that the rate at which energy flows into the material on one side of S equals the rate at which it flows out of the material on the other side. Physical intuition leads us to expect that this should be true separately for the rates of flow of electromagnetic energy and of heat. Since

$$n \cdot (E^* \times H^*) \equiv n \cdot [(n \times E^*) \times (n \times H^*)] \qquad (8.28)$$

we see from (8.27) and the boundary conditions (2.28) that this is indeed the case.

The above result gives one check that our definition of heat flux is in agreement with physical expectations. Another check comes from the use of (8.22) and (4-5.8) to express (8.18) alternatively as

$$s_1^\alpha = s\rho^\alpha - T^{\alpha\beta}\Theta_\beta = s\rho^\alpha + q_h^\alpha/T. \qquad (8.29)$$

The latter expression is identical in form with (4-6.5). It shows that the heat flux vector has the expected relationship to the entropy flux in the first approximation. This result is of sufficient importance that it could have been used to define the heat flux. There are no similar requirements of physical consistency for $\sigma_V^{\alpha\beta}$ as it is not related to the entropy flux and electromagnetic stress *can* be discontinuous at a bounding surface. This is the phenomenon of radiation pressure.

8c The entropy source strength – second approximation

We saw in Chapter 4 that a consistent relativistic theory of non-equilibrium processes must take account of the finiteness of all relaxation times. The relaxation processes contribute to s^α a term such as (4-7.4) which is parallel to ρ^α and quadratic in the thermodynamic fluxes. These fluxes are the variables which measure departures from local equilibrium, which in the present situation can be taken as the tensors $\tau^{\alpha\beta}$ and $\Pi^{\alpha\beta}$. The second approximation to s^α for a polarizable simple fluid is thus to take

$$s^\alpha = s_1^\alpha + s_2^\alpha \qquad (8.30)$$

where s_1^α is given by (8.29) and

$$s_2^\alpha \equiv -\tfrac{1}{2}\rho^\alpha(a_{\beta\gamma\delta\epsilon}\tau^{\beta\gamma}\tau^{\delta\epsilon} + 2k_{\beta\gamma\delta\epsilon}\tau^{\beta\gamma}\Pi^{\delta\epsilon} + b_{\beta\gamma\delta\epsilon}\Pi^{\beta\gamma}\Pi^{\delta\epsilon}). \quad (8.31)$$

We see that the present system needs three relaxation tensors $a_{\alpha\beta\gamma\delta}$, $k_{\alpha\beta\gamma\delta}$ and $b_{\alpha\beta\gamma\delta}$ in place of the one that sufficed in §4-7a. As in §4-7a they are functions of the comparison equilibrium state, and thus of u_0, ρ^α and $F_{\alpha\beta}$. In virtue of the algebraic properties of $\tau^{\alpha\beta}$ and $\Pi^{\alpha\beta}$, $a_{\alpha\beta\gamma\delta}$ can again be chosen to satisfy (4-7.5) while $k_{\alpha\beta\gamma\delta}$ and $b_{\alpha\beta\gamma\delta}$ can similarly be chosen to satisfy

$$k_{\alpha\beta\gamma\delta} = k_{(\alpha\beta)(\gamma\delta)}, \quad v^\alpha v^\beta k_{\alpha\beta\gamma\delta} = 0 \quad\quad (8.32)$$

and $\quad\quad\quad b_{\alpha\beta\gamma\delta} = b_{[\alpha\beta][\gamma\delta]}, \quad b_{\alpha\beta\gamma\delta} = b_{\gamma\delta\alpha\beta}. \quad\quad (8.33)$

Further restrictions on the relaxation tensors can be deduced from an invariance property that we have not previously needed. This is that s^α must be invariant under the instantaneous transformation

$$T^{\alpha\beta} \to T^{\alpha\beta}, \quad \rho^\alpha \to \rho^\alpha, \quad F^{\alpha\beta} \to -F^{\alpha\beta}, \quad I^{\alpha\beta} \to -I^{\alpha\beta}, \quad J^\alpha \to -J^\alpha. \quad (8.34)$$

The origin of this invariance lies at a microscopic level, in the invariance of the Lorentz force law (2.2) under the transformation

$$F^{\alpha\beta} \to -F^{\alpha\beta}, \quad\quad e \to -e. \quad\quad (8.35)$$

Not all microscopic invariance properties are fully reflected in the macroscopic behaviour of matter, due to the statistical nature of the relationship between the microscopic and macroscopic laws. In particular, the entropy law itself specifies a preferred time-orientation that is not present in the basic microscopic laws. However, microscopic invariance properties do leave some traces of their existence in the macroscopic laws. The result stated above is one such trace. At a purely macroscopic level it is simply a new hypothesis that we now introduce. For an account of the macroscopic consequences of microscopic invariance properties, see de Groot & Mazur (1962, Ch. 7) and Callen (1974).

For non-conducting media the transformation $J^\alpha \to -J^\alpha$ corresponds to $z \to -z$, where z is the charge/mass ratio defined by (4.1). Since we have treated z as a characteristic of the material, this transformation can be applied to our theory only when $z = 0$, i.e. when the material is electrically neutral. We shall thus continue the development only for this case. For practical purposes this is a negligible specialization as almost all media of interest *are* electrically neutral.

The thermostatic theory of equilibrium states satisfies this new invariance principle identically. For such states it follows from (6.26) that u_0 and p_0 are invariant under (8.34). By solving (7.4) for s we see that the thermostatic fundamental equation can be taken in the form

$$s = s(u_0, v, \theta, \phi, \psi) \qquad (8.36)$$

where θ, ϕ, ψ are defined by (7.1). These arguments of s are all invariant under (8.34) and hence so also is s^α by (5.7), as required. Since the variables involved in (8.34) are not all independent in equilibrium, we must also check for consistency that the Gibbs relation (6.25) can be preserved. This is so provided only that T is invariant under (8.34).

These thermostatic results can be applied in the nonequilibrium theory to the comparison equilibrium state. In the notation of §8a they show that

$$\overset{\circ}{T}{}^{\alpha\beta} \to \overset{\circ}{T}{}^{\alpha\beta}, \quad \overset{\circ}{I}{}^{\alpha\beta} \to -\overset{\circ}{I}{}^{\alpha\beta}. \qquad (8.37)$$

It then follows from (8.13), (8.15), (8.22) and (8.29) that we must have

$$\tau^{\alpha\beta} \to \tau^{\alpha\beta}, \quad \Pi^{\alpha\beta} \to -\Pi^{\alpha\beta}, \quad s_1^\alpha \to s_1^\alpha. \qquad (8.38)$$

The total entropy flux s^α of (8.30) is thus invariant under (8.36) if and only if s_2^α of (8.31) is invariant. It follows that the relaxation tensors must satisfy

$$\left.\begin{aligned} a_{\alpha\beta\gamma\delta}(u_0, \rho^\alpha, -F_{\alpha\beta}) &= a_{\alpha\beta\gamma\delta}(u_0, \rho^\alpha, F_{\alpha\beta}) \\ k_{\alpha\beta\gamma\delta}(u_0, \rho^\alpha, -F_{\alpha\beta}) &= -k_{\alpha\beta\gamma\delta}(u_0, \rho^\alpha, F_{\alpha\beta}) \\ b_{\alpha\beta\gamma\delta}(u_0, \rho^\alpha, -F_{\alpha\beta}) &= b_{\alpha\beta\gamma\delta}(u_0, \rho^\alpha, F_{\alpha\beta}) \end{aligned}\right\} \qquad (8.39)$$

in addition to their algebraic properties (4-7.5), (8.32) and (8.33).

The entropy source strength in the second approximation is easily evaluated from (8.21) and (8.31) by the method used to obtain (4-7.22). It can be expressed as

$$\sigma = \tau^{\alpha\beta} Y_{\alpha\beta} + \Pi^{\alpha\beta} Z_{\alpha\beta} \qquad (8.40)$$

where
$$Y_{\alpha\beta} = X_{\alpha\beta} - \rho a_{\alpha\beta\gamma\delta}\, D\tau^{\gamma\delta}/D\tau - \rho k_{\alpha\beta\gamma\delta}\, D\Pi^{\gamma\delta}/D\tau$$
$$- \tfrac{1}{2}\rho\tau^{\gamma\delta} D a_{\alpha\beta\gamma\delta}/D\tau - \tfrac{1}{2}\rho\Pi^{\gamma\delta} D k_{\alpha\beta\gamma\delta}/D\tau, \qquad (8.41)$$

$$Z_{\alpha\beta} = -\tfrac{1}{2}\Theta D F_{\alpha\beta}/D\tau - \rho k_{\gamma\delta\alpha\beta}\, D\tau^{\gamma\delta}/D\tau - \rho b_{\alpha\beta\gamma\delta}\, D\Pi^{\gamma\delta}/D\tau$$
$$- \tfrac{1}{2}\rho\tau^{\gamma\delta} D k_{\gamma\delta\alpha\beta}/D\tau - \tfrac{1}{2}\rho\Pi^{\gamma\delta} D b_{\alpha\beta\gamma\delta}/D\tau, \qquad (8.42)$$

and $X_{\alpha\beta}$ in (8.41) is still given by (4-6.25). With the aid of (4-6.28) and (4-7.7), the tensors $Y_{\alpha\beta}$ and $Z_{\alpha\beta}$ can be seen to satisfy

$$Y_{\alpha\beta} = Y_{(\alpha\beta)}, \quad v^\alpha v^\beta Y_{\alpha\beta} = 0, \quad Z_{\alpha\beta} = Z_{[\alpha\beta]}. \qquad (8.43)$$

9 Linearized phenomenological laws

9a *The general situation*

We saw in §4-6c that the description of material orbits by a parameter proportional to $T\tau$ is more natural than parametrization by the proper time τ itself. This was supported by the form (4-6.22) of the first approximation for σ. When an electromagnetic field is present, (4-6.22) is replaced by (8.20). This shows that $T\tau$ is still the natural parametrization of the orbits even in this more general situation.

Still following §4-6c, we see that the simplest possible phenomenological laws will now involve $F_{\alpha\beta}$ and $I^{\alpha\beta}$ and their Lie derivatives with respect to Θ^α in addition to the variables indicated in (4-6.23). Just as the arguments listed in (4-6.23) could be replaced by the equivalent set (4-6.24), so also this extended argument set is equivalent to

$$u_0, \rho^\alpha, F_{\alpha\beta}, \tau^{\alpha\beta}, \Pi^{\alpha\beta}, X_{\alpha\beta}, L_\Theta F_{\alpha\beta}, \dot{u}_0, \dot{\rho}, L_\Theta \tau_{\alpha\beta}, L_\Theta \Pi^{\alpha\beta}. \qquad (9.1)$$

The set (4-6.24) was further amended in §4-7c by the replacement of $L_\Theta \tau_{\alpha\beta}$ and $X_{\alpha\beta}$ by $D\tau^{\alpha\beta}/D\tau$ and the $Y_{\alpha\beta}$ of (4-7.23) respectively. Corresponding replacements in (9.1) give us

$$u_0, \rho^\alpha, F_{\alpha\beta}, \tau^{\alpha\beta}, \Pi^{\alpha\beta}, Y_{\alpha\beta}, Z_{\alpha\beta}, z^a, \qquad (9.2)$$

where $Y_{\alpha\beta}$ is now given by (8.41) and $\{z^a; 1 \leqslant a \leqslant 17\}$ is a set of linearly independent components of the rates of change

$$\dot{u}_0, \dot{\rho}, D\tau^{\alpha\beta}/D\tau, D\Pi^{\alpha\beta}/D\tau. \qquad (9.3)$$

The set (9.2) is the most convenient one for a study of the phenomenological laws.

The techniques of §4-6d may be applied to the expression (8.40) for σ to deduce that the phenomenological laws must include conditions of the form

$$\left.\begin{aligned} Y_{\alpha\beta} &= Y_{\alpha\beta}(u_0, \rho^\alpha, F_{\alpha\beta}, \tau^{\alpha\beta}, \Pi^{\alpha\beta}, z^a) \\ Z_{\alpha\beta} &= Z_{\alpha\beta}(u_0, \rho^\alpha, F_{\alpha\beta}, \tau^{\alpha\beta}, \Pi^{\alpha\beta}, z^a). \end{aligned}\right\} \qquad (9.4)$$

Because of the constraints (8.43) there are just 15 independent equations in this set. The universal laws (4.4) contain 11 equations governing the evolution of the system together with two equations which are constraints on the allowed initial data. In a given inertial frame these constraint equations are the equations for $\nabla \cdot \mathbf{B}$ and $\nabla \cdot \mathbf{D}$ in the three-dimensional form (1.1) and (1.2) of Maxwell's equations. Altogether we thus have 26 evolution equations. There are also just 26 linearly independent variables amongst the components of $T^{\alpha\beta}$, ρ^α, $F_{\alpha\beta}$ and

$I^{\alpha\beta}$. When the thermostatic fundamental equation (8.3) is given, the other variables such as Θ become known functions of these 26 variables. We thus again have as many evolution equations as we have variables, just as in § 4-6 d. We shall suppose that there are no further constraints on the initial data other than those implicit in Maxwell's equations. The equations (9.4) must then be the only phenomenological laws for the system. As the analogue of (4-7.27) the entropy law requires the functional forms (9.4) to satisfy

$$\partial Y_{\alpha\beta}/\partial z^a = 0, \quad \partial Z_{\alpha\beta}/\partial z^a = 0 \quad \text{when} \quad z^a = 0. \tag{9.5}$$

Close to equilibrium the laws (9.4) can be linearized in the variables $\tau^{\alpha\beta}$, $\Pi^{\alpha\beta}$ and z^a which vanish in equilibrium. For logical consistency this linearization should also omit the final two terms in each of the expressions (8.41) and (8.42) which define $Y_{\alpha\beta}$ and $Z_{\alpha\beta}$. These are quadratic in variables which vanish in equilibrium, and similar quadratic terms are being neglected in the Taylor expansion of (9.4). The linear terms in z^a must be absent by (9.5). The linearized laws thus have the forms

$$\rho a_{\alpha\beta\gamma\delta} \frac{D}{D\tau} \tau^{\gamma\delta} + \rho k_{\alpha\beta\gamma\delta} \frac{D}{D\tau} \Pi^{\gamma\delta} + c_{\alpha\beta\gamma\delta} \tau^{\gamma\delta} + l_{\alpha\beta\gamma\delta} \Pi^{\gamma\delta} = X_{\alpha\beta} \tag{9.6}$$

and

$$\rho k_{\gamma\delta\alpha\beta} \frac{D}{D\tau} \tau^{\gamma\delta} + \rho b_{\alpha\beta\gamma\delta} \frac{D}{D\tau} \Pi^{\gamma\delta} + m_{\gamma\delta\alpha\beta} \tau^{\gamma\delta} + d_{\alpha\beta\gamma\delta} \Pi^{\gamma\delta} = -\tfrac{1}{2}\Theta \frac{D}{D\tau} F_{\alpha\beta} \tag{9.7}$$

where the impedance tensors $c_{\alpha\beta\gamma\delta}$, $d_{\alpha\beta\gamma\delta}$, $l_{\alpha\beta\gamma\delta}$, $m_{\alpha\beta\gamma\delta}$ introduced by the linearization are all functions of u_0, ρ^α and $F_{\alpha\beta}$. The tensor $c_{\alpha\beta\gamma\delta}$ may again be chosen to satisfy (4-6.44), while the other three may similarly be chosen such that

$$d_{\alpha\beta\gamma\delta} = d_{[\alpha\beta][\gamma\delta]}, \tag{9.8}$$

$$l_{\alpha\beta\gamma\delta} = l_{(\alpha\beta)[\gamma\delta]}, \quad v^\alpha v^\beta l_{\alpha\beta\gamma\delta} = 0 \tag{9.9}$$

$$m_{\alpha\beta\gamma\delta} = m_{(\alpha\beta)[\gamma\delta]}, \quad v^\alpha v^\beta m_{\alpha\beta\gamma\delta} = 0. \tag{9.10}$$

When (9.6) and (9.7) hold, the expression (8.40) for σ can be put in the form

$$\sigma = \tau^{\alpha\beta}\tau^{\gamma\delta}(c_{\alpha\beta\gamma\delta} - \tfrac{1}{2}\rho Da_{\alpha\beta\gamma\delta}/D\tau) + \tau^{\alpha\beta}\Pi^{\gamma\delta}(l_{\alpha\beta\gamma\delta} + m_{\alpha\beta\gamma\delta} - \rho Dk_{\alpha\beta\gamma\delta}/D\tau)$$
$$+ \Pi^{\alpha\beta}\Pi^{\gamma\delta}(d_{\alpha\beta\gamma\delta} - \tfrac{1}{2}\rho Db_{\alpha\beta\gamma\delta}/D\tau). \tag{9.11}$$

The considerations which show the invariance of s^α under the instantaneous transformation (8.34) show also that the lowest order terms in

σ are also invariant under it. It is only for these lowest order terms that the instantaneous transformation is well defined, since the rates of change z^a are absent only at this order. These terms are the terms in (9.11) which involve the impedance tensors. Since (8.34) induces (8.38), their invariance shows that

$$c_{\alpha\beta\gamma\delta} + c_{\gamma\delta\alpha\beta} \quad \text{and} \quad d_{\alpha\beta\gamma\delta} + d_{\gamma\delta\alpha\beta} \qquad (9.12)$$

are even functions of $F_{\alpha\beta}$ while

$$l_{\alpha\beta\gamma\delta} + m_{\alpha\beta\gamma\delta} \qquad (9.13)$$

must be an odd function of $F_{\alpha\beta}$.

The rates of change which occur in (9.11) can give contributions to σ of either sign. The entropy law thus requires

$$c_{\alpha\beta\gamma\delta}\tau^{\alpha\beta}\tau^{\gamma\delta} + (l_{\alpha\beta\gamma\delta} + m_{\alpha\beta\gamma\delta})\tau^{\alpha\beta}\Pi^{\gamma\delta} + d_{\alpha\beta\gamma\delta}\Pi^{\alpha\beta}\Pi^{\gamma\delta} > 0 \qquad (9.14)$$

whenever $\tau^{\alpha\beta}$ and $\Pi^{\alpha\beta}$ do not both vanish. This will ensure that σ is non-negative for all near-equilibrium states which are not changing too rapidly.

9b The weak-field case

In the absence of an electromagnetic field we were able to give representations of the relaxation and impedance tensors in terms of a small number of scalar parameters. These were given by (4-7.6) and (4-6.45) respectively. Similar representations are possible when an electromagnetic field is present, but they are considerably more complicated. As an example, we saw in §4-6e that viscous phenomena could be described by two coefficients ζ and η when the relaxation terms are negligible. It is shown in the book of de Groot & Mazur (1962, Ch.12 §2) that seven coefficients of viscosity are needed in the presence of a strong static magnetic field. Even more would be required if the electric field were also strong.

To avoid these complications let us consider the phenomenological laws in more detail only for the simpler case in which the electromagnetic field is weak. Although this excludes the behaviour of fluids in strong static fields, it is a valid approximation for another interesting situation. Except under extreme conditions such as can be produced by lasers, the propagation of electromagnetic waves through a fluid is a weak-field phenomenon. We shall examine (9.6) and (9.7) further with this particular application in mind.

For weak fields the dependence of the relaxation and impedance tensors on $F_{\alpha\beta}$ can be neglected. The tensor (9.13) then vanishes as it is an odd function of $F_{\alpha\beta}$, as does $k_{\alpha\beta\gamma\delta}$ by (8.39). We thus have

$$k_{\alpha\beta\gamma\delta} = 0, \quad m_{\alpha\beta\gamma\delta} = -l_{\alpha\beta\gamma\delta}. \qquad (9.15)$$

The tensors $a_{\alpha\beta\gamma\delta}$ and $c_{\alpha\beta\gamma\delta}$ will again have the forms (4-7.6) and (4-6.45) respectively, with the parameters a_i and c_i being scalar functions of u_0 and ρ. The method of derivation of these equations can also be applied to $b_{\alpha\beta\gamma\delta}$, $d_{\alpha\beta\gamma\delta}$ and $l_{\alpha\beta\gamma\delta}$ to show that they must have the forms

$$b_{\alpha\beta\gamma\delta} = b_1 h_{\alpha[\gamma} h_{\delta]\beta} + b_2 v_{[\alpha} h_{\beta][\delta} v_{\gamma]} \qquad (9.16)$$

$$d_{\alpha\beta\gamma\delta} = d_1 h_{\alpha[\gamma} h_{\delta]\beta} + d_2 v_{[\alpha} h_{\beta][\delta} v_{\gamma]} \qquad (9.17)$$

and

$$l_{\alpha\beta\gamma\delta} = l v_{(\alpha} h_{\beta)[\gamma} v_{\delta]} \qquad (9.18)$$

in virtue of (8.33), (9.8) and (9.9) respectively. Here b_1, b_2, d_1, d_2, l are new scalar functions of u_0 and ρ, and $h_{\alpha\beta}$ is defined by (4-5.5). Another simplification for weak fields is that the comparison equilibrium state will satisfy the linear constitutive relations (7.7). Since the thermostatic polarization tensor $\tilde{P}^{\alpha\beta}$ of (8.5) is simply the polarization tensor of this comparison state, we see from (6.21) and (7.7) that it must be given by

$$\tilde{P}^{\alpha\beta} = \chi B^{\alpha\beta} - 2\kappa v^{[\alpha} E^{\beta]}. \qquad (9.19)$$

The dispersive polarization tensor $\Pi^{\alpha\beta}$ defined by (8.6) is thus given by

$$\Pi^{\alpha\beta} = (M^{\alpha\beta} - \chi B^{\alpha\beta}) - 2v^{[\alpha}(P^{\beta]} - \kappa E^{\beta]}). \qquad (9.20)$$

These results enable (9.6) and (9.7) to be decomposed by the methods applied to (4-6.43) in §4-6e. The equations obtained for the viscous stresses are identical with (4-7.31) and (4-7.32) and so need not be given again here. However, the equation for the heat flux q_{h}^{α} gains extra terms and there are new equations governing P^{α} and $M^{\alpha\beta}$. To replace (4-7.33) we thus need three equations, which are found to be

$$\tau_{\mathrm{t}} \frac{D}{D\tau} q_{\mathrm{h}\alpha} + q_{\mathrm{h}\alpha} = -\lambda[B_\alpha^\beta \partial_\beta T + T a_\alpha + \nu T(P_\alpha - \kappa E_\alpha)], \qquad (9.21)$$

$$\tau_{\mathrm{D}} \frac{D}{D\tau} (P_\alpha - \kappa E_\alpha) + (P_\alpha - \kappa E_\alpha) = -\kappa_{\mathrm{D}} \tau_{\mathrm{D}} \left(\frac{D}{D\tau} E_\alpha - 2\nu q_{\mathrm{h}\alpha} \right) \qquad (9.22)$$

and

$$\tau_{\mathrm{L}} \frac{D}{D\tau} (M_{\alpha\beta} - \chi B_{\alpha\beta}) + (M_{\alpha\beta} - \chi B_{\alpha\beta}) = -\chi_{\mathrm{L}} \tau_{\mathrm{L}} \frac{D}{D\tau} B_{\alpha\beta}. \qquad (9.23)$$

These equations involve the electric and magnetic susceptibilities κ and χ respectively, and seven parameters given by

$$\left. \begin{array}{l} \kappa_{\mathrm{D}} = 1/(2T\rho b_2), \quad \chi_{\mathrm{L}} = 1/(2T\rho b_1), \quad \lambda = 1/(T^2 c_3), \quad \nu = Tl, \\ \tau_{\mathrm{D}} = \rho b_2/d_2, \quad \tau_{\mathrm{L}} = \rho b_1/d_1, \quad \tau_{\mathrm{t}} = \rho a_3/c_3. \end{array} \right\} \qquad (9.24)$$

We have come across two of the constants (9.24) before, namely the thermal conductivity λ and the thermal relaxation time τ_t. The thermal conductivity was denoted by κ in Chapter 4, but λ has been used here to prevent confusion with the electric susceptibility. We shall call κ_D the Debye susceptibility,

χ_L the Langevin susceptibility,

τ_D the Debye relaxation time,

τ_L the Langevin relaxation time,

ν the cross-coupling coefficient.

The constant ν introduces into the equations a feature that was not present in §4-7c. There, the three variables p_v, q_α and $\pi_{\alpha\beta}$ were found to be governed by independent equations. This is easily seen to be a consequence of the differing tensorial natures of these variables. In the present situation we have *two* vector variables, the heat flux $q_{h\alpha}$ and the dispersive electric polarization $(P_\alpha - \kappa E_\alpha)$. These are governed jointly by (9.21) and (9.22) which are coupled together via the coefficient ν. One consequence of this coupling is that a temperature gradient will generate an electric polarization even in the absence of an applied electric field. This cross-coupling of variables with similar tensorial natures is a general feature of thermodynamics.

The middle term of (9.14) is absent in the weak-field case by (9.15). In consequence it is easily found that this condition is equivalent to the positivity of the parameters $(3c_1 + c_2)$, c_2, c_3, d_1 and d_2. It was seen in §4-7c how similar conditions on the relaxation times can be deduced from the fact that equilibrium states are stable with respect to small disturbances. This shows immediately that τ_b and τ_s of (4-7.34), and τ_L of (9.23), are all positive but the coupled equations (9.21) and (9.22) need a more subtle treatment. When the thermodynamic forces vanish and the matter is at rest in some inertial frame, these become a pair of linear homogeneous equations in the variables $q_{h\alpha}$ and $(P_\alpha - \kappa E_\alpha)$. Their general solution is a linear combination of eigensolutions with the form

$$q_{h\alpha} = Q_\alpha e^{-\omega\tau}, \quad P_\alpha - \kappa E_\alpha = R_\alpha e^{-\omega\tau} \qquad (9.25)$$

where Q_α and R_α are constant vectors and ω is the eigenfrequency. Substitution from (9.25) into the equations shows that there is a nontrivial solution only when ω is a root of the equation

$$\omega^2 \tau_t \tau_D - \omega(\tau_t + \tau_D) + 1 + 2\lambda\nu^2 T\kappa_D\tau_D = 0. \qquad (9.26)$$

For stability, both roots must have positive real parts. Since

$$2\lambda\nu^2 T\kappa_D\tau_D = l^2/c_3 d_2 > 0 \qquad (9.27)$$

by the above, this is true if and only if τ_t and τ_D are both positive. By combining all these results together, we find that the five relaxation times τ_b, τ_s, τ_t, τ_D, τ_L and the corresponding impedance parameters $\zeta, \eta, \lambda, \kappa_D, \chi_L$ all have to be positive. Only the cross-coupling coefficient ν is unrestricted.

9c Electromagnetic dispersion

Let us apply (9.22) and (9.23) to the propagation of electromagnetic radiation through a fluid at rest. For simplicity suppose that the cross-coupling coefficient ν is negligible. Use of (6.2) and (6.22) then enables us to express (9.22) in terms of the three-dimensional vectors $\mathbf{P}$ and $\mathbf{E}$, and similarly (9.23) can be expressed in terms of $\mathbf{M}$ and $\mathbf{B}$. Radiation of angular frequency ω (with respect to proper time) can be described by the real parts of complex field vectors whose time dependence is given by a multiplicative factor $e^{i\omega\tau}$. We may thus set

$$d\mathbf{P}/d\tau = i\omega\mathbf{P}, \quad d\mathbf{E}/d\tau = i\omega\mathbf{E} \qquad (9.28)$$

with the understanding that the physical fields are the real parts of the complex vectors $\mathbf{P}$ and $\mathbf{E}$. The scalar parameters κ, κ_D etc. are independent of the field vectors and so can be treated as constants. They may vary slowly as the fluid heats up through absorption of energy from the radiation, but this will be on a very long time-scale compared with ω. Substitution from (9.28) into (9.22) then shows that

$$\mathbf{P} = \tilde{\kappa}(\omega)\,\mathbf{E} \qquad (9.29)$$

where
$$\tilde{\kappa}(\omega) = (\kappa - \kappa_D) + \frac{\kappa_D}{1 + i\omega\tau_D}. \qquad (9.30)$$

The complex value $\tilde{\kappa}(\omega)$ defined by (9.29) is known as the complex electric susceptibility at frequency ω. It determines both the amplitude and the phase of $\mathbf{P}$ relative to those of $\mathbf{E}$. The dependence of this susceptibility on frequency is the phenomenon known as *dispersion*. It is known from electromagnetic theory that dispersion is necessarily also accompanied by absorption. For liquids the form of frequency dependence given by (9.30) is in agreement with experiment for radio-frequency radiation, but this form breaks down as infrared frequencies are approached. For water at room temperature, τ_D is of the order 10^{-11} seconds and the dimensionless constants κ_D and $(\kappa - \kappa_D)$ are roughly 6 and 0·06 respectively.

Statistical physics can explain both the microscopic origin of (9.30) and its breakdown at sufficiently high frequencies. The theory is due to Debye (1929, Ch. 5). The term $\kappa_D/(1 + i\omega\tau_D)$ occurs only in liquids whose molecules have a permanent electric dipole moment. Such liquids are said to be *polar*. In the absence of an applied electric field the directions of these molecular dipoles will be random and no macroscopic polarization will result. When an electric field is applied, the molecules tend to line up with it. This tendency is resisted by the effects of molecular collisions. The equilibrium value of the macroscopic polarization induced by a static electric field is determined by the balance between these two opposing influences. For weak applied fields it is proportional to the strength of the field, the coefficient of proportionality being κ_D, although saturation occurs when the applied field is strong. Molecular collisions also prevent equilibrium from being reached instantaneously. Instead, the approach to equilibrium takes a characteristic time τ_D. When a field is applied which oscillates with a frequency of the order of $1/\tau_D$, there is insufficient time for equilibrium to be reached. This reduces the amplitude of the induced polarization and causes it to lag behind the applied field. These phenomena are described quantitatively by the Debye term $\kappa_D/(1 + i\omega\tau_D)$ in (9.30).

The remaining contribution $(\kappa - \kappa_D)$ to $\tilde{\kappa}$ is due to the deformation of the molecules in an applied electric field. At low frequencies this deformation is effectively instantaneous and so it gives a contribution to $\tilde{\kappa}$ that is independent of frequency. However, molecules have natural resonant frequencies in the infra-red and optical regions of the spectrum due to their own internal structure. As the frequency of the applied field approaches these resonant frequencies, the deformation susceptibility will itself become frequency dependent and (9.30) will cease to hold. This shows clearly that the breakdown of our thermodynamic results occurs when internal molecular degrees of freedom have to be taken explicitly into account. Under these circumstances knowledge of $T^{\alpha\beta}$, ρ^α, $F_{\alpha\beta}$ and $I^{\alpha\beta}$ is no longer enough information to specify completely a macroscopic state of the system. The matter then ceases to be a polarizable simple fluid in the idealized sense defined in §4a, and our theory naturally ceases to be applicable.

Magnetic analogues of (9.29) and (9.30) can be deduced from (9.23) in a similar manner. We find that

$$\mathbf{M} = \tilde{\chi}(\omega)\,\mathbf{B} \qquad (9.31)$$

with
$$\tilde{\chi}(\omega) = (\chi - \chi_L) + \frac{\chi_L}{1 + i\omega\tau_L}. \qquad (9.32)$$

The explanation of these results by statistical physics is also analogous.

It is based on the theory of magnetic susceptibilities due to Langevin. The term $\chi_L/(1 + i\omega\tau_L)$ occurs only in materials whose molecules possess a permanent magnetic dipole moment. The only significant difference between the electric and magnetic situations does not show up in the thermodynamic results. We have deduced that κ_D and χ_L are both positive, but thermodynamics says nothing about the signs of the deformation contributions $(\kappa - \kappa_D)$ and $(\chi - \chi_L)$. Statistical physics shows that $(\kappa - \kappa_D)$ is positive but that $(\chi - \chi_L)$ is negative. The contributions $(\chi - \chi_L)$ and χ_L to χ are thus of opposite sign. They are also known as the diamagnetic and paramagnetic contributions. When the paramagnetic contribution is nonzero, i.e. when the molecules of the material have a nonzero permanent magnetic dipole moment, it is generally much larger than the diamagnetic (deformation) contribution. The overall sign of χ is thus dependent on whether or not χ_L is nonzero. No such distinction occurs for κ. Its two contributions are additive and we always have $\kappa > 0$.

Our investigation of the idealized 'polarizable simple fluid' defined in § 4a is now complete. We have seen how the theory develops naturally, from very few assumptions, to include all the main phenomena exhibited by real fluids in the presence of electromagnetic fields. Its most serious flaw is its failure to describe correctly electromagnetic dispersion at frequencies above the radio spectrum. We have traced this to a failure not of the method but of the basic idealization, which remains self-consistent but ceases to be realistic under these conditions. The ambiguities commonly associated with the energy–momentum tensor, and with the concept of pressure, when electromagnetic fields are present have been avoided completely. There is some freedom in the choice of variables used to describe a given situation, but this does not represent any physical ambiguity in the theory. The consistent use of a relativistic theory throughout produces simplifications, not complications, since distinct three-dimensional phenomena such as heat conduction and viscous stress are then represented by a single tensor. Only at the end, for the sake of interpretation, does one need to decompose such compact tensor descriptions into their irreducible constituents. These successes of a theory based almost entirely on the relativistic laws governing inert mass, momentum, energy and entropy do indeed show that special relativity is the foundation of macroscopic physics.

Appendix

Vector and dyadic notation in three dimensions

In three dimensions it is often convenient to restrict attention to coordinate systems that are right-handed Cartesian. It was seen in § 2–4a that the distinction between covariant and contravariant tensors then disappears, so that only the total valence remains significant. The most common tensors are those of valences 1 and 2. Three-dimensional vector and dyadic notation provides a compact method of writing such tensors, and the most common operations which involve them, in an index-free manner. Vectors are tensors of valence 1 and are usually denoted by boldface type, e.g. $\mathbf{a}$, $\mathbf{P}$. Dyadics are tensors of valence 2 and are usually denoted by sans-serif boldface type, e.g. T, U. Their components are denoted by the corresponding lightface symbol with appropriate suffixed indices.

Three products are defined between two vectors $\mathbf{a}$, $\mathbf{b}$. These are

the scalar product $\quad \phi = \mathbf{a} \cdot \mathbf{b} \quad$ if $\quad \phi = a_i\, b_i$

the vector product $\quad \mathbf{c} = \mathbf{a} \times \mathbf{b} \quad$ if $\quad c_i = \epsilon_{ijk}\, a_j\, b_k$

the dyadic product $\quad \mathsf{U} = \mathbf{ab} \quad$ if $\quad U_{ij} = a_i\, b_j$.

Here ϵ_{ijk} is the three-dimensional Levi–Civita symbol, defined in the notation of (2–5.7) by

$$\epsilon_{ijk} = \epsilon_{[ijk]}, \quad \epsilon_{123} = 1.$$

As usual there is an implied summation over repeated indices. Scalar and vector products involving dyadics are also defined. Only the dyadic index adjacent to the product symbol is involved. Examples are

$$\mathsf{T} = \mathsf{U} \times \mathbf{a} \quad \text{if} \quad T_{ij} = \epsilon_{jkl}\, U_{ik}\, a_l$$

$$\phi = \mathbf{a} \cdot \mathsf{U} \cdot \mathbf{b} \quad \text{if} \quad \phi = a_i\, U_{ij}\, b_j.$$

The unit dyadic I is the unique dyadic which satisfies

$$\mathsf{I} \cdot \mathbf{a} = \mathbf{a}$$

for all vectors $\mathbf{a}$.

Between two dyadics there is also the possibility of forming double products. In particular the double scalar product is defined by

$$\phi = \mathsf{T} : \mathsf{U} \quad \text{if} \quad \phi = T_{ij}\, U_{ij}.$$

In the product symbol ' : ' used here, the upper dot indicates the scalar product over the first index of each multiplicand and the lower dot indicates that over their second indices. Other double products such as mixed scalar–vector products can be defined and denoted in a similar manner.

The transpose of a dyadic $\mathbf{T}$ is the dyadic $\mathbf{T}'$ which satisfies

$$\mathbf{U} = \mathbf{T}' \quad \text{if} \quad U_{ij} = T_{ji}.$$

Derivatives of vectors and dyadics are denoted by the symbol ∇ together with any appropriate product notation, thus

$$\mathbf{U} = \nabla\mathbf{a} \quad \text{if} \quad U_{ij} = \partial_i\, a_j$$
$$\mathbf{a} = \nabla \times \mathbf{b} \quad \text{if} \quad a_i = \epsilon_{ijk}\, \partial_j\, b_k$$
$$\mathbf{a} = \nabla \cdot \mathbf{T}' \quad \text{if} \quad a_i = \partial_j\, T_{ij}$$

where ∂_i is an abbreviation for the operation $\partial/\partial x_i$. Note that the transpose operation in this final example causes the contraction to be between the index of ∇ and the *second* index of $\mathbf{T}$. The scalar differential operator $\nabla \cdot \nabla \equiv \partial_i\, \partial_i$ is denoted by ∇^2.

Publications referred to in the text

The pages on which reference is made to these publications are listed in brackets at the end of each reference.

Abraham, M. 1909 'Zur elektrodynamik bewegter Körper', *Rend. Circ. Mat. Palermo* **28**, 1–28. [*225*]

Arzelies, H. 1965 'Transformation relativiste de la température et de quelques autres grandeurs thermodynamiques', *Nuovo Cimento* **35**, 792–804. [*163*]

Batchelor, G. K. 1967 *An Introduction to Fluid Dynamics*, Cambridge: University Press. [*143, 186*]

Belinfante, F. J. 1940 'On the current and the density of the electric charge, the energy, the linear momentum and the angular momentum of arbitrary fields', *Physica* **7**, 449–74. [*128*]

Braginsky, V. B. & Panov, V. I. 1971 'Verification of the equivalence of inertial and gravitational mass', *Soviet Physics – JETP* **34**, 464–6. (Russian original in *Zh. Eksp. & Teor. Fiz.* **61**, 873–9.) [*32*]

Callen, H. B. 1974 'A symmetry interpretation of thermodynamics', pp. 61–78 of *Foundations of Continuum Thermodynamics*, eds. J. J. Delgado Domingos, M. N. R. Nina & J. H. Whitelaw, London: Macmillan. [*233*]

Coleman, B. D. 1964 'Thermodynamics of materials with memory', *Arch. Rat. Mech. Anal.* **17**, 1–46. [*147*]

Coleman, B. D. & Noll, W. 1960 'An approximation theorem for functionals, with applications in continuum mechanics', *Arch. Rat. Mech. Anal.* **6**, 355–70. [*186*]

Debye, P. 1929 *Polar Molecules*, New York: Chemical Catalog Co. [*241*]

de Groot, S. R. & Mazur, P. 1962 *Non-equilibrium Thermodynamics*, Amsterdam: North-Holland Publ. Co. [*147, 233, 237*]

de Groot, S. R. & Suttorp, L. G. 1972 *Foundations of Electrodynamics*, Amsterdam: North-Holland Publ. Co. [*200, 201, 216, 217, 220, 223, 226, 227*]

Eddington, A. S. 1948 *Fundamental Theory*, Cambridge: University Press. [*30*]

Ehlers, J. 1971 'General relativity and kinetic theory', pp. 1–70 of *General Relativity and Cosmology, Proc. Int. School of Physics 'Enrico Fermi', Course XLVII*, ed. R. K. Sachs, New York: Academic Press. [*121*]

Einstein, A. 1905 'Zur elektrodynamik bewegter Körper', *Ann. Phys. (Leipzig)* **17**, 891–921. [*3, 27*]

Einstein, A. 1911 'Über den Einfluss der Schwerkraft auf die Ausbreitung des Lichtes', *Ann. Phys. (Leipzig)* **35**, 898–908. [*34*]

Eötvös, R. von 1889 'Über die Anziehung der Erde auf verschiedene Substanzen', *Math. nat. Ber. aus Ungarn* **8**, 65–8. [*32*]

Hakim, S. S. & Higham, J. B. 1962 'An experimental determination of the excess pressure produced in a liquid dielectric by an electric field', *Proc. Phys. Soc. London* **80**, 190–8. [*224*]

Inonu, E. & Wigner, E. P.   1953   'On the contraction of groups and their representations', *Proc. Nat. Acad. Sci. U.S.* **39**, 510–24. [*67*]

Israel, W.   1972   'The relativistic Boltzmann equation', pp. 201–41 of *General Relativity – Papers in Honour of J. L. Synge*, ed. L. O'Raifeartaigh, Oxford: Clarendon Press. [*158*]

Israel, W.   1976   'Nonstationary irreversible thermodynamics: a causal relativistic theory', *Ann. Phys. (N.Y.)* **100**, 310–31. [*176*]

Israel, W. & Stewart, J. M.   1976   'Thermodynamics of nonstationary and transient effects in a relativistic gas', *Phys. Lett.* **58A**, 213–5. [*147, 186*]

Kranyš, M.   1972   'Kinetic derivation of nonstationary general relativistic thermodynamics', *Nuovo Cimento* **8B**, 417–41. [*183*]

Landsberg, P. T. & Johns, K. A.   1970   'The Lorentz transformation of heat and work', *Ann. Phys. (N.Y.)* **56**, 299–318. [*163*]

Michelson, A. A. & Morley, E. W.   1887   'On the relative motion of the Earth and the Luminiferous Ether', *Am. J. Science* **34**, 333–45. [*10*]

Minkowski, H.   1908*a*   Address delivered at the 80th Assembly of German Natural Scientists and Physicians, Cologne. English translation, 'Space and Time', in *The Principle of Relativity*, New York: Dover Publications Inc. [*3*]

Minkowski, H.   1908*b*   'Die Grundgleichungen für die elektromagnetischen Vorgänge in bewegten Körpern', *Nach. Ges. Wiss. Göttingen, math.–phys. Klasse*, 53–111 (1908). [*225*]

Onsager, L.   1931*a*   'Reciprocal relations in irreversible processes. I.', *Phys. Rev.* **37**, 405–26. [*146*]

Onsager, L.   1931*b*   'Reciprocal relations in irreversible processes. II.', *Phys. Rev.* **38**, 2265–79. [*146*]

Ott, H.   1963   'Lorentz-Transformation der Wärme und der Temperatur', *Z. Phys.* **175**, 70–104. [*163*]

Penfield, P & Haus, H. A.   1967   *Electrodynamics of Moving Media*, Cambridge, Mass.: M.I.T. Press. [*220*]

Planck, M.   1907   'Zur Dynamik bewegter Systeme', *Sitz. preuss. Akad. Wiss. Berlin*, 542–70 (1907). [*163*]

Prigogine, I.   1974   'Microscopic aspects of entropy and the statistical foundations of nonequilibrium thermodynamics', pp. 81–109 of *Foundations of Continuum Thermodynamics*, ed. J. J. Delgado Domingos, M. N. R. Nina & J. H. Whitelaw, London: Macmillan. [*148*]

Roll, P. G., Krotkov, R. & Dicke, R. H.   1964   'The equivalence of inertial and passive gravitational mass', *Ann. Phys. (N.Y.)* **26**, 442–517. [*32*]

Rosenfeld, L.   1940   'Sur le tenseur d'impulsion-énergie', *Mem. Acad. Roy. Belg.* **18**, fasc. 6, pp. 1–30. [*128*]

Saletan, E. J.   1961   'Contraction of Lie groups', *J. Math. Phys. (N.Y.)* **2**, 1–21. [*67*]

Schouten, J. A.   1954   *Ricci Calculus. An Introduction to Tensor Analysis and its Geometrical Applications*, 2nd. edn., Berlin: Springer-Verlag. [*35, 44*]

Stewart, J. M.   1971   'Non-equilibrium relativistic kinetic theory', *Lecture Notes in Physics* **10**, Berlin: Springer-Verlag. [*121, 186*]

Stewart, J. M.   1977   'On transient relativistic thermodynamics and kinetic theory', *Proc. Roy. Soc. London* A. **357**, 59–75. [*147, 186*]

Walker, G. B. & Lahoz, D. G.   1976   'Experimental observation of Abraham force in a dielectric', *Nature* **253**, 339–40. [*225*]

Weyl, H.   1946   *The Classical Groups. Their Invariants and Representations*. Princeton: University Press. [*70, 174*]

Summary and index of symbols
and conventions

The symbols and conventions summarized below do not form an exhaustive list, but they give the main notation that is often used without explicit reference to the relevant definitions. Further details can be found on the pages whose numbers are given in square brackets. Suffixes f. and ff. denote reference also to the following page or pages respectively. The items are listed in logical, rather than alphabetical, order.

Conventions

Kernel–index convention: The coordinate system to which a set of tensor components is referred is indicated by the alphabet (latin, greek, primed greek, etc.) used for the indices [35ff.]. This convention enables two coordinate systems to be used simultaneously for the same tensor, to form its *mixed components* [48].

Summation convention: There is an implied summation over every index that occurs twice in the same term, once as a superscript and once as a subscript [43f.].

Raising and lowering conventions: In an R_n, i.e. an n-dimensional affine space with a nonsingular metric tensor $g_{\alpha\beta}$, indices are raised and lowered thus:

$$t^{\alpha\beta}_{..\,\gamma} = g^{\beta\delta} t^{\alpha}_{.\,\delta\gamma} = g_{\gamma\delta} t^{\alpha\beta\delta}.$$

These are reversible and self-consistent operations [60f.]. Exceptions to this convention are the unit tensor A^{α}_{β} [61] and the fundamental alternating tensors $\eta^{\alpha\beta\gamma\delta}$ and $\eta_{\alpha\beta\gamma\delta}$ [73f.]. Care must be taken when Lie derivatives are used as the raising and lowering processes do not commute with Lie differentiation [170], e.g. $L_U t^{\alpha}_{.\,\beta} \neq g^{\alpha\gamma} L_U t_{\gamma\beta}$. In the spacetime of Newtonian physics the fundamental tensor $g^{\alpha\beta}$ is singular. It is used to raise indices, e.g.

$$t^{\alpha}_{.\,\beta} = g^{\alpha\gamma} t_{\gamma\beta},$$

but this process is irreversible and so there is no corresponding lowering convention [62].

Bracket notations for indices: Symmetrization and antisymmetrization of indices is denoted by enclosing the relevant indices between round and square brackets respectively [68ff.]. Exclusion of indices from these operations is indicated by vertical lines [70], e.g.

$$t_{[\alpha|\beta|\gamma]} = \tfrac{1}{2}(t_{\alpha\beta\gamma} - t_{\gamma\beta\alpha}).$$

Symbols from tensor analysis and geometry

E_n n-dimensional affine space [44].

R_n n-dimensional affine space with nonsingular metric tensor $g_{\alpha\beta}$ [60].

$\delta_{\alpha\beta}$, δ^α_β, $\delta^{\alpha\beta}$ Kronecker symbols, taking the value 1 if $\alpha = \beta$ and 0 if $\alpha \neq \beta$ [57f.].

A^α_β Unit tensor [45, 48].

$A^{\alpha'}_\alpha$ Mixed components of the unit tensor, used as transformation coefficients and defined by $A^{\alpha'}_\alpha = \partial x^{\alpha'}/\partial x^\alpha$ [37, 45].

$A^{a...b\gamma...\delta}_{\alpha...\beta c...d} \equiv A^a_\alpha...A^b_\beta A^\gamma_c...A^\delta_d$ Abbreviation for repeated product of transformation coefficients or unit tensors [48].

$g_{\alpha\beta}$ Nonsingular metric tensor [60f.] of an R_n, with inverse $g^{\alpha\beta}$ so that

$$g^{\alpha\gamma}g_{\gamma\beta} = A^\alpha_\beta.$$

The three- and four-dimensional cases of Euclidean geometry [57] and special relativity [58] are occasionally distinguished by prefixed superscripts (e) and (r) respectively. The signature of $^{(r)}g_{\alpha\beta}$ is $+2$ [60].

$^{(n)}g^{\alpha\beta}$ Positive semi-definite symmetric fundamental tensor of Newtonian spacetime [59f.], denoted simply by $g^{\alpha\beta}$ when the Newtonian situation is clear from the context.

t_α Fundamental vector of Newtonian spacetime [58], such that

$$^{(n)}g^{\alpha\beta} t_\beta = 0.$$

$\epsilon^{\alpha\beta...\gamma}, \epsilon_{\alpha\beta...\gamma}$ (with n indices)   Levi-Civita symbols [71], defined by

$$\epsilon_{\alpha\beta...\gamma} = \epsilon_{[\alpha\beta...\gamma]}, \quad \epsilon^{\alpha\beta...\gamma} = \epsilon^{[\alpha\beta...\gamma]}, \quad \epsilon_{12...n} = \epsilon^{12...n} = 1.$$

$u^{\alpha\beta...\gamma}, u_{\alpha\beta...\gamma}$ (with n indices)   Fundamental alternating tensors of an oriented E_n [71f.].

$\eta^{\alpha\beta...\gamma}, \eta_{\alpha\beta...\gamma}$ (with n indices)   Fundamental alternating tensors of an oriented R_n [73], given by

$$\eta_{\alpha\beta...\gamma} = |g|^{\frac{1}{2}} \epsilon_{\alpha\beta. .\gamma}, \quad \eta^{\alpha\beta...\gamma} = |g|^{-\frac{1}{2}} \epsilon^{\alpha\beta...\gamma},$$

where $g = \det g_{\alpha\beta}$.

$^{(n)}\eta^{\alpha\beta\gamma\delta}, ^{(n)}\eta_{\alpha\beta\gamma\delta}$ Fundamental alternating tensors of oriented Newtonian spacetime, such that $\eta_{1234} = \eta^{1234} = 1$ in positively oriented Galilean coordinate systems [74].

$g_{\alpha\beta\gamma\delta}$ Bivector metric tensor of relativistic [92] or Newtonian [96] spacetime, given by

$$g_{\alpha\beta\gamma\delta} = -\tfrac{1}{4}\eta_{\alpha\beta\kappa\lambda} \eta_{\gamma\delta\mu\nu} g^{\kappa\mu} g^{\lambda\nu}$$

in both cases and simplifying to

$$g_{\alpha\beta\gamma\delta} = g_{\alpha[\gamma} g_{\delta]\beta}$$

in the relativistic case.

B^α_β Projection operator in relativistic or Newtonian spacetime, projecting orthogonally to some specified future-pointing timelike unit vector v^α. Defined by

$B^\alpha_\beta = A^\alpha_\beta + v^\alpha v_\beta$ in relativistic spacetime [131], where $v^\alpha v_\alpha = -1$,

$B^\alpha_\beta = A^\alpha_\beta - v^\alpha t_\beta$ in Newtonian spacetime [132], where $v^\alpha t_\beta = 1$.

$h_{\alpha\beta}$ Covariant symmetric tensor related to the projection operator B^α_β by

$g^{\alpha\gamma}h_{\gamma\beta} = B^{\alpha}_{\beta}$, given in both Newtonian [95f.] and relativistic [164] spacetimes by

$$h_{\alpha\beta} = 2\,g_{\alpha\gamma\delta\beta}\,v^{\gamma}v^{\delta}$$

and simplifying [98] to

$$h_{\alpha\beta} = g_{\alpha\beta} + v_{\alpha}v_{\beta}$$

in the relativistic case.

$u = (\mathbf{u}, \phi)$ Three-dimensional decomposition of a four-vector with respect to a particular inertial reference frame [64]. A similar matrix notation is used for tensors of valence 2.

∂_{α} Abbreviated notation for partial derivatives, $\partial_{\alpha} \equiv \partial/\partial x^{\alpha}$ [49].

L_{U} Lie derivative with respect to the vector field U^{α} [168ff.]

$dS^{\alpha\cdots\beta}$ (with p indices)   Contravariant surface element [77] for an inner-oriented p-surface.

$dS_{\kappa\cdots\lambda}$ (with $(n-p)$ indices)   Covariant surface element [83, 85] for an outer-oriented p-surface in an E_n. When the E_n itself is oriented, we have

$$dS_{\kappa\cdots\lambda} = \frac{1}{p\,!}\,u_{\kappa\cdots\lambda\alpha\cdots\beta}\,dS^{\alpha\cdots\beta}, \quad dS^{\alpha\cdots\beta} = \frac{1}{(n-p)\,!}\,u^{\kappa\cdots\lambda\alpha\cdots\beta}\,dS_{\kappa\cdots\lambda}.$$

Symbols from relativistic continuum dynamics

v^{α} Material velocity [128].

ρ^{α} Flux vector of inert mass [133].

ρ Density of inert mass [133], with $\rho^{\alpha} = \rho v^{\alpha}$.

$v \equiv 1/\rho$ Specific volume [158].

$T^{\alpha\beta}$ (Ch. 3 §4 only) Nonsymmetric energy–momentum tensor [123]. When this is used, the spin tensor $U^{\alpha\beta\gamma}$ [125] is also required.

$T^{\alpha\beta}$ (Ch. 3 §5 onwards) Belinfante–Rosenfeld symmetric energy–momentum tensor [126], denoted in its definition in equation (3-4.29) by $*T^{\alpha\beta}$.

μ Density of rest mass [132]

u Specific internal energy [134]

q^{α} Energy flux vector [132]

$\sigma^{\alpha\beta}$ Stress tensor [132]

p Kinetic pressure [164]

$\pi^{\alpha\beta}$ Shear stress tensor [165]

related to $T^{\alpha\beta}$ by the decompositions

$$T^{\alpha\beta} = \mu v^{\alpha}v^{\beta} + 2q^{(\alpha}v^{\beta)} - \sigma^{\alpha\beta},$$
$$\mu = \rho(1+u), \quad \sigma^{\alpha\beta} = \pi^{\alpha\beta} - ph^{\alpha\beta}.$$

a^{α} Acceleration

$\omega_{\alpha\beta}$ Vorticity

$\theta_{\alpha\beta}$ Strain rate

θ Expansion rate

$e^{\alpha\beta}$ Shear rate

[165f.] related to the velocity gradient $\partial_{\alpha}v_{\beta}$ by the decompositions
$$\partial_{\alpha}v_{\beta} = \theta_{\alpha\beta} + \omega_{\alpha\beta} - v_{\alpha}a_{\beta},$$
$$\theta_{\alpha\beta} = e_{\alpha\beta} + \tfrac{1}{3}\theta h_{\alpha\beta}.$$

In the decompositions given above, $\sigma^{\alpha\beta}$, $\pi^{\alpha\beta}$, $\theta_{\alpha\beta}$, $e_{\alpha\beta}$ are symmetric, $\omega_{\alpha\beta}$ is antisymmetric, $\pi^{\alpha\beta}$ and $e_{\alpha\beta}$ are tracefree, and the vectors and tensors defined by these decompositions are all orthogonal to v^{α} on each index. The tensor $h_{\alpha\beta}$ is given by $h_{\alpha\beta} = g_{\alpha\beta} + v_{\alpha}v_{\beta}$.

$D/D\tau$ Corotating rate-of-change operator [177ff.].

Symbols from Newtonian continuum dynamics

v^α Material velocity [128].

ρ^α Mass flux vector [133].

ρ Mass density [133], with $\rho^\alpha = \rho v^\alpha$.

$T^{\alpha\beta}$ (Ch. 3 §4 only) Nonsymmetric mass–momentum tensor [123]. When this is used, the spin tensor $U^{\alpha\beta\gamma}$ [125] is also required.

$T^{\alpha\beta}$ (Ch. 3 §5 onwards) Belinfante–Rosenfeld symmetric mass–momentum tensor [126], denoted in its definition in equation (3-4.29) by $*T^{\alpha\beta}$.

$\sigma_{\dot\alpha}{}^\beta$ Stress–energy tensor [140].

u Specific internal energy [142].

q^α Energy flux vector [141].

$\sigma^{\alpha\beta}$ Stress tensor [132, 141].

These tensors obey the algebraic restrictions

$$q^\alpha t_\alpha = 0, \quad \sigma^{\alpha\beta} t_\beta = 0, \quad \sigma^{[\alpha\beta]} = 0,$$

and are mutually related by the formulae

$$\sigma^{\alpha\beta} = g^{\alpha\gamma}\sigma_{\dot\gamma}{}^\beta, \quad T^{\alpha\beta} = \rho v^\alpha v^\beta - \sigma^{\alpha\beta},$$
$$\sigma_{\dot\alpha}{}^\beta = t_\alpha(\rho u v^\beta + q^\beta) + h_{\alpha\gamma}\sigma^{\gamma\beta},$$

where $h_{\alpha\beta} = 2g_{\alpha\gamma\delta\beta} v^\gamma v^\delta.$

The conservation laws of four-momentum and of energy are given respectively by

$$\partial_\beta T^{\alpha\beta} = 0, \quad v^\alpha \partial_\beta \sigma_{\dot\alpha}{}^\beta = 0.$$

Symbols from thermodynamics

u Specific internal energy [134].

s^α Entropy flux density [150, 153, 205].

s (Ch. 4 §2 only) Entropy density [148].

s (Ch. 4 §4 onwards) Specific entropy [157, 164, 211], with $s^\alpha = s\rho^\alpha$.

σ Entropy source strength [151], defined by $\sigma = \partial_\alpha s^\alpha$.

Θ^α Inverse temperature vector [154, 159, 205].

Θ Inverse temperature [157, 159, 211], with $\Theta^\alpha = \Theta v^\alpha$.

T Absolute temperature [158, 164, 211, 229], with $\Theta = 1/T$.

Φ Thermal potential [154, 158, 205, 215].

p Equilibrium pressure [156f., 208]; kinetic pressure [164] for non-equilibrium states.

p_0 Thermostatic pressure [164]; Kelvin pressure [215f., 229] when an electromagnetic field is present.

u_0 Kelvin specific internal energy [216f., 228].

f_0 Specific free energy [223].

f^0, p^0, s^0, u^0 Helmholtz variables [223].

$\tau^{\alpha\beta}$ Thermodynamic flux tensor [164, 176, 231].

$X_{\alpha\beta}$ Thermodynamic force tensor [171, 176, 234].

$c_{\alpha\beta\gamma\delta}$ Impedance tensor [174, 176, 236].

$a_{\alpha\beta\gamma\delta}$ Relaxation tensor [177, 236].

In the absence of an electromagnetic field, the above four tensors satisfy [180] the linearized phenomenological law

$$\rho a_{\alpha\beta\gamma\delta} D\tau^{\gamma\delta}/D\tau + c_{\alpha\beta\gamma\delta}\tau^{\gamma\delta} = X_{\alpha\beta}.$$

q_h^α	Heat flux vector	defined [164f., 231] from $\tau^{\alpha\beta}$ by analogy
$\sigma_v^{\alpha\beta}$	Viscous stress tensor	with the relativistic decomposition of
p_v	Viscous pressure	$T^{\alpha\beta}$ given above.

Symbols from electrodynamics

$F^{\alpha\beta}$	Electromagnetic field tensor [191, 199]	
$I^{\alpha\beta}$	Electromagnetic induction tensor [199]	in Gaussian
$P^{\alpha\beta}$	Polarization tensor [201], with $I^{\alpha\beta} = F^{\alpha\beta} - 4\pi P^{\alpha\beta}$	units [x].
J^α	Charge-current vector [193]	
E^α	Rest-frame electric field [214]	all orthogonal to v^α and
$B^{\alpha\beta}$	Rest-frame magnetic induction [214]	defined by the decompo-
D^α	Rest-frame electric displacement [214]	sitions
$H^{\alpha\beta}$	Rest-frame magnetic field [214]	$F^{\alpha\beta} = B^{\alpha\beta} + 2v^{[\alpha}E^{\beta]}$
P^α	Rest-frame electric·polarization [216]	$I^{\alpha\beta} = H^{\alpha\beta} + 2v^{[\alpha}D^{\beta]}$
$M^{\alpha\beta}$	Rest-frame magnetization [216]	$P^{\alpha\beta} = M^{\alpha\beta} - 2v^{[\alpha}P^{\beta]}$.
g^α	Auxiliary field variable [214], defined by $g^\alpha = (1/2\pi\rho)F^{[\alpha}_{\cdot\gamma}I^{\beta]\gamma}v_\beta$.	
κ	Electric susceptibility [223], with $P^\alpha = \kappa E^\alpha$.	
χ	Magnetic susceptibility [223], with $M^{\alpha\beta} = \chi B^{\alpha\beta}$.	
$\Psi^{\alpha\beta\gamma}, \Xi_\alpha$	Electrodynamic equilibrium variables [205, 207].	
$\overset{*}{P}{}^{\alpha\beta}$	Thermostatic polarization tensor [229].	
$\Pi^{\alpha\beta}$	Dispersive polarization tensor [229], with $\Pi^{\alpha\beta} = P^{\alpha\beta} - \overset{*}{P}{}^{\alpha\beta}$.	

Subject index

In the alphabetical ordering of subdivisions, words such as 'of' and 'in' have been disregarded. Italicized page numbers refer to definitions. Suffixes f. and ff. denote reference also to the following page or pages respectively. For page references to authors cited, see pp. 245–6.

Absolute space, 3, 8
  abandoned as an *a priori* assumption, 13
Absolute time
  abandoned as an *a priori* assumption, 11f.
  deduced as one theoretical possibility, 25
  incompatible with special relativity, 3
  as Newtonian limit of proper time, 67f.
  in Newtonian physics, 8, 67, 93, 101
  operational definition of, 11
Absorption, electromagnetic, 240
Acceleration
  as cause of heat flux, 175, 182
  uniform, 162
  used in construction of $D/D\tau$, 178f.
Acceleration vector a^α, *165f.*
  in Newtonian limit, 184f.
Action-at-a-distance, significance of for construction of inertial frames, 9
Admissible parametrization, *54*
Aether, velocity of Earth through, 10
Affine scalar, *46*
Affine space E_n, *44*, Ch. 2
  basis of, *51*
  Euclidean, *60*
    tensors in, 62ff.
  geometry in an, 51ff.
  with metric, R_n, *60*
  oriented, *72*
    fundamental tensor of, 72ff.
Affine subspace, *52*
Affine tensor, *47* (*see also*: Tensors)
Affine transformation, *43*
Affine vector
  component of, *46*
  contravariant, *46*
  covariant, *46*
  (*see also*: Vectors)

Angular momentum, 103ff., 124ff.
  of continuum
    conservation of, 121, 124f.
    flux of, 124f., 127f.
    total $J^{\alpha\beta}$, 127
  dependence on origin of, 104, 124
  of single particle $j^{\alpha\beta}$, *103*
  of system of particles
    conservation of, 104, 121
    flux of, *104*, 122
    orbital $L^{\alpha\beta}$, *106*
    spin $S^{\alpha\beta}$, *106*
    total $J^{\alpha\beta}$, *104*
  (*see also*: Spin tensor; Spin vector)
Angular velocity, 162, 179 (*see also*: Vorticity)
Antiparallel, *53*
Area: *see* Hyperplane, vector area of

Barium titanate, Walker–Lahoz experiment with disc of, 225
Basis, of affine space, *51*
Basis vectors, *47*, 51, 53, 87, 94, 222
Bivector metric tensor $g_{\alpha\beta\gamma\delta}$
  in Newtonian limit, 97f., 119, 144
  in Newtonian spacetime, *96*, 111ff., 141f.
  in relativistic spacetime, *92*
Boost, *28*
  in Newtonian limit, 64, 67
Boundary ∂S, oriented, *80f.*, 85, 131 (*see also*: Orientation; Stokes' Theorem)
Boundary conditions, electromagnetic, 196ff., 212
Bracket notations for indices, *69ff.*, 75f.

Cartesian coordinates and tensors: *see under* Coordinate systems; Tensors
Centre-of-mass line, *107f.*, 112
Centroid line, 104, *106f.*

Newtonian dynamics (*cont.*)
  Third Law, 10, 111, 131
  underlying assumptions of, 8ff.
Newtonian limit, 64ff., 96ff., 115, 118ff.,
  143ff., 161ff., 183ff.
  convention concerning overbars in, 97
Normal vector, *90*
Null, *27, 87, 90*

Observers
  preferred, 1
    lack of in gravitational field, 34
Orbit, material, *129*, 168f., 170, 235
Orientation, *72ff.*
  of curve, 78
  inner, *78*, 102
  outer, *78*, 85, 102
  of spacetime, 89f., 93
  of surface, 76ff.
  time-, *88f.*, 93
Orthogonality of vectors, *87, 94*, 96
Outer product, *49*, 169f., 174, 178

Parallelism
  of *r*-planes, *52*
  of vectors, *53*
Paramagnetism, 242
Particle
  composite, *101*
  concept of, 100f.
Particle dynamics, 99ff.
  compared with continuum dynamics,
    99, 121
PCT theorem, 89
Phenomenological laws for fluids, 166f.
  linearized, 174ff., 180f.
  in electromagnetic field, 231, 235ff.
  in Newtonian limit, 185f.
  in weak electromagnetic field, 237ff.
  (*see also:* Constitutive laws; Con-
    stitutive relations)
(*r*-) plane, *52f.*
  inner and outer orientation of, 78ff.
  tangent, *55*, 77f.
Poincaré transformations, *28*
Polarizable material, 195ff.
  measurement of fields within, 198, 201
Polarization tensor $P^{\alpha\beta}$, *201*, 216
  decomposition into P^{α}, $M^{\alpha\beta}$, 216ff.
  dispersive $\Pi^{\alpha\beta}$, *229ff.*, 233ff.
  thermostatic $\overset{*}{P}{}^{\alpha\beta}$, *229*, 238
Polarization vector P, *195ff.* 219f., 240
Position vector, *46*
Poynting vector S, *203*
Pressure p, *156ff.*, 163, 208ff., 212ff., 242
  conventional nature of in polarizable
    fluid, 220
  Helmholtz $\overset{*}{p}{}^{0}$, *223ff.*

Kelvin, p_0, *216ff.*, 220, 224, *229ff.*
kinetic p, *164*
  in Newtonian limit, 184f.
  in perfect fluid, 187f.
  radiation, 232
  thermostatic p_0, *164*, 215
  viscous p_v, *164f.*, 175, 239
Principle, use of the term, 2
Principle of Equivalence, 32, 34f.
  Dynamical, 34f.
Principle of Inertia, 8, 33, 101, 192
  Extended, 14, 16, 33, 35
Principle of Special Relativity, 23
Principle of Uniformity, 21ff., 35
Projectivity, 16
Proper distance, *27*, 87
Proper time, *27*, 87, 101, 114
  Newtonian limit of, 67f.
  product with temperature as natural
    thermodynamic parameter, 170,
    235

Quadratic invariant Φ, *26f.*, 38f., 58, 91
Quotient rule, *49f.*, 191

Radiation, dispersion of electromagnetic,
  240ff.
Raising and lowering: *see under* Indices
Rank
  of a matrix, *41*
  of a tensor, *60*
Rate-of-change operator $D/D\tau$, *177ff.*, 231
  Newtonian limit D/Dt of, 184f.
Reference frame: *see* Frame of reference
Region, *54*
Relation
  binary, 72
  equivalence, *72*, 88
Relativity, special and general: *see*
  General relativity; Special rela-
  tivity
Relaxation phenomena, 170, 176ff., 185f.
Relaxation tensor $a^{\alpha\beta\gamma\delta}$, *177*, 179ff.,
  233ff.
Relaxation times, 181ff., 232, 239ff.
Rest, standard of, 4ff., 7, 14
Rigid rods, 8, *11*
Rigid rotation, 162

Scalar product, *243* (*see also:* Inner
  product)
Screwthread, used to describe orientation, 79f.
Section, spacelike, *103*
Separation, spacelike, null and timelike
  27, 87
Sheare rate $e_{\alpha\beta}$, *165f.*, 175
  in Newtonian limit, 184f.